विज्ञान-सारथिः

VIJÑĀNA-SĀRATHIḤ

Proceedings of the National Seminar

on

SCIENCE AND TECHNOLOGY

in

Sanskrit Śāstraic Traditions

held on

1-3 Sept., 2005

Organised by

Department of Sanskrit, Pali & Prakrit

Kurukshetra University, Kurukshetra (Haryana)

Chief Editor

Prof. Dr. (Mrs.) VIJAYA RANI

Director of the National Seminar

PARIMAL PUBLICATIONS

DELHI

Published by
PARIMAL PUBLICATIONS
Office : 27/28, Shakti Nagar, Delhi-110007 (INDIA)
Retail Outlet : 22/3, Shakti Nagar, Delhi-110007 (INDIA)
ph. : +91-11-23845456, 47015168
e-mail : order@parimalpublication.com
url : http://www.parimalpublication.com

Edition : Year 2020

ISBN : 978-81-7110-287-7

Price : ₹ 750.00

Printed at
Vishal Kaushik Printers
Jagatpuri Extension, Delhi (India)

Gram : Dharmakshetra
Fax : 01744-238277
Phones : Office : 238039 & 238417
Phones : Res. : 01744-238021
E-mail : vc.kuk@rediffmail.com

कुरुक्षेत्र विश्वविद्यालय, कुरुक्षेत्र - 136119 (भारत)

KURUKSHETRA UNIVERSITY, KURUKSHETRA - 136119 (INDIA)

कुलपति
Vice-Chancellor

FOREWORD

India surely enjoys a place of pride as a cradle of abiding world cultures and civilizations. Its ancient literature, mostly rendered in Sanskrit, is extremely rich and original in philosophical thought. It is equally opulent in scientific enquiry. To be precise, ancient India was a knowledge society and a leader in many intellectual pursuits in the fields of mathematics, medicine, metaphysics, and astronomy. The Sanskrit Shastraic literature has been a copious repository of scientific and technological studies since the Vedic times. A notable feature of the scientific temper of those days was its integration with the spiritual and the cultural aspects of life. Thus was science and technology devoted to the nobler cause of human happiness.

It may well be perceived that sciences in our modern world are not born of a vacuum; they had already been there either in their rudimentary form or in their developed form in ancient and medieval India, Sanskrit Shastraic traditions abound in the knowledge of such subject as Astronomy, Astrology, Minerology, Atmospheric Science, Biological Sciences, Physical, Chemical and Medical Sciences and Aeronautical Science.

Nonetheless, the fact remains that such vast knowledge has been neither full exploited nor full reserched. The National Seminar on '**Science and Technology in Sanskrit Shastraic Traditions.**' organised by the Department of Sanskrit, Pali, and Prakrit of Kurukshetra University needs to be lauded as a serious attempt in this direction. The event, I learn, saw a confluence of Sanskrit Shastriac scholars and the scientists, who presented insightful discourses underscoring the unification of science and spirituality to advance human welfare. It is through such forums, I believe, that we shall be able to explore the full potential of our ancient wisdom and garner it for the development of the society. This is how we can succeed in establishing a link between our great past and our great future.

I am glad that the Department of Sanskrit, Pali and Prakrit is going to publsih the proceedings of the National Seminar into the form of the present manual. I wish the Department success in its venture, I also believe that this manual will work as a catalyst of growth both of knowledge and research.

(R.P. Hooda)

'श्री'

Editor's note

It is a matter of great pleasure and privilege that the Department of Sanskrit, Pali and Prakrit, Kurukshetra University, Kurukshetra, has organised a National Seminar on a very important and abstruse topic, entitled 'Science and Technology in Sanskrit Śātraic Tradition.' on 1-3 Sept, 2005. It is a very relevant topic for contemporary Society, as we all need to know what treasure of knowledge is lying hidden in our old Sanskrit Śāstras.

To-day a general impression about Sanskrit is that it is a language only of rituals, ethics, religion and philosophy and it has no concern with other important fields of life such as science, technology, Astrology, Astronomy, Mathematics etc.; but it is a wrong conception. After a deep insight and a wider observation of the old Sanskrit Śāstras, scholars of India and abroad have realised that Indian Literature in Sanskrit (such as Vedas, Brāhmaṇas, Upaniṣads, Vedāṅgas, Epics, Purāṇas, Śulva Sūtras, Philosophical Sūtras and soon) has a great potential of knowledge- many it be religious, philosophical, Scientific or technological. In Nārad Saṁhitā, names of eighteen earliest Scientists of India have been mentioned, whose contribution in the field of Physics, Chemistry, Astronomy and Mathematics coincides with the discoveries of modern Science. Name of these Scientists are:

''ब्रह्माचार्यो वसिष्टोऽत्रिमनुः पौलस्त्यल्लोमसो
मरीचिरंगिरा व्यासो नारदः शौनको भृगु।
च्यवनो यवनो गर्गः कश्यपश्च पराशरः
अष्टादशैते गंभीरा ज्योतिः शास्त्रप्रवर्तकाः॥''

Apart from this, Prof. Macdonell observes about Indian wisdom: " The debt of Europe to India has been considerable. There is in the first place, the great, fact that the Indians invented the numerical figures, used all over the world. The influence which the decimal system of reckoning, dependent on those figures has had, not only on mathematics but on the progress of civilisation in general, can hardly be over-estimated. During the eighth and ninth centuries, the Indian became the teachers in arithmetic and Algebra of the Arabs, and through them of the nations of the west."

(A History of Sanskrit Literature, p. 424.)

It is, thus, evident that the scholars of west were quite acquainted of the fact that many of the Scientific theories and technologies were already known to ancient Indian Scholars. The Śāstraic tradition commencing with the Vedas and Vedāṅgas and branching out into the eighteen and sixty-four Vidyās was not only religious, but also practically important for the technologies prevalent at that time in the society. In fact, in India, religion and science co-existed in mutual support. Even the great modern scientist Albert Einstein believed that 'Religion with out science is blind and science without religion is lame.' It is noteworthy here that the Scientific inventions, innovations and technological discoveries made by the Indian Scientists of the past were wedded with the thought of human-welfare and prosperity of the society. (बहुजनहिताय बहुजन सुखाय)

To-day we are living in a high-tech age and it is a high time for all of us to take a stock of 'what we were in the past and what we are to-day in the field of scientific advancement.' For this very purpose, Department of Sanskrit, Pali and Prakrit of Kurukshetra University organised a National level Seminar on the subject cited above. Through this seminar, scholars of Sanskrit Śāstras and modern Scientists of various streams were able to come at one common plateform to discuss different problems with each other and to explore possibilities of further areas of research in that direction.

In this seminar, apart from the Inaugral and valedictory, sessions, there were total Seven academic sessions, in which about 50 scholars presented their research papers of different themes and in different languages (i.e. in Hindi/ English/ Sanskrit). Among them, only those, papers could be included in the proceedings, which were received from the authors within the stipulated time frame. These papers have been divided into seven sections, which are as follows:

Section I : Aeronautics and Metallurgy

Section II : Computer Science and Mathematics

Section III : Physics and Electronics

Section IV : Medical Science and Genetics

Section V : Different Healing Therapies in Sanskrit Śāstras.

Section VI : Miscellaneous

Now it is my foremost duty to express my deep sense of gratitude to all those who directly or indirectly have contributed in the publication of these proceedings of the seminar.

First of all I express my gratefullness to Prof. R.P. Hooda, the Vice-chancellor, Kurukshetra University, Kurukshetra, who has very kindly, agreed to give financial assistance to the Department of Sanskrit for the publication of the proceedings of the Seminar. My sincere thanks are also due to him for providing us with a thoughtful 'foreword' for this volume.

I also express my indebtedness to all those scholars, who have enriched this volume by their valuable contributions.

It is my pleasant duty to extend my thanks to M/s Parimal Publications, Delhi, who have given a final and printed shape to the proceedings of the seminar on a short notice.

In the last I can not forget to express my heart felt feelings of gratitude to Dr.V.K.Goyal, under able guidance and management, the publication of these proceedings could take place in a finer way.

March 14, 2006

Vijaya Rani
(Editor)
Professor & Chairperson
Department of Sanskrit, Pali and Prakrit
Kurukshetra University
Kurukshetra (136119)

CONTENTS

Section I

Aeroeutics and Metallurgy

1

AERONAUTICS IN SANSKRIT

C.S.R. Prabhu

1. Introduction

This report aims at presenting the preliminary results of the study and investigation on a few rare ancient Indian Scientific Śāstras in Sanskrit. While there is a general opinion that at some point in the prehistoric or protohistroic times India had a highly developed technological society, there has been till now no real corroboration of this opinion.

The recent study and experimental investigations of a few rare and obscure Sanskrit works of Scientific nature has provided some basis, especially in the fields of Chemistry, Metallurgy and material science that there could have indeed existed an advanced technological status of the society, atleast in these fields, at some point of time in India's pre or protohistory.

This preliminary report attempts to present these initial results which include the procurement and decipherment of manuscripts and also the experimental laboratory preparation of the deciphered materials.

2. History and Background

During January 1991 a set of Sanskrit manuscripts have been procured from a source at Bangalore. While most of the manuscripts procured were never published, limited publication was done for (a) Vimana Śāstra and (b) Aṁśu Bodhinī, both ascribed to Mahāriṣi Bharadvāja.

Based on the study and investigation performed till now, the following history, background and origin of the system of Ancient Sanskrit Śāstras have been identified :

(i) An integrated system of ancient scientific shastras has been identified. This system is based on an integrated framework of concepts and principles unique and characteristic to ancient Indian literature including Vedic, Puranic and Ayurvedic components. These concepts have not much in common with modern science. However, the concepts can be tested only by independent physical verification. Since theoretical models of physical phenomena can be varied with time, the conceptual framework or model may not affect the physical phenomena themselves or the experimental results thereby achieved on empirical basis.

The Shastras covered in this system run into over a hundred in number with subjects as wide ranging as Chemistry, Metallurgy, Engineering, Architecture and Medicine (human, veterinary and plant). We will be presently dealing with only the specific set of shastras which have been located by us during 1991, though there may be many other sources of shastras traditionally or otherwise which have not been dealt by us.

(ii) While only a few of the total system of Shastras have actually been orally delivered and subsequently recorded into writing around 1912 A.D., more than twenty of them are supposed to have been actually available physically with the main source, namely the late Pandit T. Subbaraya Sastry of Anekal (near Bangalore) during the period of 1875 to 1930 A.D.

(iii) Pandit Subbaraya Sastry was apparently only a medium for oral delivery from his memory (which he reportedly acquired from his Guru) of the Shastras which were written down either in parts or in whole by others (as G. Venkatachala Sharma of Bangalore) who acted only as scribes.

About twenty Shastras were orally delivered by him in parts and hand-written manuscripts (on old paper) were produced during 1911-1940 as indicated by the dates recorded by the scribe (though he himself had access to them even during 1865-1911).

Some extracts of the Shastras were also printed in a few articles in journal 'Bhoutica Kalanidhi' by late Shri B. Suryanaraya Rau (grandfather of Shri B.V. Raman of Bangalore who displayed them also).

(iv) While reportedly Pandit Subbaraya Sastry had access to these Shastras through his Guru Maharaja (who was also a Yogi) the exact means and channels of acquisition have not yet been determined clearly. Possibilities of Yogic meditation, trance or other extra normal states being the cause for this cannot be ruled out, apart from regular memorization for oral delivery.

(v) The small set of Shastras actually delivered have quoted extensively from various other texts, including dictionaries, attributed to various authors of Vedic, Post Vedic periods on varied subjects of scientific nature. None of these texts have been located from any other sources, though the subject matter covered in them can be traced to be common to many Sanskrit works on related subjects e.g. Ayurvedic texts and Nighantus (Dictionaries).

(vi) Notwithstanding the unclear origin of these texts (for which reason many persons claimed that these shastras are not authentic or genuine) some of the contents of these texts have been investigated in terms of physical experimentation by trying out in the laboratory, the given formulae for the preparation of materials as alloys, glasses, ceramics, etc., as described in these texts. The decipherment process was primarily involved in tracing the synonyms of the words used for the input ingredients for making various materials by using Ayurvedic sources as Dictionaries.

List of procured manuscripts of Ancient Scientific Shastras in Sanskrit.

	Title	Author ascribed
(i)	Vimana shastra (or Vaimanika Prakaranam)	Maharishi Bharadwaja
(ii)	Amsu Bhodhini	-do-
(iii)	Kritakavajra Nirnaya (of Ratna Pradeepika)	Not clear
(iv)	Jalatatwa Prakarana	-do-
(v)	Apatatwa Upanyasa	-do-
(vi)	Rajya Tantra	Maharishi Yajnavalkya

3. The Decipherment Process

Among the various manuscripts procured, the following were mainly subject :

(i) Vimana Shastra (or Vaimanika Prakaranam a chapter of Bharadwaja's Yantra Sarvasva)

(ii) Amsu Bodhini

(iii) Kritaka Vajra Nirnaya

The decipherment process centrally concerned itself on the identification of practicable formulae for chemical preparations (in the laboratory) from the corresponding Sanskrit descriptions and recipes for the preparation of the materials as alloys, glasses and ceramics, as per the ancient methods in terms of notions, concepts and framework which were unique to the Shastras. The only common grounds with modern times were the input ingredient materials, which after decipherment could be identified either as equivalents of simple inorganic materials in the laboratory (as metals as copper, lead etc.) or alternatively complex organic materials from the nature such as herbs, roots, gums, resins, barks and mineral ores largely known to the Ayurveda system, as is practised today, both by Ayurvedic Physicians and Ayurvedic Chemists and Pharmacists.

The decipherment process which was quite tedious and complex, primarily involved finding the equivalent modern Sanskrit words for names of input ingredients for preparation of various materials. Since most of the words used in the manuscripts were quite archaic they have gone out of use in more recent (classicai Sanskrit, the Ayurvedic Nighantus or dictionaries helped to a partial extent in giving the more recent equivalents for such words. However, there are quite a number of words for which no more recent equivalents have yet been found. The search efforts are still in progress.

The second part of the effort lies in finding the equivalent Indian Language word for Sanskrit word i.e. Hindi or Telugu equivalent which can be used for actual identification of herb or mineral ore in the market or in the nature in general. In this connection substantial help was obtained from Ayurvedic Physicians and Chemists.

The third step is procuring the identified materials from natural sources (e.g. mineral ore), a task sometimes becoming a very difficult, as some of the materials are vary rare and may not be available in the modern times.

The fourth step involves the actual preparation of the described material (as an alloy or a ceramic or glass) by mixing the identified input ingredients in the proportion given in the Sanskrit original.

At this stage the modern Chemistry laboratory was utilised and the melting and cooling procedures were performed according to the directions given in Sanskrit manuscripts.

The mixing proportions were given in terms of relative units in weight. The units of temperature used were in 'Kakshyas' and the exact interpretation of 'Kakshya' is not yet known, though roughly it has been equated (as 1 Kakshya = 12.5 C) at low temperatures (this scale may not be linear at higher temperatures).

The procedure for heat treatment was also given in Sanskrit sources in terms of either sudden pouring or gradual cooling or slow pouring to produce various effects and different properties.

As regards the equipment to be used, in all experiments only the modern laboratory equipment was used, though descriptions in Sanskrit were varied as various types of crucibles (mooshas), bellows (bhastris) and furnaces (kundas) of ancient times (which are not available now).

4. Status report as of June 1999

Ancient Indian texts and manuscripts pertaining to Science & Technology have been studied with an aim to decipher and decode formulae for making new materials as alloys, ceramics and glasses. A formula for making protein rich food extract from common Indian grasses also has been deciphered.

The following twenty formulae for new materials consisting of special alloys, ceramics and glasses have been deciphered and some of them were actually produced based on the formulae from the ancient Sanskrit texts of Vimana Shastra (quoting Lohatantra), Amsu Bodhini, Kritaka Vajra Nirnaya, etc., based on the dictations of the texts and formulae by the Late Pandit Subbaraya Shastri of Anekal (1855 - 1940 A.D).

S.No.	NAME	DESCRIPTION
1	**2**	**3**
1	"Tamogarbha Loha"	A lead alloy capable of light absorption

Status :

Already produced in the laboratory, light in weight, black in colour, found to be resistant to acids.

Displayed high level of absorption for laser light (from red Ruby laser - as observed by Prof. Robert Anderson of San Jose State University during his visit to India in December 1991). Some chemical and other properties found to be unique - patentable new alloy. A laboratory test done in 1996 in Physics Department of Osmania University, Hyderabad indicated laser absorption characteristics upto 79% of incident light from a laser (this alloy was used in 'Tamo Yantra' in the Vimana Shastra for the purposes of absorption of light escaping from a photochemical reaction which resulted in absorption of light, thereby generating 'darkness').

2.	**"Pancha Loha" (not the well-known Panchaloha for making idols)**	A copper alloy which is highly malleable and also highly corrosion resistant to salt (NaCL)

Status :

Already produced and characterised to possess :

(a) Golden Yellow Colour (described in the Sanskrit text as 'Hema Varnam' or golden colour).

(b) Corrosion resistance to moisture and salt water (displayed weight loss of only about 0.00335 mg/dm^2/day in 3% NaCL solution).

(c) High machinability and on micro structure analysis found to be single phase alloy with high malleability (described in Sanskrit as 'mridulam' or 'soft').

(d) Characteristics, composition and properties found to be not listed in ASM Reference (1988) and therefore patentable new alloy.

3. **"Araara Tamra"** A copper alloy zinc, lead and iron of light absorption.

Status :

Already produced and characterised to possess :

(a) Golden Yellow to reddish tinge (described in Sanskrit text as 'Hema Varnam' or golden colour).

(b) Brittle, light and hard on micro structure analysis found to be two phase alloy.

(c) Very hard (Young's modulus 16.9) (described in Sanskrit text as 'Dridham' or 'strong').

(d) Characteristics, composition and properties found to be not listed in ASM Reference (1988) and therefore patentable new alloy.

4. **"Chapala grahaka (ceramic)"** A fine porcelain type of ceramic.

Status :

Already produced and characterised to be resistant to all acids and alkalis.

5. **"Chapala grahaka (glass)"** A soft glass (of flow temperature melt)

Status :

Already produced and characterised to be resistant to acids and alkalis.

Refractive index found to be 1.614 (highest known among soft glasses made at low temperatures).

6. **"Ravi Shakti Apakarshana darpana (glass)"** A special glass concentrating (visible) light energy in sun light.

Status :

Already produced and study of optical properties is not yet done.

7.	**"Ushna Shakti Apakarshana darpana (glass)"**	A special glass for concentrating the heat energy in sun light

Status :

Fully deciphered and to be produced in the laboratory.

8.	**"Badhira Loha"**	A sound proof alloy

Status :

Fully deciphered and to be produced in the laboratory.

9.	**"Vidyut darpana"**	A special glass that has capability to neutralize electrical discharges as lighting.

Status :

Fully deciphered and to be produced.

10.	**"Raja Loha"**	A high-heat absorbing alloy used for the bodies of various flying craft.

Status :

Fully deciphered and to be produced in the laboratory.

11.	**"Rudanti Mani"**	A special material

Status :

Fully deciphered and to be produced in the laboratory.

12.	**"Rutika Mani"**	A special material.

Status :

Fully deciphered and to be produced in the laboratory.

13.	**"Abhra Mrid darpana"**	A special mica glass.

Status :

Fully deciphered and to be produced in the laboratory.

14.	**"Sunda Mrit Kacha"**	A special glass.

Status :

Fully deciphered and to be produced in the laboratory.

15.	**"Pingala Adarsha"**	A special glass.

Status :

Fully deciphered and to be produced in the laboratory.

16.	**"Somanka Loha"**	A special alloy.

Status :

Fully deciphered and to be produced in the laboratory.

17.	**"Ravi Shakti Apakarshana darpana"**	A special glass with solar heat collecting properties.

Status :

Fully deciphered and to be produced in the laboratory.

18.	**"Hatakasya Loha"**	A copper alloy with golden appearance.

Status :

Fully deciphered and to be produced in the laboratory.

19.	**"Vata Stambhana Loha"**	A copper, iron, lead alloy.

Status :

Fully deciphered and to be produced in the laboratory.

20.	**"Ghantarava Loha"**	An alloy that has high sensitivity to different types of sounds.

Low cost Protein rich food from Grass

In addition to the above materials, a formula for producing a protein rich food extract (powder) from specific Indian grasses is also deciphered. The Central Food Tech. Research Institute, Hyderabad has certified that the powder extracted from the specified grasses shows about 13% protein content. Other tests from nutrition and medical angles are to be taken up. This activity is aimed at producing low cost protein rich food products (as powder, biscuits, malt etc.) based on this formula.

Agni Sthambhana or Fire Resistance

In additional to materials produced or deciphered as above, a technique for preventing and resisting fire and burning is also developed. Two techniques solutions have developed for Agni Sthambhana or Fire resistance :

(a) for preventing burning of inflammable objects as paper, cloth and wood

(b) for preventing burning of human body.

It has been noted that in both the above cases the fire will not be allowed to be caught (or burning to start) even after continuous exposure to flame for upto 30 seconds (Normally fire catches any inflammable material with 0.5 seconds and any moist material within 3dampened with this liquid do not catch fire and can also be used to put off fire or escape unburnt in fire even after long exposure to flame).

5. Anahara or avoiding food

A recipe for a special type of biscuit has been developed. this biscuit when consumed upto 50 g dose can help overcome hunger and skip a meal for about 3 - 4 hours. Upon medical trial, was found very useful in obesity and diabetic cases.

(Note : items 4 & 5 are developed independent of Vimana Shastra)

6. Details of individual and organisations involved & acknowledgements

Acknowledgements with deep gratitude are due to the following individuals and organisations who have been contributing in Aarious capacities and roles for the success of this project :

	Organisation	Person involved	Role
1.	Birla Science Centre, Hyderabad	Dr. B.G. Siddharth Director, Birla Science Centre	Committee member (of the three member committee for investigation into Ancient Indian Science & Technology) and had played crucial role in arranging various resources for the project.
2.	Birla Institute of Scientific Research (BISR), Hyderabad	1. Late Dr. M.C. Ganorkar Director, BISR, Hyderbad and 2. Mr. Ranga Madhavan (then Doctoral student)	Committee member (of the three member committee for investigation into Ancient Indian Science & Technology) and had played a crucial role in terms of conducting various experiments and all activities related to Chemi-try in terms of

			preparation of materials as per formulae in the Shastras, etc., and also testing for Chemical properties of materials produced in the laboratory BISR, Hyderabad.
3.	BHEL-CTI, Bangalore	Dr. B.K. Chandrasekhar Sr. Manager, CTI	Assistance in melting at high temperature.
4.	D.M.R.L.	Sri R.B. Subramanian Project Director	For assisting with providing facilities in DMRL.
5.	Punarvasu Arogya Kendra, Secunderabad	Dr. Ram Niwas Sharma Dr. Surendra Sharma DR. K.G. Sharma	For assisting with Ayurvedic information in decipherment of various formulae.
6.		Dr. B.V.S. Subba Rao	For assisting with ceramics related information and also interacted with Indian Ceramics Society.
7.	N.G.R.I.	Dr. Y.V. Ramana	For performing mechanical tests on alloys produced at NGRI, Hyderabad and producing valuable data
8.		Sri S.R. Sharma, (Retired Director of Mines & Geology, Govt. of A.P.)	For advice from time to time in the process of decipherment.
9.	Maharaja Palace Manuscript Library, Jaipur	Head	For permitting to use the library of Ancient manuscripts in Sanskrit.
10.	I.R.D.E., Dehradun	Head	For offering test facilities.
11.	Shri Khuleesh Kothari, Jaipur		For offering support to this work.
12.	Prajna Bharati, Hyderabad		For sponsoring and arranging resources.

Conclusion

This investigation provides the basis for identifying the veracity of the description given in source of the Sanskrit texts dealing with materials. While it is not clear how these texts originated or whether they are authentic, the experimental approach only attempted in finding out the validity of the description of preparation of some special materials such as alloys, glasses, ceramics, etc., Even though some person have hinted at the whole text being a modern work, composed by the oral deliverer himself, the experimental results uphold the veracity of the textual contents and also indicate lack of availability of these materials in modern times, thereby hinting at an ancient historical origin.

* * *

2

Metallurgy as envisaged in Sanskrit texts with special reference to Arthaśāstra, Manusmṛti and Vimānaśāstra

Prafulla Kumar Mishra

The thrust of this paper is to highlight some of the striking source of metallurgy in Sanskrit studies.

I. Introduction-

Like all other sciences metallurgy, the study of matter, mines and jewels in the early days was very common. As a result of which metals were very much in use. Therefore lifting, processing and purifying were taking place from time immemorial which runs back to Vedic period as follows.

Rig-veda:

In Rig Vedic period, there are reference to words like *nidhi, khanitram, musala* and *ulukhala* etc. which are used in activities of mines

Yajurveda

Following mantras are quite objective on the mineral activities.

1. *namas takṣebhyo rathakārebhyaś ca...* 16.27
2. *namaḥ kulālebhyaḥ karmārebhyaś ca* 16...27
3. *aśmā ca me mṛttikā ca me girayaś ca me parvatāś ca me sikatā ca me vanaspatayaś ca me hiraṇyam ca me 'yaś ca me śyāmaṁ ca me lohaṁ ca me śisaṁ ca me trapu ca me yajnena kalpantam* VS 18.13

Daśakumāracarita refers to *asura-vivara ṣvyasani* who is an expert of exploring mines.

Modern scholars fix the date of earliest knowledge of metallurgy as 2500-1800 B.C which was famous in north-west India and Rajasthan. (see the table attached.]

The text on metallurgy

It is difficult to find out texts on metallurgy. But indirect references are many.

II. References of smṛtis-

Basically reference to purification of *yajna patra* or utensils in sacrifice and others are the occasion for metallurgy. In this connection all the metallic utensils made of bell metal, copper, silver etc. But many more authors like Parasara etc give only short remark on purification. The context of the utensils are only for sacrificial purpose. But *dravyasuddhi adhyaya* is a common phenomenon in *dharmaśāstras*. Mention bellow are some of the references of from Manu and Yājñavalkya.

1. 2500-1800 BC	Copper-Bronze technology Cire-perdue Process Copper mines found	North-west India Rajasthan Many parts of India
2. 1800-1000 BC	Use of Copper tools Copper-hoards	Some parts throughout India
3. 1400 BC	Discovery of the Earliest iron object	Ahar in Rajasthan
4. 1200-1100 BC	Copper rich areas with Copper smelting Traditions	Ahar, Nho in Rajasthan also in U.P. and Bihar
5. 4th Century BC	King Purushottam presents Indian made steel to Alexander	Taxila
6. 4th Century BC	Making of pure zinc by Distillation Process	Rajasthan
7. 370-375 AD	(Delhi) Iron pillar fabricated. It was shifted to Delhi in 1050 A.D.	Mathura (UP)
8. 5th Century AD	Huge Copper Statue of Buddha 7ft. 6 inches High & Weighing 1 ton, Fabricated in two Layers	Sultanganj (UP)
9. 8th Century AD onwards	Wootz Steel for making high quality swords	Export to Damascus, Syria
10. 13th to 18th Century AD	Large scale production of Zinc. Transfer of technology	Zawar-Rajasthan To England

The references are as follows:

eṣa saucasya vaḥ proktaḥ śarīrasya vinirṇayaḥ

*nānāvidhānam dravyānām śuddheḥ sruṇuta nirṇayam.*5/110

taijasānām manīnām ca sarvasyāsmamayasya ca

*bhasmanādbhirmṛdā caiva śuddhirukta manīṣibhiḥ.*5/111

apāmagnesca samyogāddhaimam roupyam ca nirbabhou

*tasmāttayoḥ svayonyaiva nirneko guṇavattaraḥ.*5/113

tāmrāyāḥ kaṁsyaraityānām trapunaḥ śisakasya ca

śoucam yathārha kartavyaṁ kṣārāmlodakavāribhiḥ. 5/114

Yājñavalkya holds:

souvarna-rajatābjānām-urdhvam pātragrahātmanām.

śākarajjumulaphalavasovidalacarmanām //182 p-254

There within three karikas (ch. v) Manu could speak of purification process of gold, silver, cupper, lead, bell metal, and other alloys. The use of purification process is quite known to the society which is indicated by Manu. It means the primary processing and purifying the metals by different rays. As per example, gold and other stones/ores are purified by fire. Secondly there are processes of purification through use of water. There are different types of clays used for the glaze of metals.The application of salt,acids are used for the purpose which testifies the knowledge of chemical processing of metals. Yājñavalkya endorses the view of Manu in this regard.

III. Kauṭilyas' Arthaśāstra

Kauṭilya refers to mines and mineral products in chapter of *adhyakṣa pracāraḥ*. Kauṭilya in the context of examination of the precious articles to the received in to the treasure describes the qualities of precious stones and jewels in details which testifies the use of different valuable gems in a particular order and for which they are named differently. They are as follows

(Section 29 examination of the precious articles received in to treasury II.11.)

A. Precious stones and gems.

Pearls were collected from Tamraparne, Pandyaka vata, Pasika, Kula, that from the Curni, (Mt.) Mahendra, Kardama, Srotasi, Lake, and Himavat. He is of opinion (see sec.4 of II.11),

"A pearl that is lentil-shaped, triangular, tortoise shaped semi-spherical, with a layer, coupled, cut up, rough, spotted, gourd-shaped, dark, blue, and badly perforated, is defective.

(that which is) big, round, without a flat surface, lustrous, white, heavy, smooth and perforated at the proper place is excellent." He classifies the types of necklace made out of it:

1. *śirsaka, upaśirsaka, prakāndaka, avaghataka,* and *taralapratibaddha*,-these are varieties of (pearl) strings.
2. One thousand and eight strings make the *indracchanda* (necklace).
3. Half that (number) make the *vijayacchanda.*
4. Sixty-four make the *ardha-hāra.*
5. Fifty-four make the *rasmi-kalāpa*
6. Thirty-two make the *guccha.*
7. Twenty-seven make the *nakṣatra-mālā.*
8. Twenty-four make the *ardha gucha.*
9. Twenty (string) make the *mānavaka.*
10. Half that number makes the *ardha mānavaka.*
11. The same with a gem in the centre, it is the *mani-sopānaka.*
12. A necklace of only *sirsakas* is an unmixed *hāra.*
13. With a gem in the centre, it is an *ardhamāṇavaka.*
14. One with these strips is a *phalaka hāra* or one with five strips.
15. A single strings is the unmixed *ekāvalī*
16. The same with a gem in the centre, it is *yaṣṭi.*
17. Variegated with gold and gems it is the *ratnavali.*
18. With gold gems and pearls at intervals, it is *apavartaka.*
19. With stringing in a gold thread it is the *sopānaka.*
20. With a gem in the centre it is the *maṇi-sopānaka.*
21. By that are explained varieties of strings and net-work for the head, hands, feet and waist.

The gems from oversea and their identification are quite interesting.

1. Gems come from Koti, from the Mālā, and from beyond the sea.
2. The ruby, of the colour of the red lotus, of the colour of saffron, of the colour of parijata flower, and of the colour of the morning sun.
3. The beryl, of the colour of the blue lotus, of the *sirisa* flower, of the colour of water, of the colour of green bamboos, of the colour of the parrot's wing, yellow coloured, of the colour of cow's fat.
4. The sapphire, with blue lines like the kalaya flower, deep blue, having the lustre of the jambu fruit, having the lustre of the dark cloud, the delighter and the streaming interior.
5. The pure crystal, of the colour of mulata flower shedding a cool shower and the sun stone. These are the gems.
6. Hexagonal, square or round of a flashing colour, having a suitable form, clear, smooth, heavy, lustrous, with lustre inside and imparting lustre these are the excellences of gems.
7. With a dull colour and lustre, with grains, with a hole in the bloom, broken, badly bored, and covered with scratches, these are blemishes.
8. The *vimalaka*, the *śasyaka*,the *añjanamulaka*,the *pittaka*, the *sūlabjaka*, the *rohitākṣa*, the *mṛgāsmaka*, the *jyotirāsaka*, the *māleyaka,* the *ahicchatraka,* the *kūrpa*, the *pratikūrpa*, the *sugandhikūrpa*,the *kṣiravaka*, the *śūktivurnaka*, the *śilapravālaka,* the *pulaka*, the *śūklapulaka*, these are subsidiary types of gems.
9. The rest are glass crystals.
10. Diamond comes from Sabha-rastra, from Tajjama-rastra, from Kastira-rastra, from (Mt.) Srikatanaka, from (Mt.) Manimanta and from Indravana.
11. The mine, the stream, and miscellaneous are the sources.
12. Like the cat's eye the sirisa flower, cow's urine, cow's fat, pure crystal, the mulata flower, and of the colour of any of the gems, these are the colours of diamonds.
13. A diamond that is big, heavy, capable of bearing blows, with symmetrical points, capable of scratching a vessel, revolving like a spindle and brilliantly shining is excellent.
14. That with points lost, without edges and defective on one side is bad.
15. The coral from Alkanda and from Vivarna red and the colour of lotus is excellent.

B. Mines and ores

With above mention of precious gems there is detailed description of mines and affairs of metallurgy in the next chapter under *akaradhyksa,*. The director of mines is well-versed in the concerned discipline. Kauṭilya says,

"The Director of Mines being conversant with the science of (metal) venue in the earth and metallurgy, the art of smelting and the art of colouring gems, or having the assistance of experts in these, and fully equipped with workmen skilled in the work and with implements, should inspect an old mine by the marks of dross, crucibles, coal and ashes, or a new mine, where there are ores in the earth, in rocks or in liquid form, with excessive colour and heaviness and with a strong smell and taste".

C How to identify a mine or mineral products:

Kauṭilya quite naturally discusses the importance of the superficial identification of mines. Mentioned bellow is the salient points.

1. Those liquids that flow inside a hole, a cave, a table land at the foot a rock cut cave, or a secret dug out in mountains whose regions are known (to contain gold-ore),that are of the colour of the jambu, the mango or the palm fruit or of a cross section of ripe turmeric or of jaggery of orpiment or red arsenic or honey or vermilion or white lotus or of the feather of a parrot or a peacock, that have in their environs water and plants of the same colour, and that are viscous, clear and heavy, are gold bearing liquids.

2. If when thrown in water, they spread on the surface like oil and absorb mud and dirt, they are capable of transmuting copper and silver up to one hundred (times their own weight.)

3. Exactly similar to that in appearance, (but) with a strong smell and taste, he should know to be bitumen.

4. Ores in earth or rocks, which are yellow or copper coloured or reddish yellow, which, when broken, show blue lines or are of the colour of the midge or mesa bean or kraal, which are variegated with spots or lumps as of curds, which are of the colour of turmeric or microbalance or lotus leaf or moss or liver or spleen or saffron, which when broken, show lines, spots or swastikas of fine sand, which are possessed of pebbles and are lustrous, which, when heated, do not break and yield plenty of foam and

smoke are gold ores, to be used for insertion, as transmutes of copper and silver.

5. (Ores) of the colour of a conch-shell or camphor or crystal or butter or a dove or a pigeon, or vimalaka(gem) or the peacock's neck, Or of the colour of assay(gem) or comedic gem or jiggery or unrefined sugar, or of the colour of the flower of caviar or lotus or petal-flower or Kalama flower or the flower of flax or of linseed, and containing lead, containing antimony smelling like raw flesh, when broken either black with a white shimmer or white with a black shimmer or all variegated with lines and spots, soft, which when being smelted do not split and yield plenty of foam and smoke are silver ores.
6. In the case of all ores, when there is increase in heaviness there is increase in metal content.
7. Of these, those that are impure or dim in the interior flow in a pure form when infused in strong urine and caustic, when formed into lumps with (a paste of) the remarks, the banyan tree, the pilu, cow's bile, pigment and the urine and dung of buffalo the donkey and the young camel, either mixed with this(during boiling) or smeared with this(paste)
8. Ore from rocks or a region of the earth, which is heavy, unctuous and soft (and which is) tawny, green, reddish or red (in colour) is copper ore.
9. That which is crow black or of the colour of the dove or yellow pigment or studded with white lines and smelling like raw flesh, is lead ore.
10. That which is grey like saline earth or of the colour of a baked lump of earth is tin ore.
11. That which is made up mostly of smooth stones, is whitish red or of the colour of sinduvara flower is iron ore.
12. That which is of the colour of *kakanda* (row's egg) or birch leaf is *vaikrntaka* ore.
13. That which is clear, smooth, lustrous, possessed of sound, cold, hard, and of a light colour is gem ore.(ch, II.12)

D. Process of softening and hardening:

Kauṭilya was aware of the process of hardening and softening by application of some liquid mixture prepared in the countryside. This knowledge could enable the sculptors to carve small and thin lines on the hard granite stone which is evident from the temple walls of different famous temples. As for example in the Sun temple of Konark the granite images of god Surya wears cloth where small motives are engraved like a thin silken cloth.This is not possible by any instrument unless it is soften

Kauṭilya mentions the process bellow.

1. The insertion (in the ore) of the bulbous roots of the kadali and the vajra, along with the caustic (of the ashes) of barley, masa-been, sesamum, palasa and pilu or along with the milk of the cow the goat, is productive of softness.
2. Honey and liquorices, goats milk with sesamum-oil, mixed with ghee, jaggery and fermenting stuff, together with the kandali, with only there infusions in this, a (metal ore) that may have been broken even a hundred thousand fold becomes soft.
3. The insertion of (the power of) the teeth and horns of the cow is the remover of softness(in metals)

It testifies that some detailed process was probably available at his time.

E. Lifting and utilization of ores

1. What is produces from ore he should put to use in factories for the respective metals.
2. He should establish trade in manufactured gold in a single place, and (lay down) a penalty for those who manufacture, purchase or sell elsewhere.

F. Protection and law against theft and rubbery of mines.

1. He should make a miner who robs pay eight fold, except in the case of jewels.
2. The thief and the person who lives (by mining) without permission, he should bind and force to work (in the mines) the person who offers personal labour in place of a fine.

G. Policy of mines to lease, lift and mint.

1. He should let for part share or on lease a mine that is burden some in point of expenses or working; a light one he should work himself.
2. The Director of Metals should establish factories for copper, lead, tin, *vaikrntaka*, brass, steel, bronze, bell metal and iron, also (establish) trade in metal ware.

3. The Mint Master should cause to be mined silver coin with one forth part copper and containing a hardening alloy one masa (in weight) of one of the following, (viz)iron, tin, lead, and antimony,(of the denomination of) one pana, a half pana, a quarter pana, and a one eight pana ;(further) copper coins with one quarter sustenance (of a n alloy),(of the denominations of) one masaka, a half masaka, a kakani and a half kakani.
4. The Examiner of Coins should establish the currency of panas, for trade and for receipts in the treasury.
5. (He should also fix) a coining fee of eight per cent, a commission of five per cent, an inspection fee of one eighth per cent, and a penalty of twenty five panas for those who manufacture, purchase sell and examine in other places.
6. The Superintendent of Mines should establish factories for (articles of) conch shells, diamonds, gems, pearls, corals, and caustics as well as commerce in them.
7. The salt Commissioner should collect at the proper time the share of salt as released after crystallization as well as the lease rent, also the price, the inspection fee and the surcharge from the sale.
8. Thus from the mines he should collect the price, the share the surcharge the monopoly tax, penalty, duty, compensation, fine, inspection fee and manufacturing charges as well as the twelve kinds of metals and commodities(made from them) In this way he should fix the collection (of income) under various heads in the case of all commodities.

Kauṭilya is quite objective of the use of mines and explore them in proper way and solves the legal intricacies occurred there of. He attaches importance to the treasury which can be nourished and full by the revenue of mines.So he says;

"The treasury has its source in the mines; from the treasury the army comes into being. With the treasury and the army, the earth is obtained with the treasury as its ornament." (ch 12)

H. Identification and manufacture of gold and silver.

1. In the middle of the market highway state by superintendent of gold should establish the goldsmith, skill in his profession, of noble birth and trustworthy.

2. That form the Jambu river, that from (Mt.) Satakumbha, that from hataka, that from venu, that produced in srngasukti that found in a natural condition, that transmuted by means of liquids and that produced from mines, these are types of gold.
3. Gold of the colour of the lotus filament, soft, smooth, not producing a sound and lustrous is best, reddish yellow is of a middling quality, red of the lowest quality.
4. Of the best varieties, the pale yellow and the white are impure.
5. He should cause that because of which it is impure to be removed by means of lead four times that quality.
6. If it becomes brittle by the admixture of lead, he should cause it to be smelted with dried lumps of cow dung.
7. If it is brittle because of its own roughness, he should cause it to be infused in sesamum oil and cow dung.
8. Gold produced from the mines, becoming brittle by the admixture of lead, he sold turn into leaves by heating and causes them to be pounded on wooden anvils, or should cause it to be infused in the pulp of the bulbous roots of the kadali and the vajra plants.

H.1 Silver

1 That origination in Tuttha, that from Gauda, that from Kambu and that from Cakravala, these are types of silver.

2 Silver which is white, smooth, and soft is best..

3 The opposite kind and that which tends to burst is bad.

4 He should purify that with one-forth part of lead.

H.2 Purification:

This knowledge is also popular even today which it appears older than Kauṭilyas period.

1 That in which a crest has appeared at the top, which is clear, lustrous and of the colour of curds is pure.

2 One suvarna of pure, turmeric colored gold is the standard.

3 Thence by the substitution of one kakani copper onward (in a suvarna of gold) up to the limit of four (masakas), sixteen standards (are obtain)

4 After first rubbing the gold to be tested on the touchstone, he should afterwards rub the standard gold on it.

5 That with a streak of the same colour as the standard on places on the stone that are neither depressed nor elevated, he should know as properly tested; what is over rubbed or lightly rubbed or powdered over with red chalk from underneath the finger nail, he should know as deception.

6 Gold touched with the forepart of the hand smeared with jati vermilion or sulphate of iron dissolved in cow's urine, be comes white.

7 The streak (of the gold) in the touchstone, that has filaments, is smooth, soft and lustrous if best.

8 Stone from the Kaliṅgas or from the Tapi which is or the colour of the mudga-bean is the best touch stone.

9 That which shows the exact colour is advantageous for either sale or purchase.

10 That which has the colour of an elephant, with a green tinge, is over sensitive to colour and is advantageous for sale.

11 That which is farm, hard and of uneven colour, is not sensitive to colour and is advantageous for purchase.

12 The cut of gold that is strictly, of even colour, smooth, soft and lustrous is best.

13 The heating that is uniform outside and inside, and has the colour of the lotus filament or the colour of kurandaks flower, is best.

14 That which is dark or blue shows impurity.

Needless to mention how weights and measures and the law there of is laid down in detail. Besides the process of ornaments and the varieties are all comes under this context.

However metallurgy with all its aspects is known even before Kauṭilya which he collects them in Arthaśāstra.

Kangle quotes many texts like Bṛhat-samhitā, Kalāvilasa of kṣemendra, Gajakumbhādikam, Vārhaspatya of Bhattasvāmin etc. for the references. (see the respective footnotes of R.P.Kangle's edition of Arthaśāstra vol II, Univ. of Bombay, 1972)

IV. Use of metallurgy from *vimana„Èstra*

The references of texts are quite objective in *vimanśāstra* of Bharadvāja. There are as good as 97 types of texts referred by him on flying and its allied fields. In the connection to *vimanas* only, the necessary objectives are shorted out of their well-known *sutras*. There is a list under this head:

"*vaimanika prakarane nirdiṣṭa purātana grantha-suci :*

A. For iron:-

1. *Loha tantram*
2. *Loha prakaranam*
3. *Loha sarvasvam*
4. *Lohatattva prakaraṇam*
5. *Loharatnākaraḥ*
6. *Lohadhikaraṇam*
7. *Loharahasyam*
8. *Lohaśāstram (śākaṭāyana kṛta)*

B. Other allied texts

1. *Darpaṇa pakarana*
2. *Darpaṇaśāstram*
3. *Mukura kalpaḥ*
4. *Dhatu sarvasvam (bodhāyana kṛta)*
5. *Kunda kalpaḥ*
6. *Kunda nirnayaḥ*
7. *Bhastrikā nibandhanam*
8. *Maniratnākara*

Besides, authorities are quoted and texts are referred from the texts. In this connection only loha and allied texts for metallurgy are taken.

C. Texts:

1 Saunakīya sūtram

2. Manibhadra kārikā
3. Loha tantram
4. Lohakalpaḥ
5. Loha rahasyam
6. Nāmārtha kalpaḥ
7. Viśvambhara kārikā
8. Samskāra darpaṇaḥ
9. Darpaṇa prakaraṇam
10. Loha ratnākaraḥ
11. Nirṇayādhikāraḥ
12. Mūsā kalpaḥ
13. Kunda kalpaḥ
14. Kunda nirṇayaḥ
15. Bhastrikā nibandhanam
16. Mukura-kalpaḥ
17. Kriya-sāraḥ
18. Darpaṇa-kalpaḥ
19. Sammohana kriyā kāṇḍam
20. Aṁśubodhinī
21. Prapañca sāraḥ

D. Authors:

1 Śaunaka
2 Maṇibhadra
3 Sāmba ācārya
4 Lalla
5 Nārāyana
6 Dhundinātha parāṅkuśa
7 Visvambhara

It would be interesting therefore to search out some of the important facets on use of metals and its purification there of.

E. The sutras of *vimana„Èstra* are as follows:

1. *athā yāna lauhāni* I.13.

2. *tatcchuddhir yathāṁśo dhanādhikāre* I. 14

3. *uṣmapās trilohamayaḥ* II.1

4. *melanāt* II. 2

5 *pañcamāt dvitiye* I.

6. *atha saptame nava* II. 4

7 *syād bhastrikā 'ṣṭame ṣodasi* II.4 (I. 12)

8. *darpaṇas ca* III.1.

F. Types of iron

saumaka *sauṇḍalika* *maurtvika*

These three constitute of usma loha which is 16 types They are like :usnam bhara, usnapa, unṣa-dduna, rājāmatrt, vīrahā, pañcaghna, agnitṛt, bḥārahan, śītahan, garalaghna, amlahan, viṣambhara, viṣalyakṛt, dvijamitra etc.

1. The above three types of iron mixed together form a *usma* type of iron that holds good due to its light weight,for making of air ṣcraft.

2. Types of iron and its location in the mines:

Regarding the identification of the good iron the opinion of Samba acharya is quite informative.

Identification of saumaja type of iron Out of 7 layers of iron ore the third one is saumaja. The opinion of Samba and Lohatantra are one.

Samba:

bhūgarbha sthitakhānijarekhapanktiṣu saptame // 83

tṛtiyakhanijasthā je te lohās saumajātayaḥ //

Bodhananda taking the help of *Lohatantra* says that,

rekhā saptamasya tṛtīya khanijalohāḥ pañcaśaktimayās saumajātīyas te vījalohā iti //

Again he takes recourse to *Lohakalpa* for types of iron. He is quite sure of the nature of the mineral deposits. He holds,

bhugarbha khanija rekhā tri-sahasrādhikā smṛtā /

trimsatottara sahasra rekhās steṣuttamaḥ kramāt // 85

There are more than 3000 layers of ore of which up to 1300 layers are good. Out of them in the seventh layer there are 27 minerals. Out of them three mineral / ores are the *garbhakoṣa or the core mines having* 5 *sakti*s. The iron with5 saktis are named as sauma, sandala and maurtika. This origin is as per *Loharahasyam*. Atri holds that there are 14 *vija-loha* originated from the earlier one prospectively. They are:

Sauma- saumyaka - sundāsya - soma- pañcanana- usnari - uṣmapa- sṛṅga- saunḍira - lāghava - urmiśa - prāṇana - śaṅkha - kapila

saumya produces *saumyaka* and *saumyaka* produces *sundasya* etc.

All the seed iron or *vijaloha* has powers of four types. The divisions of power on each division group/ varga are of 167768 types. This has been determined in *viśvambhara kārikā* (102- 106) on the basis of *vālmikigaṇita*. Subsequently the force of each alphabet sa, u, and ma are calculated which have force of *sūrya, varuṇa, indu* and *candramā*. Martaṇḍa and the bhūtas produce 71th force. A force of *bhuta* named *rucikā* is 106th., Similarly the force of *nakṣatra* named as *bhauma* is 101th. Again among *dhruva* group in soma vadava force named as indu-candramā is 109th. (see kārikā 107 - 116)

Loha rahasyam describes the śaktis which speak of the ingredients respectively. The words are used in a symbolic way as follows;

kūrma- centrifucal force./ centre of gravitational pull.

kaśyapa- outer orbital force.

mārtaṇḍa- the flow of solar rays.

*bhūta - tanmātrā*s- specially the flow of wind.

bhā - graha śakti/ planetary force.

arka - internal attractive force of sun.

iṅdu - moon

vādava - The holding force in between the moving time or the sun and the earth etc.

All of them have mineral portions in their force in the proportion of 3, 8,11,5,2,6,4 and 9 respectively. Again a conglomerated force called *vādava-śaktimelana* is of 14th part. Both of them mix together. Thus 4 groups of forces being inter mingled in the core of the mine (garbha-koṣa) [due to natural pressure] in *kālapāka* or being heated in time becomes *soma* type of iron. Similarly 8 forces are thus visible being intermingled.

evam catvāri vargastha śaktayas tāḥ parsparam /

khanijanām garbhakoṣe militvā kālapākataḥ // 113

After the description of *soma* type of iron he then describes the process of other types like *saundala* [*kārikā* 127-133]etc.. The number of *sakti* stands for a unit which needs further study.

G. The process of purification

The purification process is spelt out on the basis of *samskāra darpaṇavidhiḥ*. The aphorism says,

tatsuddhir yathāṁśodhanādhikāre I.14

***sauma* iron**:

saumaloham samāhṛtya pācake samprapūrayet /

saptavimsati kakṣosṇavegāt sampācayet dravat (drutāt?)// 139

jambira-likuca-vyaghra-cinca-jambu-rasais tatha /

vistrtasyena nalayantre pacayed divasavadhi // 140

tat saṁgṛhyātha vidhivat kṣālayitvā tataḥ param /

pancatailais catur dravai kasais saptabhis tatha // 141

pṛthak pṛthak gālayitvā loham paścāt samāharet /

Ore put in the woven needs 27 *kaksa* of heat speed. / degrees {the unit of degree is not known again the speed of heat }.The juice of different fruits are applied there and heated for the whole day. They are:

jambira (orange), *likuca* (soft sour fruit),*vyāghra (karanja* or red *eranda caster), cinca*, (turmeric) and *jambu*.

All of them are applied in a wide faced pot Then being washed apply 5 oils and 4 *drava* like tanka etc. and gradually put out in different pots and get different *lohas* accordingly.

The 5 oils of gunjjā, kañjala / kañjara / amvala, cañcū / eraṇḍa, kuñjara / pippala/ kantha kucai and karanja fruits.

The 4 dravas of pranaksara / nausadara, viranci / sajjiksara, kancuki ṣyava / yavaksara/ khuraksara

Purified by hingu, parpati, -parpatipadmāvati sugandhadravya, ghontikā,- supāriphala, jatāmāmsi- bālāchada, vidārānginī- kanigara gaṇdha bṛkṣa or vidāri kanda, matsyākṣi maccecci, raktakaṇḍha kuvari red coloured thuhar thorns

Similarly the process of purification of saundāla and maurtvika are explained.

The process of purification is organic and nature friendly.

5. Conclusion

It is difficult to conclude on the topic as the information is not exhaustive. Yet it could be safely deducted from the above description that metallurgy was a science which was not only developed but also applied thoroughly by different authorities.

References:

- Arthaśāstram, University of Bombay, 1972
- Arthaśāstram, Bharatiya Vidya prakashan, Delhi, 1984
- Cultural Heritage of India, vol.VI, Calcutta, 1986
- Manusmṛtiḥ, - MLBD, Delhi 2000
- Samskṛta-vijñānavaibhavam-Rastriya Samskṛta Vidyapitha, Tirupati, 2001
- Samskṛtame Vijñāna, Vidyadhara Sharma Guleri, Samskṛta Bhāratī, Delhi, 2000
- Vṛhat- vimāna- śāstra, Sarvadeshika Arya Pratinidhi-sabha, Delhi, 1959
- Yājñavalkya smṛtih- Parimal Publications, Delhi, 2000

3

Metallurgy in Sanskrit Shastric Literature

Dr. V.K. Didolkar
and
Mr. M.K.Kawadkar

1. Introduction

The existence and flowering of a great civilization, know as Sindhu civilization is acknowledged as the genesis of Indian culture. This civilization is well known to have possessed the knowledge and the skills of the processing of materials. This civilization flourished in the hunge land mass of India and was responsible for the innovations in the development of materials of metals. The publication of De re Metallica by Geogius Agricola in 1556 A.D. is a major landmark as it represents of the first acknowledged treatise on the metal extraction and processing. The existence of various ancient Indian texts predating this publication, is only now gaining recognition.

These are numerous fascinating monuments of stones in India such as Kailasa at Ellora, Cave temples, in Western India at Ajanta, Ellora, Elephanta etc. Its is interesting to think of the infrastructure that think of the infrastructure that was necessary to carve or build these magnificent works of art. It will not take us long to realize that our cultural heritage reflects the strong base our ancient technology. Indian expertise in gold, silver, copper, brasses and bronze, zinc, lead, iron, and steel etc. is acknowledged by one and all.

2. Decorative applications of Metal Powders since Vedic Period:

The earliest application of very fine metal powder was in the manufacture of various colours for painting. Metals like gold, silver, copper tin etc. were used for making of dyes of various shades and colours [(1)] It is quoted in the text that metal powders were produced by thinning the metal pieces into leaves/ foils. The later texts Manasollasa [(2)] and Silparatna[(3)] give full details for the production of metal powders by comminution techniques. Panning technique was repeatedly used for the purification of metal powders and were then used as a paint after mixing with suitable binders. The morphological properties of the superfine and flaky metal powders are noteworthy from the viewpoint of their application as colourants in paints and dyes.

Filing was another method of making metal powders by mechanical comminution Rasaratnasamuccaya[4] gives the details of producing iron/ steel powders for the manufacturing of 'Iron Bhasma.'

Grinding in a mortar with a pestle was also a method for making metal power. Rasendrasar Sangraha[5] states 'Thin leaves of iron/ steel were ground in an iron/ steel mortar by an iron/ steel pestle.'

"Lauhe drsadi lauhanca mudgarena hatam muhuh" This iron/ steel powder was used for manufacturing 'Iron Bhasma.'

3. Applications of Metal Powders of Making Explosives:

'Sukraniti'[6] gives a formulation for making explosive mixtures. These were prepared by mixing charcoal, sulphur, potassium nitrate, mainsila, haratala, lead, oxide, hingula, steel powder, camphor, lac, rala and gum.

Explosive mixtures required for guns were discussed in details in the text giving various combinations of the above said materials.

There are few available in Atharvaveda[7] wherein the reference is made to lead. Though, there is no clear mention of the techniques of producing lead granules / shots, there is a evidence, suggesting the use of lead granules/ shorts in meeting the punishment to the criminals. In the documentary text 'Ain-I-Akbari' (Sixteenth Century A.D.), there is a clear evidence of the use of atomization technique to produce lead granules/shots, which is the earliest reference of atomization of liquid metals.

4. Medicinal Applications of Metal Powders:

The most important application of metal powders/ Bhasma was in the preparation of various types of ayurvedic medicines,. Kasyapa Saṁhitā[8] give the formulation for medicine to treat eye diseases, as follows–

' Equal amounts of pippali, adraka, maschi, anjana, triphala, sankhanabhi, saindhava and copper powder were mixed and formed into tablets which were used to treat eye- diseases.

Aswalayan Grihya Sutra [9] (1.15.1) mentions when a chilled is born, the father should feed him from a golden vessel with clarified butter and honey with which he has ground the gold dust. This rite was performed for the sake of longevity and the general health of the child.

A similar reference is also available in Manusmriti [10] as follows–

" Before the navel string is cut, Jatkarma ceremony must be performed for the newborn feeding it with gold (dust), honey and clarified butter."

Susruta (Sarir- 10.12)[11] also prescribes similar procedure for the newly born child.

Susruta (Sarir- 10.72-74)[12] discusses four formulations for tonics given to the child as follows–

1. Gold powder (pure), kustha, honey, clarified butter and vaca
2. Matsyaksaka, sankhapuspi, honey, clarified butter and gold powder.
3. Arkapuspi, honey, clarified butter, gold powder and vaca.
4. Gold powder, kaidarya, svetadurva, clarified butter and honey.

These tonics are useful for promoting physical strength and intelligence in the child. Each of these formulations contains well- prepared fine, pure gold powder (sufficiently fine to be suitable for oral application).

Another very interesting medicinal application of metal powders in ancient India, was in the treatments of poisoning. Caraka (cikitsa 23.10- 14, 234-236)[13] mentions the effect GARA poison on human body as producing variety of ailments like high fever, jaundice, weakness, increased palpitation, poor digestion etc. The text describes a treatment of feeding such patients with a small amount of very fine copper powder alongwith honey so that undesirable materials are moved by vomiting. Then these patients were fed with small amount of very fine (pure) gold powder with honey. Thus, use of gold powder eliminated the effect of poisons.

Some verses of Sukla Yajurveda and Atharvaveda [M] imply that the use of gold powder to promote longevity and better physical strength. The next describes gold powder giving longevity; imparting glow, beauty and better physical strength; uplift and leads to victory.

5. Use of Metal Powders as a Currency:

A very interesting application of meta powders in India, was it use as a currency. The documentary evidence by Traill clearly mentions this fact.

Traill G.W, [14] quoted in his report (prepared in 1830 A.D.) on the Bhotia Mehals of Kumaon (Himalayan range)–

"Gold is calculated (measured) by the 'Sarswo' or 'Phetang', equal to seven and half Masas. Gold dust separated into phetangs, each tied up in a bit of cloth, is used as a coin (currency). One phetang equals to eight rupees."

It must be realized that the use of any loose powder as a currency material is rather inconvenient as the chances of wastage during handling is high. A more convenient method would have been to used to compacts made from the powder, instead of loose powder as such. One of the reasons for using loose powder may be the availability of alluvial placer gold which was found in abundance in the river-beds of India.

6. Summary and Concluding Remarks :

The earliest reference to metal in the form of lead granules/ shots in ancient Indian texts dates, back to the period of Atharvaveda (C. 2400-2500 B.C). The use of fine powders of gold, silver, copper, tin etc:, as paint materials can be traced to as early 400-500 A.D., which perhaps represents in the very first application of the fine metal powders in India (and perhaps in the world).

From the literary evidences of vedic period, it appears that the foremost application of fine metal powder/dust, was in medicine.

A detailed study of the vedic literature indicates that alluvial placer gold/ stream gold, was found in abundance in the river–beds of Indians know the technique of washing the auriferous sands of these rivers and heavier gold powder was separated from the lighter sand and clay particles by using gravity separation techniques such as panning, jigging etc.

Fine gold powder or gold dust was an important ingredient of medicinal preparation for oral consumption for longevity and better physical strength and for improving intellect and memory as quoted in the above said texts.

The details of the manufacturing process of such fine metal powders are available in Indian texts composed in early ninth century A. D. onwards. The metal powder manufacturing techniques based in mechanical comminution was filing, grinding etc. During the same period, fine metal powder of gold, silver, copper, iron, (steel) were produced in a reactor by an amalgamation technique, to be immediately converted in situ into a very fine metal compound powder, known as 'Bhasma'. It is apparent that Indian were well conversant with the amalgamation technique and the superior reactivity of the fine powder thus produced by 9th century A.D. Indian texts in this period illustrate various alternative method of producing fine metal powders to be used for producing 'Bhasma', which was the basis of important medicines. One such techniques was the direct reaction with sulphur or sulphide of thin metal leaves, in the absence of air.

Although the concept of atomization of liquid metal is embodied in Atharvaveda, the description of making metal granules / shorts by atomization appears in the Indian exists of 16th Century A.D.

Indian texts composed in 9th century such as Rasaratnasamuçcaya, Rasendrasara Sangraha etc., mention various aspects of meal powder production and the characterization of powders such as classification of powders preparation of various types of sieves/ screens etc.

Documentary evidence shows that a weighed quantity of gold powder or dust in a piece of cloths was also in use as a currency.

The literary and documentary evidences can be made use of establishing chronological development of powder technology in ancient India. More over, it is apparent from these evidences that this technology has flourished in India through many ages. Certain aspects of powder technology were in existence as early as vedic period.

Refferences

1 Visnudharmottara Purana- III 40. 25-26 (400-500) Venkateshvara Press, Bombay, 1913.

2 Manasollasa- Vol. 2, 1.3.171-174 (1131 A.D) Oriental Institute, Baroda, 1967

3 Silparatna- 46. 124-128 (16th Century A.D.) Trivendrum Sanskrit Series, No. 98, Trivendrum, 1921.

4 Rasaratnasamucchya- Chater 5.106 (9th Century A.D) Meharchand Lachhamandas, Delhi 1982.

5 Rasendrasara Sangraha- Cited in Rasaratnasamucchya Meharchand Lachhamandas, Delhi 1982.

6 Sukranitit (4.7.194. 196) Risidevi Rooplal Kappor Trust, Bahalgarh, 1983.

7 Atharvaveda (1.16.2-4) Swadhyaya Mandal Paradi, 1958.

8 Kasyapa Samhita Page 238 Chaukhambha Varanasi, 1986.

9 Aswalayana Grihya Sutra Motilal Banarasidasss, Delhi. 1984

10 Manusmriti (2.29) Chaukhambha Sanskrit Series, Varanasi, 1986.

11 Susruta Samhita Meharchand Lakshmandas, Delhi 1989.

12 Caraka Saṁhitā Chaukhambha Bharati Academy, Varanasi, 1984.

13 Sukla Yajurveda & Atharvaveda (Vol. 1-5) Swadhyaya Mandal, Paradi, 1985.

14 Traill G.W. 'Asiatic Researchers' Vol. 17, pp. 24 (first published in 1832.) Reprint Cosmo Publication, Delhi.

4

Technical Sciences in Sanskrit

Prof. Rajendra Mishra

It would be really unjust to think that the divine language Sanskrit is deprived of technical Sciences and its illumines the grandeur of poetry only. Actually, right from the hymns of Ṛgveda to the 11th century A.D. Sanskrit has produced hundreds and thousands authentic works, related with various disciplines, other than *belles letters.* like the original streams of philosophical systems, drama and rhetoric etc. we can very easily trace the origin of all technical sciences in the Ṛgveda alone. In the context of India's battles with vṛtra, Namuci, Śambara, Bala, Rauhiṇa and Bṛngada etc, the war strategy has been highlighted along with the names of various weapons like Vajra and Dhanuṣ etc. India सरमा as his spy in order to know the where about of paṇis who had stolen the cows. Thereafter, the spoilt as iron-forts of his enemies. Moreover, he helps Sudāsa in the fierce battle, namely Dāsarājña.[1]

The medicinal value of Atharvaveda also cannot be denied. Our great Ṛṣis and Maharṣis, dwelling in dense forests, know the medicinal quality of each and every herb. Eventually, they became successful in developing the discipline known as आयुर्वेद, which even today proves to be the best pathy of curing any disease.

The Śatapatha Brāhmaṇa explains the constituents of clouds and elaborates the symptoms of the rain clouds. When the great poet Kālidāsa describes the cloud saying: धूमज्योति: सलिलमरुतां सन्निपात: क्व मेघ: the seems to borrow this concept from Śatapatha which holds:

Further, we can gather some other information from these hymns which prove the antiquity of our technical sciences. Ṛgveda delineates the boat of Bhujyu which could be waded about in the deep water through one hundred helms - शतारित्रां नावम् (ऋग्वेद १.११६.५) The Aryas of the Ṛgvedic era were frequent users of planes. The king Tugru sends his son Bhujyu through a cosmic vessel which was controlled by the अश्विनौ. In Ṛgveda (1.34.2) a chariot of Aśvins has been narrated which had three wheels and three pillars. It was a horse-less chariot and could float in air very easily.

---

1. शतमश्मन्मयीनां पुरमिन्द्रो व्यास्यत् दिवोदासाय दाशुषे। ऋग्वेद ४.३०.२०.

प्रति मदस्य वज्रं वाह्नोर्धुर्हत्वो दस्यून् पुर आयसीर्नि तारीत्। ऋग्वेद २.२०.८

Swami Brahmamuni, who edited the मंत्रसर्वस्व of महर्षि भरद्वाज has clearly stated in his preface to the वैमानिकप्रकरण that the work of Bharadvāja is based upon the sole testimony of the Vedic Mantras. Satyavrata Samashrami also holds the view that almost all the technical sciences are described in Vedas indirectly as well as quite minutely. The American writer Wheeler Villaxe has written in her book- *Sublimity of the Vedas* that Vedic Ṛṣis were absolutely adept in Aeronautics, navigation, electrolysis, Radiology and so on, so forth.

It is a well known historical fact that the Talapade brothers of India could achieve success in preparing their aeroplane in 1895 in complete 8 years prior to Right-brothers. But due to our political slavery the credit could not be given to Talapade-brothers and it is accepted quite ignorantly that the aeroplane could be originated only in 1903 by the Right-brothers. Shivaji Bapuji Talapade achieved this knowledge from the great scholar Maharṣi Sulbarai Shastri. He lived in the Chirabazar area near Dukkarwadi and later on was shifted to Mulad with his son विनायक who breathed his last in the sixth decade of the 20th century. The wife of S.B. Talapade was also a great vedic scholar. With the help of his wife and younger brother, Talapade prepared his plane which could fly in the sky of the chowpatty area of Bombay, continuously for five minutes on the hight of 1500 feet. B. Jaikar, N. Ranaday, Sayaji Rao Gaekward, the Maharaja of Baroda and Walji Narayanaji and thousands were eye witness on this historical moment. The authorities of the Hindustan Aeronautics Ltd. have still preserved the original frame and other fabrics of the plane for their historical record. In Vedas and Purāṇas, eight varieties of the aeroplanes have been discussed of which the 6th is called पुष्पक and the 8th मरुत्सखा. The plane, prepared by Talapade was based upon the synthesis of Puṣpaka and Marutsakhā from the technical point of view.[1]

The vedic evidences of aeronautics are seen more effectively practised in the Puranic era. The reference of the Puṣpaka in the Vālmīki Rāmāyaṇa is well known. It was an automatic plane and could move freely as per traveller's with (कामग) that techonology is still undeveloped in the history of aeronautics. In Śrīmadbhāgavatam also we have two references: the Saubha Vimāna of Sālva and the Vimāna of Maharṣi Kardama. The Saubha of Sālvarāja was a cruiser type of plane. It could create dense clouds around itself in order to conceal its identity. The plane of Maharṣi Kardama was fully furnished with seasonal views. The pigeons,

1. Times of India, 6.9.95 issue.

erected on the windows of the plane were so alive and concrete that the wild pigeons could join them fearlessly. It was also a pilotless plane and could move automatically.[1]

Now, to prove the superb development of Science and technology in ancient India, I would like to assess them one by one. Astrology, astronomy and mathematics are considered to be interlinked. The Indian astronomy undoubtedly touches the horizen of Vedic hymns when the Ṛṣis utters in a Mantra.

येनोदितेन दिनम् अस्तमितेन रात्रिः

The rising sun brings the day and disappears leaving behind the night. The vedic Ṛṣis thought that the sun plunges into the water of ocean and the other day rises up from the top of the mountain. The Tattiriya Brāhmaṇa says:

य उदगात् महतोऽर्णवात् विभ्राजमानः सलिलस्य मध्यात्।। तै०ब्रा० २.१.

The receding and increasing Kalas of the Mooṇ in the two fortnights of a month respectively, have been examined by our Ṛṣis in their own way. It occurs in the Ṛgveda 10.85.19

नवो नवो भवति जायमानोऽन्हां केतुरुषसामेत्यग्रम्।

भागं देवेभ्यो विदधाति आयन् प्र चन्द्रमास्तिरते दीर्घमायुः।।

This mantra seems to be quite allegorical in nature but the later commentators have unfolded the mystery, lying in it. Likewise, several other Mantras delineate some significant scientific truth. For example:

अन्तरिक्षं नाम पक्षिमार्गः (ऋग्वेद १.२५.६)

सूर्यः द्युलोके सर्वोच्चः (ऋग्वेद १.५६.११)

चन्द्रः नक्षत्रमण्डले सञ्चरति (तै०सं० ७.५.२३.१)

पृथ्वी लघुतमा विश्वमापारम् (ऋग्वेद १.५२.११)

ऋतून् प्रशासद् विदधौ सूर्यः (ऋग्वेद १.९५.३)

Shri S.B. Velankar, in his article published in the Girvaṇa Sudhā has given an exhaustive account of the Vedic astronomy.[2] The geometry of Vedic Shulvasutras, along with decimal-base numerals, squaring a circle and algebra etc. prove the antiquity of Indian astronomy and Mathematics. Lagadha's Vedānga jyautiṣa stands between Vedic and Traditional astronomy. Although, the

1. Śrīmadbhāgavatam, skandha 5th and 10th.
2. प्राचीनभारते खगोलशास्त्रज्ञानम्— एस्० वी वेलङ्कर। गीर्वाणसुधा—अक्टूबर/नवम्बर १९९५ ई० अङ्कः।

astronomers of the later period have oftenly quoted the authority of Brahmā, Sūrya, Vasiṣṭha, Manu, Soma, Lomasha, Marichi, Angiras, Vyāsa, Nārada, Shaunaka, Bhṛgu, Chyavana, Garga, Kashyapa and Parāshara etc. But only a few Saṁhitās are available now like Nārada Samhitā, Parāshara Saṁhitā and Kāshyapa Samhitā. Nevertheless, we can imagine the prosperity and splendour of our technical science even before Christian era.

Then, we find a galaxy of great astronomers and mathematicians. Aryabhat I, the writer of आर्यभटीयम् (476 A.D.) Varāhamihira, the writer of बृहज्जातक (6th C.A.D) Bhāskarācārya I, the writer of महाभास्करीयम् (end of the 6th C.A.D) Brahmagupta, the writer of ब्रह्मस्फुटसिद्धान्त and खण्डटवाद्य (628 A.D) भास्कराचार्य II, the writer of Lilāvati and Siddhāntashiromani (12th C.A.D) Ganesh Daivajna, the writer of ग्रहलाघव etc. enriched Indian astronomy. Their original contributions cannot be highlighted here, thoroughly in this papers but some salient features are to be quoted.

1. The theory of gravitation has been dealt with by Bhaskarācārya II in the Golādhyāya of his Siddhānta shiromani.

आकृष्टशक्तिश्च मही। गोला० ३.६

Newton, the so called inventor of this theory was born in the 16th century A.D. It is to be noted.

2. Āryabhaṭa I depicted in his Āryabhaṭīyam that the earth revolves around the Sun, which is absolutely constant. But the same truth could be known and declared by Galilio in the 17th century A.D.

3. It was Brahmagupta (7th C.A.D) who introduced यन्त्राध्याय in his ब्रह्मस्फुटसिद्धान्त comprising 24 chapters. In his observatory, Brahmagupta installed various yantras (technical devices) for the astronomical observations of planets, stars and the solar as well as lunar eclipses.

Bhāskarācārya, following the tradition of Brahmagupta explained the techniques of ताम्रघटीयंत्र, शंकुयंत्र, यष्टियंत्र, चक्रयंत्र and चक्रार्धयन्त्र. Above all he established the धीयंत्र i.e. the human mind itself. Bhāskarācārya wrote a fullfledged text on these techincal devices, namely सर्वतोभद्र which was followed by the Yantrarāja of Mahendra Suri, Dhruvabhramayantra of पद्मनाभ, yantra chintāmaṇi of Chakradhara, Pratodayantra of Gaṇesh Daivajna and Golānanda of Chintāmaṇi Dikshita. But all these yantras were confined to the text only and therefore, the ruler of Amer, Mahārāja Jaisingh got built a new observatory at Jaipur in 1730, wherein the Samrāt, Nādivalaya, Diganṣh, Bhitti and Shanku Yantras are still preserved.

Regarding the process of eclipse an exhaustive account is available in the 8th chapter of Bhāskarācārya's Siddhānta shiromani. The solar and lunar eclipses are never due to Rāhu and Ketu, as described in the Puranic texts. Actually it happens so when the moon comes in between the Sun and the earth. It is called solar eclipse. According to Bhāskarācārya:

पश्चाद्भागाद् जलदवदधः संस्थितोऽभ्येत्य चन्द्रो
भानोर्बिम्बं स्फुरदसितयाऽच्छादयत्यात्ममूर्त्या।
पश्चात्स्पर्शो हरिदिशि ततो मुक्तिरस्यात एव
क्वापि च्छन्नः क्वचिदपिहितो नैष कक्षान्तरत्वात्।।

However, the Ancient Indian astronomy and mathematics have hundreds and thousands subjects to explain and they have been elaborated as well. Right from the 5th to 18th century A.D. the Indian astronomy championed in the field of astronomical observations and inventions but due to our political subordination and lack of self-respect, the crown of supremacy is on the head of the western scientists.

Astronomy and mathematics are the two branches of the same tree. The Mathematics of today was known as Rāshividyā, Dhūlikarma and Parilekha etc. in old days. Bhāskarācārya II for the first time, divided mathematics in three parts, Gaṇita, Bījaganita and Pātigaṇita in his Siddhanta Shiromaṇi. The Gaṇita Sūtras were formulated by Aryabhata I in the 5th century A.D. For several reasons the Ankavidyā or Ankagaṇita was considered to be the back bone of astrology:

यथा शिखा मयूराणां नागानां मणयो यथा।
तद्वत् वेदाङ्गशास्त्राणां गणितं मूर्ध्नि संस्थितम्।।

Varāhamihira also divided astrology in three branches सिद्धान्त, होरा and संहिता. The measurement of time, the movement of planets and eclipses etc. come under the preview of Siddhānta. The Horāskandha deals with Jātaka in the birth of a child along with his life circle and the auspicious- inauspicious effects of various Grahas (planets) on his life etc. The Samhitā skandha has another name also,. that is Muhūrta Skandha. In this part the Acharya describes various yogas and Muhurtas (auspicious occasions) and this good and bad result. In Nārada Samhitā, these three Skandha have been given much significance as they are just like the vision of the Veda.

सिद्धान्तसंहिताहोरारूपस्कन्धत्रयात्मकम्।
वेदस्य निर्मलं चक्षुः ज्योतिश्शास्त्रमनुत्तमम्।।

Varāhamihira, repeats the above, in other words:

गणितं जातकं शाखां यो वेत्ति द्विजपुङ्गवः।
त्रिस्कन्धतो विनिर्दिष्टः संहितापारगश्च सः।।

Bhāskarācārya elaborates two separate phase of Mathematics Vyakta and Avyakta.

द्विविधगणितमुक्तं व्यक्तमव्यक्तमुक्तं
तदवगमननिष्ठः शब्दशास्त्रे पटिष्ठः।
भवति यदि तदेदं ज्यौतिषं भूरिभेदं
प्रपठितुमधिकारी सोऽन्यथा नामधारी।।

If any astrologer is not adept in mathematics, how can be explain the Science of influence of stars of human affairs? Specially the Gaṇita Jyautiṣa is totally based upon the mathematical calculations only. Such pseudo-astrologers are really नामधारी only and they are destined to be discarded. As I have stated earlier, Aryabhata I formulated several Gaṇita sūtras like the system of finding the vargamūla.

भागं हरेदवर्गान्नित्यं द्विगुणेन वर्गमूलेन।
वर्गाद् वर्गे शुद्धे लब्धे स्थानान्तरे मूलम्।।

This is one example. Many other examples can be had in connection with triangles, quadrilaterals, squares, rectangles, parallelograms, rhombus (quadrilateral figure with equal sides but unequal angles, असमकोणचतुर्भुज) Diagonals (straight line joining opposite angles of rectilineal figure कर्ण-विकर्ण) area and altitude etc. Addition, substraction, multiplication, division, big numbers, negative numbers, zero and decimal notations have been dealt with in ancient Indian Arithmetics at full length. Hypotenuse (side opposite to right angled triangle: कर्ण) and its converse (opposite) theorem, at present, which are named after Pythagorous, have been propounded by Bothayana, much earlier. In the field of algebraic equations the contributions of Āchārya Brahmagupta, Bhāskara, Pṛthūdakasveāmī, Mādhava and Nārāyaṇa can never be forgotten. They have proved to be the milestones in Indian algebra.

However, the roots of Indian mathematics have to be traced back to the Vedas just like the astronomy. The Sulva Sutra, which provide various formulas for constricting यज्ञभूमिका in sacrificial alters is the oldest mathematical text ever known to the world. Baudhāyana and Apastamba sutras give very good rational approximation for the irrational number √2 in the form: 1+ 1/3+1/3.4– 1/1.3.4.34 which is correct to 5 decimal places.[1]

How, the Indian mathematics is proved to be the pre-eminent and predominant in nature and how it acquires the precedence also is evident from the following statement which I borrow from the script for Dr. Kota Venkatachalam Rao. Putting forth the mathematical rediscoveries in Europe he wirtes:

1 Gregory (1671) rediscovers Mādhava's (14th century) series for ten =1(0)

2 Eular (1732) solves a problem of Fermat (1657) which was solved much earlier by Bhāskara (1150 A.D) what is nowadays known as Pell's equation.

3 Whwiller (1782) rediscovers the formula of Papameswara (1430) for the Circumradius (अर्धव्यास) of a cyclic quadrilateral.

4 Snell (1690) rediscovers Brahmagupta's theorem (7th century) on diagonals of a cyclic quadrilateral.

5 Herigone (1634) rediscovers Mahāvira's (9th century) formula for nC.

Now, a few words about the ancient Indian Medical system. It is evident from various (internal and external both) sources that the अथर्वसंहिता is more popuarly recognized as the भैषज्यवेद. India was known to be workshop of chemical technology from the time immemorial. It is evident from the archaeological excavations at Mohenjodaro and Harappa (now in Pakistan) which belongs to the Indus Valley Civilization (2500 B.C). These persons were skilled in chemical processes and used bricks, water pots, vessels, jars, terracota, jewellery, metal vessels, seals, painted pots, chrome glazed pottery and glass vessels etc. Metals like copper and bronze were also in their frequent use. They used Gold and silver also for making jewellery. All these remind us of the Vedic hymn occurring in Kṛṣ ṇa Yajurveda 4.7.5.

---

1. In Ṛgveda the Sun has been stated as the base of all the Bhuvanas.
सप्त युजन्ति रथमेकचक्रमेको अश्वो वहति सप्त नाम।
त्रिनाभिचक्रमजरमनर्वं यत्रेमा विश्वा भुवनानि तस्थुः ।।
In Atharva Jyautiṣa also the Sun etc. have been said to be the controller of this respective days:
आदित्यः सोमो भौमश्च तथा बुधबृहस्पती।
भार्गवः शनैश्चरश्चैव एते सप्तदिनाधिपाः ।।

अश्मा च मे मृत्तिका च मे गिरयश्च मे पर्वताश्च मे सिक्ताश्चमे

वनस्पतयश्च मे हिरण्यञ्च मेऽयश्च मे सीसं च मे त्रपुश्च मे श्यामं च मे लोहं च मे etc.

From the same Vedic medicinal and chemical sources the Indian Medical Science developed. In ancient texts it has been named as the रसविद्या. In the Rasaratnasamuchchaya we find the description of a full fledged of रसशाला i.e. the chemical laboratory for preparing the chemicals and drugs.

रसशालां प्रकुर्वीत सर्वबाधाविवर्जिताम्।
सर्वौषधिमये देशे रम्ये कूपसमन्विते।।
पदार्थसंग्रहः कार्यो रससाधनहेतुकः।
सत्त्वपातनकोष्ठीञ्च सुराकोष्ठीं सुशोभनाम्।।
भूमिकोष्ठीं चलत्कोष्ठीं जलद्रोण्योऽप्यनेकशः।
भस्त्रिकायुगलं तद्वन्नलिके वंशलोहयोः।।
स्वर्णायोघोषशुल्वाश्मकुण्ड्यश्चर्मकृतां तथा।
करणानि विचित्राणि द्रव्याण्यपि समाहरेत्।
कण्डनी पेषणी खल्वा द्रोणीरूपाश्च वर्तुलाः।
आयसास्तप्तखल्वाश्च मर्दकाश्च तथाविधाः।।
सूक्ष्मच्छिद्रसहस्राढ्या द्रव्यगालनहेतवे।
चालनी च कटत्राणि शलाका हि च कुण्डली।।

रसरत्न० अ०६

Actually, these exigmatic expressions are to be explained. But, one can know very easily that such was the chemical laboratory of ancient day, fully equipped with various apparatus, which were called yantras. Some of them have been narrated in the रसरत्नसमुच्चय along with their process. They are कोष्ठीमंत्र, तिर्यक्पातनयंत्र, ठेकीयंत्र. The Rasārṇava text describes the qualifications of the staff i,e. the Rasasādhakas and the servants.

रसचिन्तका वैद्या निघण्टुज्ञाश्च वार्तिकाः।
सर्वदेशजभाषाज्ञाः संग्राह्यास्त्रेऽपि साधकैः।।
यमिनः पथ्यभोक्तारो योजनीया रसायने।
तत्तदौषधनामज्ञाः शुचयो वञ्चनोज्झिताः।।
शुचीनां सत्यवाक्यानामास्तिकानां मनस्विनाम्।

सन्देहोज्झितचित्तानां रसः सिद्ध्यति सर्वदा॥

सोद्यमाः शुचयः शूराः बलिष्ठाः परिचारकाः॥

In the Charaka and Sushruta Saṁhitās not only the various diseases and their treatments have been discussed minutely, but on the other hand, the philosophical background of these diseases have also been elaborated. According to Charaka अयोग अतियोग and मिथ्यायोग of the mind, speech and action are the main cause of any disease and therefore, we must get rid of it.

Apart from the Charka and Sushruta Saṁhitās, more than 50 texts have been written on the medical discipline, which strengthen the Ayurvedic tradition of India. Some of them are the Rasaratnasamuccaya and Aṣṭāngahṛdaya of Vāgbhata, Rasaratnākara, Rasārṇavakalpa, Rasarājamṛgāṇka, Rasaratnapradīpa, Rasadarpaṇa, Rasarājamahodadhi and Rasayogamuktāvali.[1]

The ancient Indian technology has many other branches also like Indian architecture, Indian Jurisprudence, Environment and Social Science, Fine arts and Folk Culture etc. But it requires much time to shed an adequate light on all these subjects. However, I stop here, with a humble submission to my enlightened friends.

दृश्यते नैव तच्छास्त्रं यत्प्रणीतं न भाटते।

तस्माद्धि निखिलं विश्वं सततं भारतोन्मुखम्॥

1. Indian Tradition of Chemistry and Chemical Technology by Prof. A.R. Vasudevamurthy and

Section II

Computer Science and Mathematics

1

सङ्गणकीय-संस्कृतम्

डॉ. चंद्रगुप्त वर्णेकर, प्राचार्यः

प्रस्तावना :

संमेलनस्यास्य संयोजकाः! कुरुक्षेत्रविद्यापीठस्य पदाधिकारिणः! अत्र समुपस्थिताः, जेष्ठाः, विद्वांसः, प्राध्यापकाः, विज्ञानिनः, प्रतिनिधयः! मान्याः! छात्राः! श्रोतारश्च!

विदितं यत् प्रतिवर्षं श्रावणपौर्णिमायां तिथौ भारते सर्वत्र संस्कृतदिनोत्सवः प्रवर्तते, इति। दिवसेऽस्मिन् केन्द्र-राज्य-प्रशासनद्वारा संस्कृतपण्डित-विद्वज्जनानां यथोचितं सभाजनं भवति। स्थाने स्थाने संस्कृतानुरागिणः विविधान् उपक्रमान् समायोजयन्ति; सङ्गीत-भाषणादिकार्यक्रमाः प्रचलन्ति। तस्मिन्नेव श्रावणमासे समायोजितेऽस्मिन् ''संस्कृतशास्त्रपरम्परायां विज्ञानं प्रौद्योगिकी च'' विषयस्य सम्मेलने ''सङ्गणकीय-संस्कृतम्'' इति विषयमधिकृत्य भवान् भाषणं करोतु, इति संयोजकानां समुचिता सूचना मया अचिरादेव सम्प्राप्ता। तदनुसारम् एतस्य बृहद्विज्ञानविषयस्य शोधविषयस्य वा संक्षेपं विवरणमत्र मया प्रस्तूयते।

नवविज्ञानयुगस्य वैशिष्ट्यम् :

सम्प्रति विज्ञानयुगं वर्तते इति कथ्यते। यतो हि विज्ञानतन्त्रैः विकसितानां विविधानां यन्त्रोपकरणसाधनानां प्राचूर्यं सद्यस्तने जीवने पदे पदे सर्वत्र दरीदृश्यते। यथा अत्र विद्युद्दीपः, व्यजनम् अथवा अयं ध्वनिवर्धकः इत्यादि। एतादृशी भौतिक विज्ञानोपकरणमालिका प्रायः द्विशतवर्षपूर्वं प्रारब्धा, अद्य यावत् प्रचलति च (छायाचित्रम्) तत्रापि दीप-व्यजन-यानादि साधनानि जनसाधारणस्यापि परितः भृतकवदुपतिष्ठन्ति। वृद्धिङ्गता खलु सामान्यविज्ञानस्य व्याप्तिः, यतो हि ह्यः एव यत् किमपि असामान्यं विज्ञानोपकरणमभूत् तदद्य सामान्यरूपेण उपतिष्ठते; किं बहुना? अस्यां विज्ञानोपकरणमालिकायां नवविकसितपुष्परूपेण विलसति तदाधुनिकं सङ्गणकयन्त्रम्!

पुरा विज्ञानोपकरणानां विषये भाषारचनायाः चर्चा न कदापि सञ्जाता। विद्युद्दीपस्य का नाम भाषा? का वा भवति व्यजनस्य? परमस्मिन् सङ्गणकविज्ञाने तु Computer Language, Programming Language, Natural Language Processing, Speech Processing, Speaker Recognition, Phonetic input इत्येतादृशी भाषाविज्ञानचर्चा यथेष्टं प्रवर्तते। यत्र कुत्रापि तादृशं भाषारचनायाः शास्त्रीयं, विज्ञानचिन्तनं भवति, तत्र अस्माकं विज्ञानतन्त्रतः शुद्धायाः पाणिनीयव्याकरण-नियम-सुबद्धायाः संकृतभाषारचनायाः महत्त्वं

अनिवार्यम् इत्यनेन न्यायेन सङ्गणक-विज्ञाने संस्कृतरचनानां चर्चा देशविदेशेषु वैज्ञानिकैः अङ्गीकृता। तेषु प्रमुखः वैज्ञानिकाः सन्ति अमेरिकादेशीयः नासा-वैज्ञानिकः डॉ. रिक् ब्रिग्जः कॅनडादेशीयौ आर्मस्ट्रॉङ्गदम्पती, ब्रिटिशमायक्रोसॉफ्ट-संशोधकः डॉ. टोनी होअर, भ्रमणशीलः विज्ञानी डॉ. कॉनेन् मैश्लरः आय्.आय्.टी. प्राध्यापकाः डॉ. महाबलः, सिडॅक् परमनिर्माता डॉ. विजयभटकरः, डॉ. राजीव सिङ्घलः, डॉ. विनीतचैतन्यः इत्यादि। शोधविषयमिमम् अधिकृत्य तैः नैके शोधनिबन्धाः लिखिताः सन्ति; सभासम्मेलनेषु च प्रस्तुताः सन्ति। तेभ्यः ये केचन ज्ञानविज्ञानामृतबिन्दवः दिष्ट्या मया समुपलब्धाः, ते एव अत्र प्रस्तुताः सन्ति। अतः

यदत्र सौष्ठवं किञ्चित् तद् गुरोरेव मे नहि।
यदत्रासौष्ठवं सर्वं तन्ममैव गुरोर्नहि॥

सङ्गणकविज्ञाने संस्कृतरचनानां महत्त्वं

विदितं यत् सामान्यगणितव्यवहारे यान्त्रिकता विद्यते इति। के सन्ति व्यवहाराः ? तथा हि—

संयोगश्चान्तरं भागो गुणो वर्गो घनस्तथा।
द्वे मूले चेति सामान्यमष्टाङ्गं गणितं स्मृतम्॥

इति। एतेषु या यान्त्रिकता भवति सा संस्कृत-साहित्ये विविधासु बीजाङ्करेखागणितप्रक्रियासु वर्णिता अस्ति। आधुनिके सङ्गणकविज्ञाने तादृशी प्रक्रिया Algorithm इत्युच्यते। किं नाम अल्गोरिदम्? कुतः आयातः अयं शब्दः? पुरा किल ईस्वीयाम् अष्टम्यां शताब्द्याम् अरबदेशीयः कश्चन् अल्ख्वा-रिझ्मी नामधेयः गणितज्ञः स्वपुस्तके तादृश्यः प्रक्रियाः वर्णितवान्। पुस्तकस्यास्य नाम भवति ''मुकाबला-ए-हिन्दसा'' इति; यस्य आङ्ग्ल भाषान्तरम् '' Calculations with Hindu Numerals" अद्यापि समुपलभ्यते। कालान्तरेण ''Al-Khwa-Rizmi's Methods" इति क्लिष्ट-शब्दावल्याः अपभ्रंशः सञ्जातः Algorithm इति। एवम् अल्गोरिदम् शब्दस्य व्याख्या आङ्ग्लभाषायाम् इदृशी क्रियते वैज्ञानिकैः ''A step-by-step logical & mathematical procedure expressed in common natural language to solve a class of problems."

अस्यां व्याख्यायां सङ्गणकविज्ञानत्रयी एव वर्णिता अस्ति। तथाहि बीजाङ्करेखागणितम् (Algorithms), सामान्यः तर्कः (Computational Logic), भाषारचना (Computational Linguistics) चेति। सङ्गणकविज्ञानत्रयी-विषये विपुलं मौलिकं संस्कृतसाहित्यं समुपलभ्यते। परमुचितज्ञानां निमित्तनामभावेन उपेक्षिताः खलु तादृश्यः ग्रन्थगतारचनाः। नुनमुपनतं तन्निमित्तामिदानीं सङ्गणक-विज्ञानरूपेण। अतः साहित्यस्यास्य पुनरावलोकनं सङ्गणकवैज्ञानिकैः सम्प्रति प्रारब्धम् अस्ति देशविदेशेषु।

अल्गोरिदम् इत्यस्य कृते नैके पारम्पारिकाः संस्कृत-शब्दाः सन्ति। तथाहि रीतिः, पद्धति, प्रक्रिया, गणितम्, पाटी, व्यवहारः, विधिः, विधानम्, सोपानम्, सूत्रम्, मार्गः इत्यादि। तानेव समाश्रित्य सङ्गणकस्य गणप्रणाली (Computer Program), लिख्यते। तादृशी गणप्रणाली सूत्रस्वरूपा स्यात् तथाहि—

अल्पाक्षरमसन्दिग्धं सारवद्विश्वतो मुखम्।

अस्तोभमनवद्यं च षड्विधं सूत्रलक्षणम्।। इति।

एवं सङ्गणकोपकरणस्य सम्यक् सञ्चालने वस्तुतः सूत्रस्वरूपा अतीवं महत्त्वं भजते। एकमेवेदमुपकरणं स्यात् यस्य सञ्चालनालय भाषारचना उपयुक्ता भवति। तत्रापि संक्षेपेन यानि वैशिष्यानि भाषारचनाविषये विविधेषु चर्चासत्रेषु चिन्तितानि सन्ति तानि निम्नलिखितानि भवन्ति।

सङ्गणकप्रयोगे अपेक्षितानि भाषारचनायाः वैशिट्यानि

1. संस्कृतरचनासदृशा सुनियमबद्धा सुसूत्रा शब्दरूपावलिः धातुरूपावलिश्च (Tabulted Forms of inclinations).
2. लिपेः वैज्ञानिकता/यथोच्चारणं लेखनम् (Phonetic Transcription).
3. व्याकरणसूत्राणां गणितविधिसाधर्म्यम् (Computational Linguistics).
4. यथा सामान्य-संस्कृतवाक्ये पदक्रमपरिवर्तनेन वाक्यार्थस्य परिवर्तनं न भवति तथा (Free word- order).
5. खण्डान्वय/ दण्डान्वय प्रक्रिये (Top down/ Bottom up Parsing process in compilation).
6. संख्यापद्धति-विषये मौलिक्यः रचनाः/ संख्यायाः वर्ण-चिह्नाङ्कनम् (Variable Names).
7. अक्षर-गण-वृत्तादि विषयेषु शास्त्रीयं साहित्यम् (Binary Symbol System) (छायाचित्रम्)।
8. बीजाङ्क-रेखागणितविषये विविधाः सोदाहरणरचनाः (संस्कृत Algorithms).
9. सामान्य-तर्क-विधानम् (Computational logic)
10 भौतिकविज्ञान विषये सैद्धान्तिकं विवरणम् (Theoretical Physics).
11. शाब्दबोधादि प्रक्रियाविवरणम् (Semantics).
12. ज्योतिर्गणितस्य मौलिकं साहित्यम् (Algorithms).
13. संशोधनकार्ये मार्गदर्शनक्षमता (Research Capability).

एभिः लक्षणैः संस्कृतरचनानां सङ्गणकीयता एव स्वयंसिद्धा भवति। अतः

येषां रतिः सङ्गणकीयशास्त्रे ये देववाणीमभिकामयन्ते।

तेषां मनोबुद्धिविकासकार्ये सम्मेलनेऽस्मिन् भवति प्रयासः।।

सन्दर्भ ग्रन्थावली:

1. Rick Briggs, Al Journal 1983 about Shastrik Sanskrit & Al.

2. Rajiv Singhal & Dr. Vineet Chaitnya, ''Natural Language Processing" :

3. A Paninian perspective Tata McGraw-Hill Publication.

3. भास्कराचार्य: ''लीलावती:'' (मराठी अनुवादसहिता)

4. पिङ्लाचार्य: ''वृत्तरत्नाकर:'' (छन्दस् शास्त्रम्)

5. लोगाक्षिभास्कर: ''तर्कभाषा'' (मराठी अनुवाद:)

6. सोमणशास्त्री, ''ज्योति:शास्त्रम्'' (मराठी भाषायाम्)

2

Development of Mathematics In Ancient India

B. S. Rajput

1. Introducton

In ancient India time at the early stage of their awakening, the Indian philosophers conceived the notion of ultimate harmony and symmetry of this universe and realized that it must have much richer mathematical structure such that wherever one has to analyse any of its fundamental processes upto the prescribed limit of accuracy, he will definitely have to follow certain mathematical rules. Relying upon the rational analysis free from the discipline of direct observational contents, they went off in vivid different directions in their study of concepts of indivisible parmanu to the original bootstrap model and to more abstract mathematical concepts of numbers, geometry, trigonometry, algebra, gravitations, planetary motions and symmetries. In all these developments, the mathematics was given a very respectable position as stated in Vedaṅga Jyotiṣa[1]

यथा शिखा मयूराणां नागानां मणयो यथा।

तद्वद् वेदांगशास्त्राणां गणितं मूर्धनि स्थिताम्॥

which means that just as the crest is to peacockes, as is the head-germ to snakes, so is ganita, among Vedang-śāstras, as crown of all sciences.

In mathematics the inventions of zero and other nine number, decimal placing value system and infinity were made for the first time in India;

पूर्णमदः पूर्णमिदं पूर्णात् पूर्णमुदच्यते

पूर्णस्य पूर्णमादाय पूर्णमेवावशिष्यते॥

In his book on "Foundation of Process of Mathematics" Prof. J. P. Halsted described the Indian contribution for discovering zero as the greatest discovery for the intellectual development in mathematical researches. According to Professors Hackworth and Holland, "Zero as a symbol was introduced by Hindus in India in about 500 AD. Hindu word for zero is Sunya, and the symbol O enclosing nothing in its boundary conveys that idea."[2]

In ancient India the mathematics was development as per need in understanding the Nature in terms of processes and constituents. The concepts of independent nine numbers were incorporated in Vedic versus and Maharishi Meghatithi developed the various methods of mathematical calculations. In the oldest scripture of the world, the Rigved, there are 313 versus (Richas) associated with his name. In Arithmetic the classical contribution of Meghatithis is the power of ten. This concept originated in Rik Mantras and it was fully developed in versus (Mantras) of Yajurveda. The famous mantras associated with Meghalithi, which appeared in all the four Vedas, is

अतो देवा अवन्तु नो यतो विष्णुर्विचक्रमे। पृथिव्याः सप्त धामभिः।।१६।।''

इदं विष्णुर्वि चक्रमे त्रेधा नि दधे पदम्। समूळ्हमस्य पांसुरे।।१७।।

त्रीणि पदा वि चक्रमे विष्णुर्गोपा अदाभ्यः। अतो धर्माणि धारयन्।।१८।।[३]

2. Development of Numbers

In Rigveda the independent nine numbers originated in the following versus;

For number one and two:

सप्ती चिद्घा मदच्युता मिथुना वहतो रथम्। एवेद्धूर्वृष्ण उत्तरा।।१८।।[४]

For number three

अग्ने देवाँ इहा वह सादया योनिषु त्रिषु। परि भूष पिब ऋतुना।।४।।[५]

For number four

यत् त्वा तुरीयमृतुभिर्द्रविणोदो यजामहे। अध स्मा नो ददिर्भव।।१०।।[६]

For number five:

इहि तिस्रः परावत इहि पञ्च जनाँ अति। धेना इन्द्रावचाकशत्।।२२।।[७]

For number six:

उतो स मह्यमिन्दुभिः षड्युक्ताँ अनुसेषिधत्। गोभिर्यवं न चर्कृषत्।।१५।।[८]

For number seven;

अतो देवा अवन्तु नो यतो विष्णुर्विचक्रमे। पृथिव्याः सप्त धामभिः।।१६।।[९]

For number eight and nine:

अष्टापदी नवपदी बभूवुषी सहस्राक्षरा परमे व्योमन्।।४१।।[१०]

For number ten, hundred, thousand and ten thousand:

५९८३. ये ते सन्ति दशग्विनः शतिनो ये सहस्रिणः।
अश्वासो ये ते वृषणो रघुदुवस्तेभिर्नस्तूयमा गहि।।९।।(11)

In Yajurveda different numbers were developed in the following manner:(12)

एका च मे तिस्रश्च मे	:	$1 + 2 = 3 = 2^2 - 1^2$
तिस्रश्च मे पंच च मे	:	$3 + 2 = 5 = 3^2 - 2^2$
पंच च मे सप्त च मे	:	$5 + 2 = 7 = 4^2 - 3^2$
सप्त च मे नव च मे	:	$7 + 2 = 9 = 5^2 - 4^2$
नव च मेएकादश च मे	:	$9 + 2 = 11 = 6^2 - 5^2$
....		....
....		....
चतस्रश्च मेऽष्टौ च	:	$4 + 4 = 8 = 3^2 - 1^2$
मे द्वादश च मे	:	$8 + 4 = 12 = 4^2 - 2^2$
षोडश च मे विंशतिश्च	:	$12 + 4 = 16 = 5^2 - 3^2$
मे चतुर्विंशतिश्च मे	:	$16 + 4 = 20 = 6^2 - 4^2$
....	:	....
....	:	....

where the first series (3, 5, 7, 9,.........) has been obtained by substracting the square of a number from the square of next higher number while the second series (8, 12, 16, 20,..........) has been obtained by substracting the square of a number from the square of next to next higher number. In this mantara the numbers have been constructed according to the principle of calculus of Ganita-Jyotisha.

Similarly in Atharveda the first ten numbers had been introduced in the following manner:

य एतं देवमेकवृतं वेद।
न द्वितीयो न तृतीयश्चतुर्थो नाप्युच्यते। यएतं देवमेकवृत्त वेद।
पंचमो न षष्ठः सप्तमो नाप्युच्यते। यएतं देवमेकवृतं वेद॥
नाष्टमो न नवमो दशमो नाप्युच्यते। यएतं देवमेकवृतं वेद॥

It is interesting to note that the number have been placed in definite order at definite places in the following versus of Rigveda;(13)

आ द्वाभ्यां हरिभ्यामिन्द्र याह्या चतुर्भिरा षड्भिर्हूयमानः।

आष्टाभिर्दशभिः सोमपेयमयं सुतः सुमख मा मृधस्कः ।।४।।
आ विंशत्या त्रिंशता याह्यर्वाङा चत्वारिंशता हरिभिर्युजानः।
आ पञ्चाशता सुरथेभिरिन्द्रा षष्ट्या सप्तत्या सोमपेयम् ।।५।।
आशीत्या नवत्या याह्यर्वाङा शतेन हरिभिरुह्यमानः।
अयं हि ते शुनहोत्रेषु सोम इन्द्र त्वाया परिषिक्तो मदाय।।६।।

In these versuses the following numbers have been given in perfect order: 2, 4, 6, 10,,20, 30, 40, 50, 60, 70, 80, 90, 100.

India did not only originate independent numbers for the human civilization and intellectual development but also gave names to these numbers which became popular in Europian countries (with slightly changed names) after more than ten thousand years. This fact is universally accepted as Prof. Ginsbeg has stated that;[(14)]

"In 770 a Hindu scholar Kanak of Ujjain visited Bagdad on the invitation of Abbas Sayeed Khalifa Almansoor and taught Hindu Mathematics and Astrology and Hindu system of numbers reached west from Arab Via Egypt as Arabic numbers."

3. Large Numbers In Ancient India Mathematics

For analysis the obserned behaviour of processes and constituents of nature mathematically, very often we need large numbers. The mentions of large numbers at various places in ancient Indian literature are the evidences of the fact that the processes and constant representations associated with these numbers were known to ancient Indian thinkers and seers. In ancient times Indians could count the numbers upto 10^{17}, 10^{53} and also upto 10^{62}. For instance we mention the following shlokas of Valmiki Ramayan which is at least five thousand years old:

शतं शतसहस्त्राणां कोटिमाहुर्मनीषिणः।
शतं कोटिसहस्त्राणां शकुंरित्यभिधीयते॥
शतं शकुसहस्त्राणां महाशंकुरिति स्मृतः।
महाशकुसहस्त्राणां शतं वृन्दमिहोच्यते॥
शतं वृन्दसहस्त्राणां महावृन्दमिति स्मृतम्।
महावृन्दसहस्त्राणां शतं पद्ममिहोच्यते॥
शतं पद्मसहस्त्राणां महापद्ममिति स्मृतम्।
महापद्मसहस्त्राणां शतं खर्वमिहोच्यते॥

शतं खर्वसहस्त्राणां महाखर्वमिति स्मृतम्।
महाखर्वसहस्त्रेण समुद्रमभिधीयते।।
शतं समुद्रसाहस्रं मोघ इत्यभिधीयते।
शतं मोघसहस्त्रेण महौद्य इति विश्रुतः।[15]

In other words

शतं सहस्त्राणां	=Hundred Thousand =105 = 1 Lakh
शतं शतसहस्त्राणां	=Hundred Lakh =107 = कोटि
1 लाख कोटि	$=10^{12}$ = 1 शंकु
1 लाख शकु	$=10^{17}$ = 1 महाशंकु
1 लाख महाशंकु	$=10^{22}$ = 1 वृन्द
1 लाख वृन्द	$=10^{27}$ = 1 महावृन्द
1 लाख महावृन्द	$=10^{32}$ = 1 पद्म
1 लाख पद्म	$=10^{37}$ = 1 महापद्म
1 लाख महापद्म	$=10^{42}$ = 1 खर्व
1 लाख खर्व	$=10^{47}$ = 1 महाखर्व
1 हजार महाखर्व	$=10^{50}$ = 1 समुद्र
1 लाख समुद्र	$=10^{55}$ = 1 औद्य
1 लाख मौघ	$=10^{60}$ = 1 महौघ
शतं महौघ	$=10^{62}$

It is worth mentioning that such a large number is not needed for analyzing any process or constituent representation known to modern scientist. This fact provides an evidence that during Ramayan time some processes, still unknown to modern physicists, were known for the analysis of which such large numbers were required. In the modern cosmology the largest scale of number is 10^{48}, the square of ratio of diameter of universe and the diameter of atom, which may occur in future in the theory of strong gravity. In any foreseen future theory any number large than this may not be required in modern science.

4. Small time scale in Ancient Indian Mathematics

In Varaharmihir *panch sidhantika* there are several examples of temporal measurements of very large as well small scales. For instance

रविशशिनो पंचयुगं वर्षाणि पितामहोपदिष्टानि।

अधिमासास्त्रिंशदुभिर्मा सैरवभास्त्रिंशहद्यप्तुम्।।[16]

and also

नवत्राणां स्थानधारेण =४,३२०,०००,००० सौर वर्षाणि कल्पः।

एषः कल्पः चतुर्युगरूपेण विभक्तः।

भूमौ सूर्यपरिभ्रमणकालम् =एक सौर वर्षम्

एक दिनं	=	60 घटिका
एका घटिका	=	60 विघटिका
एका विघटिका	=	60 परघटिका
एका परघटिका	=	60 सूक्ष्म घटिका
एका सूक्ष्म घटिका	=	60 क्षणं
सहत्रंश क्षण	=	10^{-3} क्षणं = 1 पल

It give 1 क्षणं = 10^{-4} सेकण्ड

And 1 पल =10^{-7} सेकण्ड।

Such a small time is not needed in any analysis of any classical (macroscopic) process. Such time scales like Pal = 10^{-7} sec. or Satanshpal= 10^{-9} sec. are needed in the analysis of atomic (microscopic) phenenomera occurring on the scale of nano-second (10^{-9}). It shows that such process (of quantum nature) were known in ancient India, for which such a small time scale was introduced.

5. Contributions of some renowned ancient Indian Mathematicians

Ancient Indians development the systems of Arithmetic, algebra, geometry, analytical geometry, diffential calculus, permutations and contributions, square root, logarithm, quadratic equations, planetary motion, gravitational force and electromagnetisms. Here we shall give a brief review of the eontributions of some of ancient Indian thinkers,

(a) Arya Bhatt

Arya Bhatt, The great astronomer and mathematician of 476 A.D., established algebraic methods of calculations for the first time and formulated very famous and fundamental thirty seven versus. In all his literature he admitted that whatever he wrote was based (with his own modifications) on the words of early writers of ancient time. His famous book Arya-Battiya was classified in four sections (pads);

(i) Gitika Pad which contains ten sutras which are called Dasgitika Sutras:[17]

दशगीतिका सूत्रमिदं भूग्रहचरितं भपंजरे ज्ञात्वा।
ग्रहम गणापरिभ्रमणं स याति मित्वा परंब्रहम।।

(ii) Ganita pad contains thirty three shlokas which deals with square, cube (Shloka-3); Square root (Shloka-4); Cube root (Shloka-5), Area of triangle and volume of parallelopiped pyramid (Shloka-6); Area of circle and volume of sphere (Shloka-10), Sine (Shloka-11); side and base of rightangle triangle (Shloka-16); Diagonal of rightangle triangle (Shloka-17); cosine (Shloka-18); tri-roots (shloka-23); Decimal value of ratio (shloka-27); value of unknown (Shloka-30); Kuttkai Ganita (Shlokas 32 and 33).

In these thirty three versuses Arya Bhatt presented very original mathematical methods. For instance the correct value of Ṇ was obtained in the following verse ;

चतुरधिकं शतमष्टगुणम् द्वाषष्टिस्तथा सहस्राणां।
अयुतद्वय विष्कम्भस्यासन्नो वृत परिणाहः।[18]

which means

[(100+4)x 8+62000]/20000 = = ð = 3.1416... $\frac{\text{वृत परिधि}}{\text{व्यास}}$

For determining the area of a triangle and volume of a solid figure, he gave the following Shloka (formula);

त्रिभुजस्य फलशरीरं समदलकोटिभुजार्ध संवर्ग।
उर्ध्वभुजा तत्संवर्गाय स धनुष षडश्रिरिति:।[19]

and

यश्चैव भुजावर्गः कोटिवर्गश्च कर्णवर्गस्य।
वृते शरसंवर्मोर्धाज्या खलुधनुषो।।[20]

Which means that in a right angle triangle the sum of squares of base and normal is the square of diagonal and area of circle is the product of square of half of diamater with the ratio of circumference and diameter.

and also,

सर्वेषां क्षेत्राणां प्रसाधय पार्श्वे फलं तदभ्यासः।
परिगधोष् षड् भाजज्या विष्क्रमममर्धेन सा तुल्या।।[21]

which means that in a circle the Jeeva of one sixth part of circumference is half of its diameter. Aryabhatt conceived the motion of sines and cosines through the following shlokas;

समवृतपरिधिवादं छिन्द्यात् त्रिभुजाच्चतुर्भुजाच्चैव।
समं चापज्यार्धानि तु विष्कम्भार्धो यथेष्टानि॥
प्रथमाच्चापज्यार्धा घैरूनं खंडित द्वितयाधर्म।
तत्प्रथमं ज्यार्धाशैस्तैरूनानि शेषाणि॥[22]

(iii) In Kalkirya Pad there are twenty five shlokas on time, its scale and its meaurement. For instance,

वर्ष द्वादशमासास्त्रिंशमद्दिवसो भवेत् स मासस्तु।
षष्टिनडियो दिवस षष्टिस्तु विनाडिका नाडि॥[23]

and

गर्वक्षराणि षष्टिर्विनाडिकाक्षी षडेव वा प्राण:।
एवं कालविभाग: क्षेत्रविभागस्तथा भगणात्॥[24]

(iv) In Golepad there are fifty shlokas with the evidences of rotating round earth, planetary motion and the details about Sun and other stars. For instance,

अनुलोम गति: नौस्य पश्यति अचलं विलोमगम्।
यद्वत अचलानि माति तद्वत् सम पश्चिमगानि॥[25]

Arya Bhatt is also recognized as founder of algebra. He also gave the following method to solve the algebraic equations:

मूलफलं सफलं कालमूलगुणधर्म मूलकृतियुक्तम्।
मूलं मूलार्धोन कालह्यतं स्यात् स्वमूलफलम्॥[26]

which gives the solution of the following type of problem;

फलं शतस्य मासे यक्षतं ततस्फलान्तरम्।
मासषटके षोडशकं जाते मूलफलं वद॥

which gives the value of x in the equation

$$c x^2 + ax - da = 0$$

Arya Bhatt gave the solution of simultaneous equations.

$$x \text{ṣ} y = a$$

$$xy = b$$

in the following from:

द्विकृतिगुणात्संवर्गाद् द्वयन्तरवर्गेण संयुतान्मूलम्।
अन्तरयुक्तहीनं तद् गुणाकारद्वयं दलितम्॥[27]

which means

$$x = \tfrac{3}{4} [((a^2 + 4b)_{.}^{\frac{1}{}} + a]$$

and $y = \frac{3}{4} [\sqrt{(a^2 + 4b)} - a]$

Arya Bhatt also gave miraculous formulae in Jyotisha. In fact the very foundation of Jyotisha was laid in Vedic period and algebra was fully developed by Aray Bhatt as the mathematical method required in Jyotisha. The Vedang Jyotisha was reformulated by Maharishi Lagadh in 1200 BC. During Mahabharat period (2000 BC.) the knowledge of planets, stars, elipses and planetry motion was very commonly known to Indians.

(b) Algebra of Bhaskara

Bhaskara acquired very high esteemed position among ancient Indian mathemathecians and Jyotishavidas. His famous contributions are Sidhant Siromani, Karan-Kutuhal, Leelawati and algebra. Karan Kutuhal was written in 1183 while Sidhant Siromani was completed in 1150. The first part of Sidhant Siromani is known as Pati-Ganita or Leelawati which contains 278 Shlokas each one of which is a classic in mathematic. The second part of Sidhant Siromani is known as algebra which contains 213 shlokas. In 1587 the Mugal emperor Akbar got the Leelawati and algebra translated into persian by Abulfazal and Atta Alla Rashudi.

(c) Varaharmihir's Panchsidhantika (490 A.D)

Another great mathematician and Astromer was Varahamihir whose famous work "Panch-Sidhantika" got universal recognition in the field of Astronomy, Astrology and Astrophysics. He also prepared Brihajjataka and then moved to Astrology. His other works include Laghujataka, Brohajjataka, Vivaha-patala,. Brihasamhita and Yogayatra. In Surya Sidhanta, the measurement of time, sinetable and cosine functions, meridiam, equinoxes and solstices; eclipses, planetry motions, inclinations and asterism to the elliptic, helical rising setting of Mars, relative motions of the Sun and moon and calender computation all are mentioned.

In Pitamaha Sidhanta the year is of 365 days and 21.5' ghatikas and in Pulisha Sidhanta, the knowledge of the oblique path and appearance and disappearance of the stars is found. Accordingly, the following Nadikas (the symbolic periods in days of planets) were mentioned;

584/11 Budha (Mercury) = 89 days and 2 hours

115/52 days of Shukra (Venus) =224 days and 17 hours

7802/45 days Mangal (Mars)= 687 days

3991/11 days Shani (Saturn) = 10765 days and 19 hours

The Romak Sidhanta gives a year of 365 days, 14 ghatika and 18 palas.

The Calculations of modern solar system are based on these results.

Varahimihir also calculated the correct sizes of Sun and Moon and their distances from the earth. There is interesting record of astronomical instruments in various Sanskrit texts on astronomy like nadika (Waterclock); Shanku (gromon); chakra (circle), chap (half circle), dhanu (arc) ; turiya (quadrant) and soon.

(d) Brahmgupta (598 A.D.)

He fixed Sayan Meshasankramana Brahmagupta did not accept Arya Bhatt's theory of rotation of earth and epicycles. His two treatises on astronomy Brahmasphuta - Sidhanta and khandan - khadyaka were translated into Arabic by Muhamad Ibu Ibrahim al- Fazari and Yaqub ibn Tariq in the late eighth century A.D. as sindhind and Arkand. It is supposed that an Indian astroner paid a visit to the court of Abbasid and the Greek thoughts were enriched by Indian thoughts in astronomy which was further carried to Arabia and later again enriched the calculations of Indian astromers. Thus ancient Indians were not that closed minded as some of the western critics would like us to believe.

(e) Shatananda

He authored Bhasvatikarana and determined the positions of Sun and the moon according to constellations. In fact the priests from the Vedic days had knowledge of the calender. The Vedas speak of seasonal sacrifices (Chaturmasyani), the full-moon sacrifies or the Soma feasts. In Chandogya Upnishad, Narada tells Sanat Kuwa [15], "I know also astronomy". In Brahmana texts, the astrologysts were known as Nakshatradarshaks and in Yajurveda there is mention of 27 or 28 stellar constellations. The reference to Rahu planet and Abhijit constellations occurs in Atharvaveda also. The Sanskrit term Nakshatra = Na + kshatra, which means which does not change or fall. The first mention of these fixed stars is in Krishna Yajurveda and in Atharva-Veda. The astronomical topics discussed is various books of Sidhantas (principles) are too technical to be discussed here. The Sanskrit texts mention the important great circles, one passing through the north-south poles and the zenith where the nadir is the meridian (Yamottara- vrtta). There are also references to horizontal, equatorial and and elliptic systems representing the positions of celestial dies.

(f) Bodhayan

Bodhayan, another great mathematician of ancient India, was known as father of geometry. Trigonometry and mensuration. He developed the principles of geometry on the basis of art of construction of altars of vedic yajna. He established the system of measurement in the term of Sulba-Sutras, constructed circular, semicular, square and more complicated altars all of equal area and discovered the so called Phythagorous theoem independently. He also invented lines, surfaces and measuring devices by using his Sulba-Sutras;

कर्णी तत्कर्णी त्रिर्यक्मानी पार्श्वमान्यक्षणया चेति पंच रज्जवः।(28)

Geometrical literature originated from Rigveda which describes three places for fire;

Circular altars (Grahayapata);

Square allars (Ahayanya);

And Semicircular altars (Dakshinaagni)

which shows that the Indian knowledge of circle, square and diagonal of square originated from Vedas. For the constructions of these altars of same area, the following geometrical processes were necessary;

(i) Construction of square on a line

(ii) Construction of circles touching the sides of a squares.

(iii) Construction of circle touching the corners of squares.

(iv) Doubling the circle, which requires the knowledge of $\sqrt{2}$.

Other altars were of still complicated shapes, like syenasit altar, which require very complicated geometrical constructions.

Bodhayan independently discovered the so called Phythagorous theorem,

दीर्घचतसृस्याक्ष्णया रज्जुः पार्श्वमाणी।
तियङ्-मानी च यत्पृथग्भूते कुरूतस्तु भयं करोति।।(29) or
समचतुस्रस्याक्ष्णया रज्जुद्विकरणी।(30)

In the light of these facts the so called Phythtagorous theorem should be recognised as Bodhayan Sulba-Sutra. There are several other names who contributed in ancient Indian mathematics, like Manjula, (who wrote Udayantara), Ganesha Daivajna (who wrote Grihalaghara) etc.

The Hindu contributed to astronomy is specifically the lunar month (derived from moon-eth) of thirty days (adhika masa). The lunar day (the tithi) is again peculiar to Indian astronomy. Paying much tribute to Indian genius is Mathematics, Laplace, the great scientist of Greece, said, "How grateful we should be to Hindus who discovered zero and decimal systems that did not occur is the minds of such mighty mathematicians as Archimedes and Apollonious".

Reference:

1. Brahmspotsiddhant 1 P.P.153
2. R. D. Hachworth and J. Howland; Modular Math. Ser. 19.76, P.14
3. Rigveda 1.22. 16
4. Rigveda 8.33. 18.
5. Rigveda 1.15. 4
6. Rigved 1.15. 10
7. Rigveda 8.32.22
8. Rigveda 1.23.15
9. Rigveda 1.22.16
10. Rigveda 1.164.41
11. Rigveda 8.1.9.
12. Yajurveda 18.24
13. Rigveda 2.18.5-6
14. Ginsberg, Bull. Am. Math. Soc. 25 (II ser) 336-369
15. Srimad Valmiki Ramayan, Astavins: Sarg 33, 37, 39
16. पंचसिद्धान्तिका - 12-1
17. Arya Bhatt- Gitika
18. आ. भ. ग.-10
19. आ. भ. ग.- 6
20. आ. भ. ग.- 17
21. आ. भ. ग.- 9
22. आ. भ. ग.- 11] 12
23. आ. भ. का. क्रि. - 1
24. आ. भ. का. क्रि. - 2
25. आ. भ. ग.- 30
26. आ. भ. ग.- 29
27. आ. भ. ग.- 24 .
28. बौ. सुल्व सूत्र - 1- 1
29. बौ. सु. सू. - 1- 48
30. बौ. सु. सू. - 2- 12

Section III

Physics and Electronics

1

Quantum Physics and Its Classical Ancient Indian Counterpart

B. S. Rajput

1. Introduction

Though most of the Indians believe that we have a great cultural heritage, only a few among them will reconcile to include science in it. The reason for this apathetic view towards science as our ancient cultural heritage lies with our defective system of science education, where no serious effort has been made, even in the post independent period, to evolve a patriotic environment to cultivate originally in thoughts on the basis of our ancient scientific achievements. Inspite of all richness of systematic classification of knowledge of Science and Mathematics in our very ancient literature like four Vedas (Rigveda, Samaveda, Yajurveda and Atharvaveda), their four upvedas (Ayurveda, Gandharvaveda, Dhanurveda and Sthapatyaveda), six vedic darshans (Nyaya, Vaisesik, Sankhya, Yoga, Mimansa and Vedanta), four heterodox darshans (Jaina, Buddhist, Tantrics and Charvak), Upanishads, Brahmanas, Aranyakas, Srimad Bhagwat Gita and Mahabharat, desired consciousness could not be generated internationally about Indian contributions in Mathematics, Astrophysics, Geometry, Material Science, Cosmology and Life Science in ancient time. For instance, in the celebrated treatise "Human Intellectual Development" there is discovery of paper in China and history of science and agriculture in China but there is no mention at all about any scientific achievement of ancient India. Similarly, a scientist and thinker like S. Hawkins has referred to Greek, Arab and subsequent Christian religious thinkers but has not given any indication pointing towards Indian thoughts. The main reason for this attitude is that most of western scientists compare the ancient thoughts with the present thinking based on evidences of books available in west and the ideas in vogue since generations. They have either no knowledge of the India Vedic principles or even if they have some, it is limited only to the physical rituals of worshipping or the abstract spirituality. The world would have been enormously benefited if the wonderful scientific researches made by the scientists during past two centuries had been based on the background of Vedic thoughts.

According to the western science only that knowledge which is acquired by direct perception of the senses or through apparatus is the true knowledge. Indian science in ancient time went much beyond this concept and touched the last periphery of omniscience according to which we must acquire knowledge of all we can see, feel and perceive and also that we can not perceive. The ancient Indian science has always held a holistic view point while the modern science considers the creation in pieces or parts. The direct and simple course as followed in ancient India, is to take the whole first and then study its parts. There as countless evidences to believe that science first flourished long ago with the quest of early Indian philosophers for an underlying unity of the apparently rich diversities and complexities of Nature. This search in understanding Nature at the fundamental level in terms of processes and constituents, carried them much beyond the sensory world of appearance. Relying upon the rational analysis free from direct observational contents, they went off in vividly different directions in their study of concepts of individual atoms, the original bootstrap model of indefinitely divisible seeds within seeds and the more general Mathematical concepts of numbers, geometries and symmetries. They included science and philosophies in Vedanga and Upanga in terms of which the Vedic Knowledge could be perceived. The meaning of word Veda is an unlimited source of all knowledge and every Vedic mantra is original and authentic in its own right. The different formulae(in terms of these mantras) expounded in a complex language in Vedas, have been explained by great seers in a simple lucid language with the help of examples of legends. These seers or rishis were most enlightened persons and knowers of the facts, who saw the philosophy of life through mantras and they were also exuded scholars of science. The scientific researches were conducted in ancient India under their guidance and directions. The Purush Sukta gives an idea of the detailed scientific analysis of the whole universe and its different units. Every where in Vedas the references to Purush Sukta is not just formal but it is used as the formulae or codes to understand the scientific definitions. It is the sublimity of the Vedic literature that the words and sequences describing in a nut-shell the summary of science of creation can also easily explain the methodology of ancient scientific research. Ancient Indian thinkers noticed that the matter and consequently the universe are connected with motion which involves the concept of space as well as time. The doctrine of change has been understood in terms of cause and effect as narrated by Vedavyas in very simple words:

व्यतिरेकानवस्थितश्च अनपेक्षत्वात्

which means that the inanimate objects do not change their position unless they are affected by external force. The same concept was given by Newton in modern science as a law of motion that every particle of matter, continues in a state of rest or motion with a constant speed in a straight line unless compelled by a force to change the state. In Vedic literature the theory of universe had been dealt with matter, space, time and causality. Our Vedic literature is replete with the description of Brahmavidya or the cosmic science which is the nucleus of all knowledge, which has been acquired and also which is yet to be acquired. In this paper the classical and quantum concepts and the theory of causality expounded in ancient Indian literature (specially in Rigveda, Vaiseshik darshan and Sankhya darshan) will be explained and the links of these concepts with the modern quantum mechanics will be established. It will also be explained how the modern physicists, facing the reality which lies beyond the opposite concepts of mutually exclusiveness and complementarily, may be guided by the ancient Indian concepts to adopt a special way of thinking where the mind is not fixed in the rigid frame of classical logic but keeps moving and changing the view points in microscopic world.

2. Quantum Concepts of Sankhya Darshan

In Rigveda (which is believed to be most ancient scripture in the world), we find the beginning of the enquiry about the mysteries of creation (Nasadiya Sukta) and in Yajurveda, there is first mention of 27 or 28 stellar constellations. In their quests for ultimate harmony in the apparent diversities of nature, Vedic rishis had realized that the universe and its objects are the manifestations of positive unitarily principle known as Viśvajyoti (Cosmic energy) with their dynamical aspects assigned to another principle called Yajña. Different seers suggested different elements (Bhutas) to be the building blocks of the universe and consequently the Sankhya system of the Vedic philosophy, evolved on account of cosmic evolutions, had both the continuum and quantum concepts. At the time of outcome of Upanishads this darshan was intrinsically mixed up with Vedanta and for the first time Maharshi Kapil established it as a separate independent darshan. It admits only two elements (Prakriti and Purush) as fundamental ones which create this universe and hence this darshan is a dualistic representation which automatically incorporates the modern theory of causality related with cause and effect. According to it, Purusa by itself is inactive and plays no part in the evolution of Prakriti, which is considered as formless, undifferentiated, limitless, indestructible and uncontrollable (continuum). In the state of equilibrium Prakriti consists of three extremely fine gunas (sattva, rajas and tamas) which are not its qualities but are the constituent principles of the universe. With the transcendental influence of Purusa, the equilibrium of Prakriti gets disturbed and it starts to evolve itself through the relative preponderance of one or two gunas over the rest and subsequently gunas starts uniting, separating and uniting again and the process of creation begins with the emergence of mahat from the Prakriti with the predominance of sattva. Through the transformation with predominance of rajas, ahankara emerges from mahat. The accretion of tamasik component of anhankara leads to the creation of five tanmatras (sabda, sparsa, rupa, rasa and gandh), product of which gives the fundamental building block of universe. In Sankhya darshan the distinctions between dik and kala (space and time) is removed, the objective existence of time apart from moment is denied and the light is regarded as a kind of substance which has velocity and energy (quantum nature of light). The Prakriti of Sankhya is said to be the primordial cause which unfolds itself and manifests effects in the form of flow of energy leading to evolution in the frame work of niyamas (laws) like conservation of energy (and of mass). through keeping the totality of sattva, rajas and tamas gunas separately constant. In the destruction process the reverse of creation takes place and every thing finally gets dissolved into original Prakriti in a state of equilibrium and then the transcendental influence of Purusa starts new cycle again and this process of creation and annihilation goes on with a systematic rhythm making this universe a dynamic one exactly in the form as described by the modern theories of evolution and different type of interactions among the fundamental particles. In Sankhya darshan the world is conceived in terms of movement, flow and change, where one sees the universe as an inseparable web whose inter-connections are dynamic and not static. Modern physics also has come to conceive the universe as such a dynamical web, where the dynamical aspect of matter arises in quantum mechanics as a consequence of wave nature of atomic particles, which implies the fundamental restlessness of matter as a characteristic of sub-atomic world.

3. Quantum Concepts in Vaisesik Darshan

All basic characteristics of Prakriti were explained at fundamental level in Vaisesik darshan (a plural representation) which is one of the well known six systems of Vedic philosophy. This darshan has played an important role in the rigorous attempts to achieve the highest goal through the true knowledge of the physical world. Its expounder, Kanad, divided all measurable things primarily into two classes i.e. the existence and non -existence where the objects of the first category are further classified as Dravya (Substance), Guna (attributes), Karma (action), Samanya (genus), Visesa (species) and Samavaya (combination). The substance, attribute and action are further subdivided in to nine, twenty four and five classes respectively, The substance possesses action and attributes and is not the combinatory cause;

क्रियागुणवत् समवायिकारणमिति द्रव्यलक्षणम्।[1]

अथवा

कारणमिति द्रव्ये कार्यसमवायात्।[2]

It is relatively eternal as ultimate paramanu and non- eternal as observable bodies in the wide sense of the terms. The matter, eternal form of substance, is again of two types i.e. parmanvic and ubiquitous. In Vaisesik, the divisions of substance are nine only; Kshiti (earth), Apah (water), Tejas (fire), Vayu (air), Akasha (ether), Dik (space), Kala (time), Manas (mind) and Atman (self), where the first five elements are Bhautik (material) and the others are Abhautik (non-material) Ether, time and space denote single reality and provide eternal (continuum) back ground of the creative (discrete) processes. Mind helps in organization of the matter produced in to structure capable of experiencing pleasure and pain in association with self. The first five elements enter as principles (building blocks) in the composition of the physical world and not as ordinary earth, water, air etc. All these elements have number, dimension, distinction, conjugation and disjunction. In the modern physics, starting with electrons, protons, neutrons and mesons as the constituent of atom, the new constituents like a large number of strange particles, several types of mesons and resonances and various other mysterious strongly interacting particles (hadrons) have been discovered. Rigorously perturbed and disillusioned in the zoo of these large number of the so called fundamental particles (building blocks) modern high energy physicists described these particles as composite ones, classified them into the category of leptons (electron, μ- meson, $\hat{o}$- meson and the corresponding neutrinos) and hadrons (strongly interacting particles) and proposed quarks (five required by experimental reasons and six by the symmetry theory) as the constituents of hadrons. This five quark model resembles with the old five elements theory of Vaisesik darshan. This assumption of five quarks (or six quarks) implies the existence of thirty (or thirty six) constituents (each quark in three colors and anti-quark corresponding to each quark) of hadrons. This number thirty (or thirty six) is also to large to be accepted as building blocks used by very intelligent Prakriti (Nature) which is always keeping many secrets hidden in spite of recent inventions of very sophisticated and sensitive instruments on the ever shrinking scales of space and time. A new concept is recently creeping in assuming these quarks also composites each carrying two fundamental constituents which are space ṣ time manifestations of certain external degrees of freedom. It resembles with the dual representation described long ago in Sankhya darshan.

Vaisesik had gone much ahead than the modern Physics and laid emphasis on the attributes which do not have separate existence but reside in the matter. An attribute is not an independent cause of conjunction and disjunction.

रूप-रस-गंध-स्पर्शसंख्याः परिमाणानि
पृथक्त्वं संयोगविभागौ परत्वापरत्वे बुद्धयः
सुख-दुखे इच्छाद्वेषौ प्रयत्नश्च गुणाः।[3]

These are twenty four in number and in terms of them five tanmatras (patrons) create matter (paramanu) with definite discrete orders(quantum numbers) from the continuum. Vaisesik believes that the paramanus, the ultimate constituent of matter, is super-sensuous, indestructible, indivisible, and un-perceivable. Its non-perceptance does not imply its non existence since according to Nyaya-Vaisesik, the presence of magnitude (mahattava) in an object is one of the conditions of its perceptions and as there is no magnitude is the paramanu it is not perceived. By themselves, paramanus can not produce any thing and yet they are the ultimate material cause of the universe. As such the vaisesic paramanu has nothing in common with the modern atom but shows exciting similarly with the fundamental building blocks (quarrels) of modern Physics. Vedant also accepts parmanuvada but it speaks of paramanus not as the ultimate constituent of matter but as the smallest conceivable quantum of matter.

The mathematical framework of modern quantum theory has passed countless successful tests and it is now the universally accepted as a consistent and accurate description of all atomic phenomena. It is based on the postulate that a wave is associated to every moving particle and vice- visa. It takes into account the interactions between the observation and the quantity being observed and shows that an experiment is a violent interference with the object being measured. These probability waves lead to the opposite concepts of existence and non- existence (described in details in Vaisesik darshan) and preclude the possibility of obtaining complementary informations about a system. This wave-particle duality led Bohr to postulate the principle of complementity according to which the claim of causality and space - time coordination are complementary to each other. It resulted into Heisenberg's uncertainty principle of mutual exclusiveness of space-temporal and casual description of microphysical processes. Thus the quantum description of any phenomena involves two concepts that are mutually exclusive but both are necessary. This complementarity of casual description and space- time coordination had been nicely explained in Vaisesik darshan where in tenth book there are a number of sutras related with the nature of cause and effect. The vivarti on one of these mantras refers to three kind of causes:

कारणं त्रिविधं समवायिकारणासमवायिकारणनिमित्तकारणभेदात्।[4]

How the cause is related to effect is a very difficult question to answer even in the most branches of modern science. On the other hand the Vaisesik gives the following important maxims relating cause and effect;

न द्रव्यं कार्यकारणं च बाधति। [5]

which means that substance in not annihilated either by itself or by cause;

क्रियागुणवत् समवायिकारणमिति द्रव्यलक्षणम्।[6]

which means that substance is characterized by action and attribute and is not an inherent cause;

कारणाभावात् कार्याभावः।[7]

which means that if cause is destroyed the effect would be destroyed,

न तु कार्याभावात् कारणाभावः।[8]

which means that the non- existence of cause does not follow from the non existence of effect,

कारणगुणपूर्वककार्यगुणो दृष्टः।[9]

which means that the attribute of the effect is seen to be preceded by cause.

These maxims of Vaisesik lead to the conclusion that matter has inherent causality which is turn in responsible for its (matter's) conservations. Vedanta rejected arambhvad and suggested vivartvad as the principle of causality. In Vaisesik the claim of causality and space- time coordination are complementary and mutually exclusive. It forms the basis of Bohr's complementary principle and Heiserberg's uncertainty principle, which are the foundation stones of modern quantum theory but the Vaisesik complementary principle is free from the wave - particle duality or the probabilitist interpretations which made Einstein most eloquent opponent of modern quantum mechanics as became clear in fifth Solvay congress held during 24-29 Oct. 1927 and in sixth Solvay congress held during 20-25 Oct. 1930 where he used the statements like "Not facing the facts", "Avoiding reality and reasons" and "Having abandoned the reasons" against the modern quantum Physics. The debates between Einstein and Bohr during these conferences were viewed by many thinkers as "Irreconcilability of materialism and idealism".[10] In 1935, Einstein, Podolosby and Rosen[11] demonstrated that the wave function of quantum equation does not provide a complete description of physical reality and used their arguments to prove that the modern quantum theory is an incomplete theory. Einstein never changed his mind on this matter. In 1950, less than five years before his death, he wrote to Schrodinger that the fundamentally statistical character of the theory was simply a consequence of incompleteness of the description of quantum mechanics where one hand does not know what does other hand do. In a letter to Born he wrote that he could not believe in dice throwing God through this quantum mechanics. Had the quantum mechanics been formulated on the basis of complementary principle of Vaisesik, these arguments could have not been raised against it.

Quantum mechanics does not generally yield with certainty the result of a given experiment performed on an individually selected system, but the statistical distribution of the results obtained when one repeats the same measurement on a very large number of independent systems represented by the same wave functions, According to this view the knowledge of wave function would not suffice to define completely the dynamical state of an individual physical system. As such the dynamical state of the physical system should be defined at each instant by a certain number of hidden variables whose evolution would be governed by some specific laws.

Bohr maintained that since the interaction between object and the instrument can not be neglected in microworld, an independent reality in the ordinary physical sense can neither be ascribed to the phenomena nor to the agencies of observation. Accordingly, on the microscope level, one can not make a sharp separation, required by the ordinary concept of observation, between the natural phenomena and the instrument with which it is observed. As such, certain cognitions which co-exist and harmonize in the classical theory are split, in quantum mechanics, into mutually exclusive views complementary to each other. In view of these aspects of quantum uncertainty, we can not say that an atomic particle exists at a certain place, nor can we say that it does not exist there. We can neither say that it changes its position nor that it remains at rest. In the words of Robert Oppenheiver this microscopic indeterminacy puts us in the following situation:

"If we ask, whether the position of an electrons remains the same, we must say no; if we ask whether the electron is at rest, we must say no and if we ask whether it is in motion, we must say no."

This statement of Oppenheiver seems to echo the words of Upanishad [12];

तदेजति तन्नेजति तद्दूरे तदन्तिके।

तदन्यस्य सर्वस्य तदु सर्वस्यास्य बाह्यतः॥

which means that

"It moves, it moves not,

It is far and it is near,

It is within all this

And it is outside of all this".

The quantum uncertainty is the helplessness of the common intelligence which has been indicated by the expression "Yo Budhheh Paratastush" in Bhagvad Gita and also by the expression "Sadasadanirvachaniya" in the Advaita Vedantic Philosophy of Adi Shankracharya, meaning that which is existence and also non - existence and therefore indeterminate.

Force and matter, particle and wave, motion and rest, existence and non-existence etc. are some of the contradictory concepts which are transcendent in modern quantum mechanics. In atomic physics we have to go beyond the concept of existence and non - existence. Ancient Indian faith deals with a reality which lies beyond existence and non - existence. This important fact has been emphasized in the following statement of Ashvaghosha;

"Suchness is neither that which is existence; nor that which is non - existence; nor that which is at once existence and non - existence; nor that which is not at once existence and non - existence"

Faced with a reality which lies beyond the opposite concepts, physicists have to adopt a special way of thinking where the mind is not fixed in the frame work of classical logic, but keeps moving and changing the view points in atomic physics as described in Mahabharat[13]

यत् किंचिन्मन्यसेऽसृटीर्ति सर्वं नास्तीति विद्धि तत्।

एवं न व्यथते प्राज्ञः कृच्छ्रामप्यापदं गतः॥

which means for any thing you think existence first understand that it is non- existence

Also

यद्धि भूतं भविष्यं च सर्वं तन्न भविष्यति।

एवं विदित वेद्यस्तव मर्घ्मेभ्यः प्रमोक्ष्यसे॥[14]

which means any thing which existed in past and will exist in future was non- existence and will remain non- existence.

In contrast to various attempts at extending the conventional formalism of quantum mechanics to incorporate a consistent theory of measurement, it has been argued that quantum mechanics should be reformulated to include altogether the notion of measurement since the orthodox quantum theory does not specify whether an object plus apparatus undergoes a determinable transformation in accordance with Schrodinger's time dependent equation or whether it undergoes a probabilistic transformation associated with the reduction of wave packets [15] It is this inability of the orthodox quantum theory to specify precisely such mutual exclusive conditions that lies at the root of insolubility of the measurement problem. Physicists have to learn how to play with two mutually exclusive pictures switching form one to other and back in order to cope with the reality. This concepts is reflected in Bodh view;

"Reality is a multidimensional impression formed from the superimposition of single impression from different point of views"

References

1. Vaisesik Sutra 1.1.15
2. Vaisesik Sutra 10.2.1
3. Vaisesik Sutra 1.1.16
4. Tarkasangrah 1.
5. Vaisesik Sutra 1.1.12
6. Vaisesik Sutra 1.1.15
7. Vaisesik Sutra 1.2.1
8. Vaisesik Sutra 1.2.2
9. Vaisesik Sutra 2.2.24
10. H. Naumann; Deutsche Zeitschreft fur Philosophie 7 (1959) 389
11. A. Einstein, B. Podolsby and N. Rosen; Phys. Rev. 47 (1935) 777
12. Ishavasopnishad-5
13. Mahabharat (Shanti Parva)
14. Mahabharat (Shanti Parva)
15. N. Maxwell; Am. J. Phys. 40(1972) 1431

2

Quantum Teleportation: A Vedic View

S P Gupta

Introduction

Technology is the offspring of science. And science is an organized knowledge. Then the present day scientific knowledge - with all its intellectual fervour - is not the complete understanding of the nature. It is at best a training, which teaches a man to be an adopt 'magician', performing wonders by taming genies, the forces of nature, and not that sublime knowledge which opens him up to a cosmic vision - the vision of harmony, peace and beatitude. Quantum teleportation is a technique discussed in Quantum information science to transfer a quantum state to an arbitrarily distant location using an enlarged state and the transmission of some classical information Eistien et al (1935). In quantum mechanics, the state of a system is described by a mathematical entity called the wave function or state vector. A system means basically any physical object, but think of particles such as photons, atoms, ions or molecules maybe more than just one.

According to Heisenberg's uncertainity principle, it is impossible to completely measure the state vector of an individual quantum system, as its different constituents (amplitudes) are complimentary: If you measure one quantity to precisely obtain its value, you necessarily disturb (change) some other quantity so that you cannot find out its value. It is impossible to get all the "quantum information", which completely describes a system, out of it. This makes it impossible to create an exact duplicate of it. Hence, no-cloning theorem asserts that this still holds if you try to find other techniques than measuring and recreating the system. However, you can transfer the complete state of a system onto a similar system which is at some other place. The original system loses its state, of course, in order to follow the no-cloning postulate. A teleportation machine would be like a fax machine, except that it would work on 3-dimensional objects as well as documents, it would produce an exact copy rather than an approximate facsimile, and it would destroy the original in the process of scanning it.

Indistinguishability

Let's say that Alice has a rubidium atom Hagley et al (1997), which is in its ground state, and Bob also has such an atom, as well in its ground state. It is important to see that these two atoms are indistinguishable; that means that there really is no difference between them.

If Alice and Bob had, say, two glass balls, which exactly look alike, and they exchanged them, then something would change. If you had a powerful techneque, you could certainly find some difference between the two balls. For atoms of the same kind and in the same quantum state, however, there really is *no* difference at all. The physical situation with Alice having the first atom and Bob the second is exactly the same as vice versa Bennett et al (1993). In a certain sense, it is even wrong to say that the two atoms have any individuality or identity. It would be more appropriate to say that the two locations in space both have the property that the fundamental quantum fields have those values which define the ground state of the rubidium atom.

In the 1964, John Bell showed that a pair of entangled particles, which were once in contact but later move too far apart to interact directly, can exhibit individually random behaviour that is too strongly correlated to be explained by classical statistics. Experiments on photons and other particles have repeatedly confirmed these correlations, thereby providing strong evidence for the validity of quantum mechanics, which neatly explains them.

Another well-known fact about EPR correlations is that they cannot by themselves deliver a meaningful and controllable message. It was thought that their only usefulness was in proving the validity of quantum mechanics. But now it is known that, through the phenomenon of quantum teleportation, they can deliver exactly that part of the information in an object that is too delicate to be scanned out and delivered by conventional methods.

In the Fig.1 and Fig.2, we compare conventional facsimile transmission with quantum teleportation. In conventional facsimile transmission the original is scanned, extracting partial information about it, but remains more or less intact after the scanning process.

The scanned information is sent to the receiving station, where it is imprinted on some raw material (e.g. paper) to produce an approximate copy of the original. In quantum teleportation two objects B and C are first brought into contact and then separated.

Object B is taken to the sending station, while object C is taken to the receiving station. At the sending station object B is scanned together with the original object A which one wishes to teleport, yielding some information and totally disrupting the state of A and B. The scanned information is sent to the receiving station, where it is used to select one of several treatments to be applied to object C, thereby putting C into an exact replica of the former state of A.

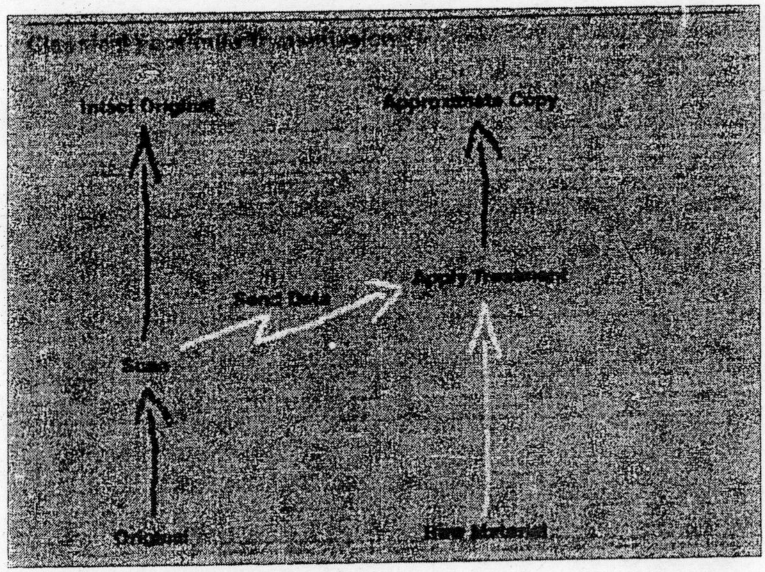

Fig. 1

In 1993, Bennett et al at I B M confirmed the intuitions of the majority of science fiction writers by showing that perfect teleportation is indeed possible in principle, but only if the original is destroyed.

Until recently, teleportation was not taken seriously by scientists, because it was thought to violate the uncertainty principle of quantum mechanics, which forbids any measuring or scanning process from extracting all the information in an atom or other object.

According to the uncertainty principle, the more accurately an object is scanned Pan et al (1998), the more it is disturbed by the scanning process, until one reaches a point where the object's original state has been completely disrupted, still without having extracted enough information to make a perfect replica.

This sounds like a solid argument against teleportation: if one cannot extract enough information from an object to make a perfect copy, it would seem that a perfect copy cannot be made. But the six scientists found a way to make an end-run around this logic, using a celebrated and paradoxical feature of quantum mechanics known as the Einstein-Podolsky-Rosen effect.

In brief, they found a way to scan out part of the information from an object A, which one wishes to teleport, while causing the remaining, unscanned, part of the information to pass, via the Einstein-Podolsky-Rosen effect, into another object C which has never been in contact with A. Later, by applying to C a treatment depending on the scanned-out information, it is possible to maneuver C into exactly the same state as A was in before it was scanned.

A itself is no longer in that state, having been thoroughly disrupted by the scanning, so what has been achieved is teleportation, not replication.

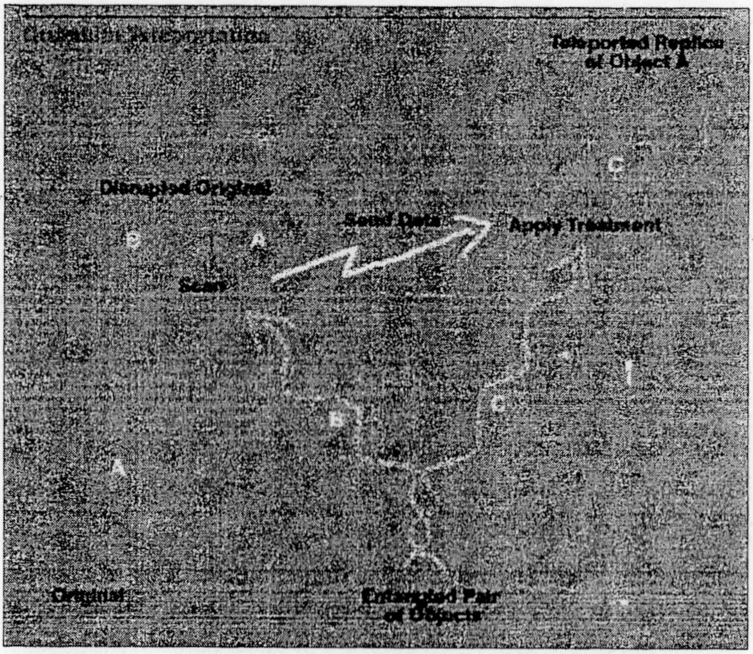

Fig. 2

The Fig.2 suggests, the unscanned part of the information is conveyed from A to C by an intermediary object B, which interacts first with C and then with A. What can it really be correct to say "first with C and then with A"?

In order to convey something from A to C, the delivery vehicle must visit A before C, not the other way around.

But there is a subtle, unscannable kind of information that, unlike any material cargo, and even unlike ordinary information, can indeed be delivered in such a backward fashion. This subtle kind of information, also called "Einstein-Podolsky-Rosen (EPR) correlation" or "entanglement", has been at least partly understood since the 1930s when Albert Einstein, Boris Podolsky, and Nathan Rosen discussed it in a famous paper.

The Method of Teleportation

As a prerequisite, Bob has produced two particles (or atoms) called I and II, which are maximally entangled, e.g. in the Bell state

$$|\phi)=\sqrt{1/2}\,(|0(I\otimes|0)\Pi+(1)I\otimes|1\rangle()$$

This means that the particles I and II are neither in the state |0> nor in |1> but rather in both simultaneously, and if you measure one of them you will find it to be |0> or |1> each with probability 1/2. But the other is then always in the same state, and this connects them by some "spooky action at a distance", as Einstein has put it.

Bob passed particle I to Alice (by moving it there through a quantum channel) and kept particle II. From now on, Alice can send a quantum state to Bob, whenever he wants (or Bob to Alice).

If Alice now has a particle which he wants to teleport to Bob, he does a so-called Bell measurment on the particle to be sent and on particle I. The Bell measurement projects Alice's two particle into one of the four Bell state. The two measured particles are afterwards in a known entangled state, specified by the measurement result, and the particle to be sent has lost its former state. That state, however, is not gone: it was teleported to Laxaman's particle II due to the previously existing entanglement.

One problem remains: The state of Bob's particle might be "rotated". This depends on the outcome of Alice's measurement, and Alice has to tell Bob this outcome by using some classical (i.e. ordinary) communication channel and Bob will then apply a unitary operation onto his particle. Afterwards, he really has the desired state.

Bell-State Measurement

In previous discussions we almost always talked about the spin state of electrons, although we regularly pointed out that the same situations exist for the polarization of light, albeit with a difference of a factor of 2 in the angles being used. Here we will reverse the situation, and mostly talk about polarization states for photons, although the arguments also apply to spin states of electrons.

The fact that we may talk about light polarization in almost the same way that we discuss electron spin is not a coincidence. It turns out that photons have spins that can exist in only two different states. And those different spins states are related to the polarization of the light when we think of it as a wave.

Here we shall prepare pairs of entangled photons with opposite polarizations; we shall call them *E1* and *E2*. The entanglement means that if we measure a beam of, say, *E1* photons with a polarizer, one-half of the incident photons will pass the filter, regardless of the orientation of the polarizer. Whether a particular photon will pass the filter is random. However, if we measure its companion *E2* photon with a polarizer oriented at 90 degrees relative to the first, then if *E1* passes its filter *E2* will also pass its filter. Similarly if *E1* does not pass its filter its companion *E2* will not.

In a number of standard textbooks, we discussed the Michelson-Morley experiment, and later the Mach-Zehnder interferometer. You will recall that for both of these we had half-silvered mirrors, which reflect one-half of the light incident on them and transmit the other half without reflection. These mirrors are sometimes called beam splitters because they split a light beam into two equal parts.

We shall use a half-silvered mirror to perform Bell State Measurements. The name is after the originator of Bell's Theorem.

We direct one of the entangled photons, say *E1*, to the beam splitter.

Meanwhile, we prepare another photon with a polarization of 45^0, and direct it to the same beam splitter from the other side, as shown. This is the photon whose properties will be transported; we label it *K* (for Kirk). We time it so that both *E1* and *K* reach the beam splitter at the same time.

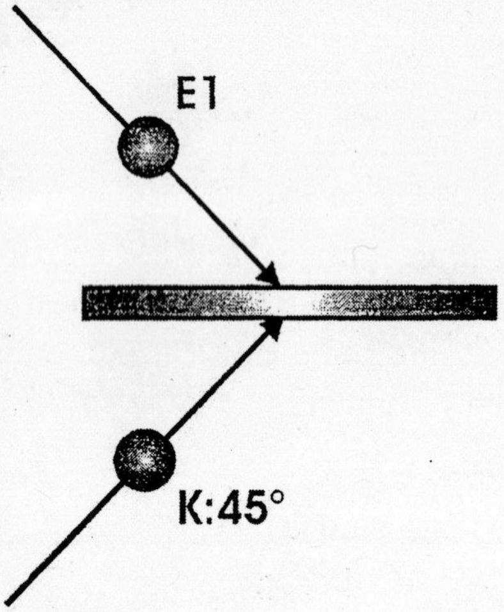

The *E1* photon incident from above will be reflected by the beam splitter some of the time and will be transmitted some of the time. Similarly for the *K* photon that is incident from below. So sometimes both photons will end up going up and to the right as shown.

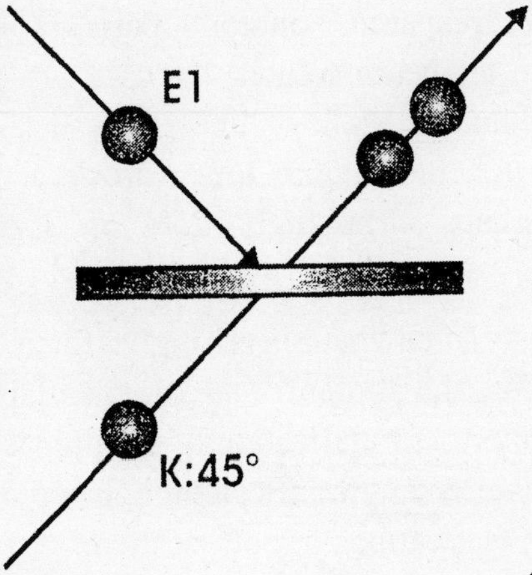

Similarly, sometimes both photons will end up going down and to the right.

But sometimes one photon will end up going upwards and the other will be going downwards, as shown. This will occur when either both photons have been reflected or both photons have been transmitted.

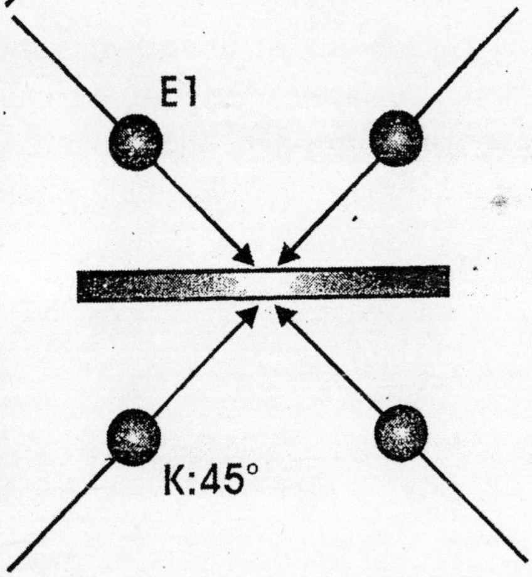

Thus there are three possible arrangements for the photons from the beam splitter: both upwards, both downwards, or one upwards and one downwards.

Which of these three possibilities has occurred can be determined if we put detectors in the paths of the photons after they have left the beam splitter.

However, in the case of one photon going upwards and the other going downwards, we cannot tell which is which. Perhaps the beam splitter reflected both photons, but perhaps both were transmitted.

This means that the two photons have become entangled.

If we have a large beam of identically prepared photon pairs incident on the beam splitter, the case of one photon ending up going upwards and the other downwards occurs, perhaps surprisingly, 25% of the time.

Also somewhat surprisingly, for a single pair of photons incident on the beam splitter, the photon *E1* has now collapsed into a state where its polarization is -45^0, the opposite polarization of the prepared 45^0 one. This is because the photons have become entangled. So although we don't know which photon is which, we know the polarizations of both of them.

The explanation of these two somewhat surprising results is beyond the level of this discussion, but can be explained by the phase shifts the light experiences when reflected, the mixture of polarization states of *E1*, and the consequent interference between the two photons.

The Teleporter

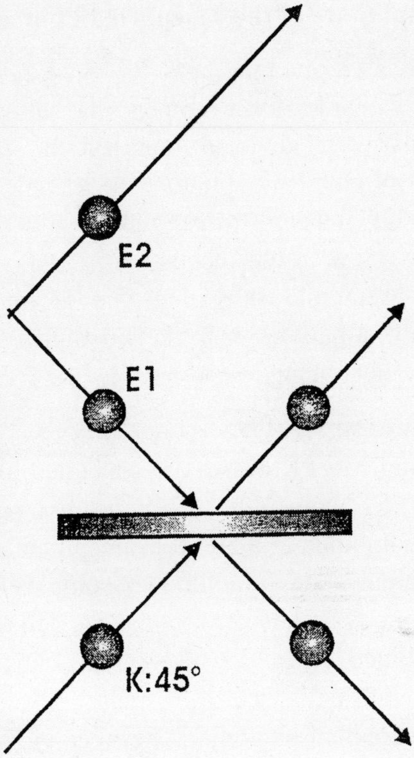

Now we shall think about the *E2* companion to *E1*.

25 percent of the time, the Bell-state measurement resulted in the circumstance shown, and in these cases we have collapsed the state of the *E1* photon into a state where its polarization is -45^0.

But since the two photon system E1 and E2 was prepared with opposite polarizations, this means that the companion to E1, *E2*, now has a polarization of +45^0. Thus the state of the *K* photon has now been transferred to the *E2* photon. We have teleported the information about the *K* photon to *E2*.

Although this collapse of *E2* into a 45^0 polarization state occurs instantaneously, we haven't achieved teleportation until we communicate that the Bell-state measurement has yielded the result shown. Thus the teleportation does not occur instantaneously.

Note that the teleportation has destroyed the state of the original *K* photon.

Quantum entanglements such as exist between E1 and E2 in principle are independent of how far apart the two photons become. This has been experimentally verified for distances as large as 10km. Thus, the Quantum Teleportation is similarly independent of the distance.

The Original State of the Teleported Photon Must be Destroyed

Above we saw that the K photon's state was destroyed when the E2 photon acquired it. Consider for a moment that this was not the case, so we end up with two photons with identical polarization states. Then we could measure the polarization of one of the photons at, say, 45^0 and the other photon at 22.5^0. Then we would know the polarization state of both photons for both of those angles.

As we saw in our discussion of Bell's State Measurement, the Heisenberg Uncertainty Principle says that this is impossible: we can never know the polarization of a photon for these two angles. Thus any teleporter must destroy the state of the object being teleported.

Teleportation Apparatus

Entangled EPR beams are generated by combining two beams of squeezed light at a 50/50 beamsplitter. EPR beam 1 propagates to Alice's sending station, where it is combined at a 50/50 beamsplitter with the unknown input state, in this case a coherent state of unknown complex amplitude. Alice uses two sets of balanced homodyne detectors to make a Bell-state measurement on the amplitudes of the combined state. Because of the entanglement between the EPR beams, Alice's detection collapses Bob's field (EPR beam 2) into a state conditioned on Alice's measurement outcome. After receiving the classical result from Alice, Bob is able to construct the teleported state via a simple phase-space displacement of the EPR field 2 (Fig 3).

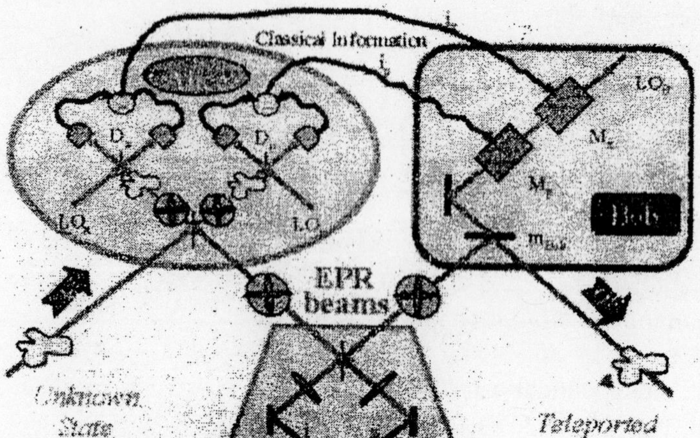

Fig. 3 Theoretical view of experiment

The first experimental demonstration of entanglement was carried out by Bouwmeester et al (1997). Their setup allowed to teleport the polarization quantum state of photons across an optical table. A slightly similar experiment by Boschi et al (1998) showed the principle of teleportation, however it involved only two particles, instead of three: one carrying the state to be teleported, and two entangled ones, which provide the "quantum channel" and must be initially independent of the one carrying the state.

Riebe et al (2004) and Barrett et al (2004) demonstrated the teleportation of atoms: The electronic quantum state of a calcium or beryllium ion was teleported to another one, with all the ions being held in a linear ion trap.

Rigolin (2005) have shown a protocol in which an arbitrary two qubits is faithfully and deterministically teleported from Alice to Bob. He constructed the 16 orthogonal general Bell states, which can be used to teleport the two qubits. While, Diaz-Caro (2005) introduced a quantum circuit to implement teleportation of N qubits with Bennet protocol. He analyzed the teleportation algorithm to simplify the operations to be implemented by Bob (receiver) in order to obtain the teleported N qubits state.

Vaidman (1998) emphasized that teleportation of people will remain a dream for future. It is fact that teleportation of an unknown quantum state has not yet been achieved but recent experiments clearly demonstrate that it can be done. I suggest the experimentalists to perform a pursuance teleportation experiment: Carol gives to Alice (single) particle in different state (unknown to Alice), Alice teleports the state to Bob, Bob gives them back to Carol who test that what she gets is which she sent before

How this is accomplished is usually not explained in detail, but the general idea seems to be that the original object is scanned in such a way as to extract all the information from it, then this information is transmitted to the receiving location and used to construct the replica, not necessarily from the actual material of the original Neves et al (2005), but perhaps from atoms of the same kinds, arranged in exactly the same pattern as the original.

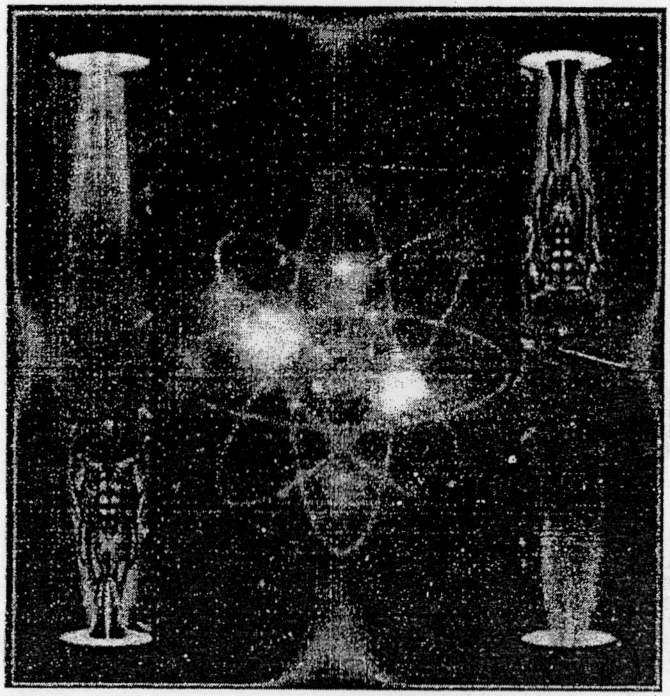

Fig. 4

It is pointless to put blame in epoch of eternal conflict between Truth and Untruth, Light and Dark and Immortality and Death (Fig.4). Finally, quantum teleportation is the beginning of axiomatic statement of *Upnisad (Kenwilber, 1981) - sarvam khalu idam brhama.*

References:

1. A. Einstein et. al.; Phys. Rev. **47** (1935) 777.
2. C. H. Bennett, G. Brassard, C. Crêpeau, R. Jozsa, A. Peres and W. K. Wootters; Phys. Rev. Lett. 70 (1993) 1895.
3. Kenwilber, Up From Eden (Routledge and Kegan Paul, 1981).
4. G. Brassard, S Braunstein and R Cleve; Physics D **120** (1998) 43
5. G. Rigolin; Phys. Rev. A **71** (2005) 032303.
6. A. Dîaz-Caro; arXiv quant-ph/0505009 (2005).
7. D. Bouwmeester et al; Nature **394** (1998) 840
8. D. Bouwmeester, J.-W. Pan, K. Mattle, M. Eibl, H. Weinfurter and A. Zeilinger; Nature **390** (1997) 575.

9. D. Boschi, S. Branca, F. De Martini, L. Hardy and S. Popescu; Phys. Rev. **80** (1998) 1121.

10. M. Riebe, H. Häffner, C. F. Roos, W. Hänsel, J. Benhelm, G. P. T. Lancaster, T. W. Körber, C. Becher, F. Schmidt-Kaler, D. F. V. James and R. Blatt; Nature **429** (2004) 734.

11. M. D. Barrett, J. Chiaverini, T. Schaetz, J. Britton, W. M. Itano, J. D. Jost, E. Knill, C. Langer, D. Leibfried, R. Ozeri and D. J. Wineland; Nature **429** (2004) 737.

12. J W Pan et al; Phys Rev Lett **80** (1998) 3891

13. D Greenberger; Physica Scripta **176** (1998) 57

14. L Neves et al; Phys Rev Lett; **94** (2005) 100501

15. L Vaidman; arXiv; quant.ph/9810089 v1 (29 Oct 1998)

16. E Hagley; Phys Rev Lett **79** (1997) 1

17. Kim et al Phys Rev Lett **86** (2001) 1370

18. Tarlton M, Amarican Institute of Physics (Private Communication, 2005).

3

The Origin and Development of Technology (Electronics) in Vedic Literature

M.K. Kawadkar
Vijay Didolkar

1. Let us face one question, and ask ourselves, why do we want to study VEDAS for TECHNOLOGY ?

A little elaboration may help us to go near the centre of the problem if not hit the bull's eye.

a. Are we short of technology? Possibly, tremendous growth has taken place in the modern times of which the Indian talent is a sizable partner. In a few cases we are also world leaders. Do these scientists and technologists need VEDIC TECHNOLOGIES ? The answer is probably obvious and negative.

b. Then who needs it and why ?

This question can be variedly answered.

1. The first suggestion may be to increase the self respect amounting to pride.
2. It may also be a significant thought that the Vedic technology being INDIGENOUS, will provide us self reliance.
3. Curiosity to find something, wonderous ? or speculative?
4. Do we want to create another universe from Vedic knowledge, which will be more favourable to us, than others ? As there is a strong belief that VEDAS contains the TOTAL knowledge of the inceptions to evolution of the universe.
5. Is it for rediscovery of the mythology?
6. Do we want to provide more firm foundations for the occult and metaphysics?
7. Do we want to discover more about such practices as Yajñas, mantras, with their stipulated magical power. The existence of Svarga, Narak, Moksha, Mukti etc ?

8. Shall we limit out studies to PRACTICAL sciences such as agriculture, irrigation, medicine, the traditional Viman shastra and the like ?

9. There can be 'n' number of such suggestions. One may have to choose his own goal, as to why he should pursue the study of Vedas for technology. More clarity is required in this matter. There is a full length of scale between planned and regulated study and the freelancing.

2. Second point to ponder may be, who is competent to take the task of study. Obviously the scholars of the respective branches of science and technology will be most suitable for this task. Will such meritorious men find time to do this job, and what is the incentive to them. Else, we have to be optimistic and hope that something will comeout.

3. It is well known that the Vedas are available to us in a language called SANSKRIT. It is fortunate that a few scholars do accept that there are some difference in Vedic Sanskrit and the modern Sanskrit. The magnitude of the difference is considered as inconsequential and not phenomenal. Considering the fact that the language undergoes a change over a period and has a SIGNIFICANT difference over different territories, a researcher may have to be a little sceptic as to whether, the difference between the modern Sanskrit and the Vedic language is SIGNIFICANT or not. Then, what role is enjoyed by modern Sanskrit scholars in study of Vedic literature, who possibly have no technological background. This may be a very crucial point to carry out a research in the Vedas because of the great difference in time and space between the present and the Vedic past.

The study of a language aims at the study of Syntax and Semantics. Here, the words have PRE-FIXED meanings with due reference to their positions in sentences. This is irrespective of whether a TEXT exists or not and every given text is interpreted with reference to this RIGID frame of syntax and sementics. How rational, will this approach be, when applied to interpretation of Vedas where, the context has little role for the meanings of words. The present picture is that the RIGID semantics control the text and vice versa is not permitted. This needs a change: Since Vedas are ANCIENT, application of RIGID semantics is likely to be perplexing and intangible. Therefore first point of research is to develop semantics fully guided and CONTROLLED by the Vedic context and not attempt an experiment of fitting a square peg in round hole. Such method are well in progress. Of course some minimum knowledge of traditional Sanskrit will be a great help along with technological background. Rigidity may be only a curse rather than a blessing. It may only be remarked that two languages i.e. modern Sanskrit and the Vedic language have great difference. It is very well known that Vedic language is an accentuated one but not the modern one. It is also well known that the accentuations makes all the difference in the meanings. However, the treatment given to this subject is more of a MUSCIAL nature and of high and low tones, rather than conducive to semantics. Thus, in Vedic Sanskrit, the word masa, masa, masa, and masa written in four different styles have different meanings, and will be heterogeneous.

Another character worth observation is, certain concepts may not have developed in a particular period, but the words existed with different meanings. To cite an example, let us consider `aśīti' which in modern language means eighty. However, a context guided study of Vedic literature, suggests the meaning as aśīti = fully grown, fully developed or matured. For test, a case study of Ādya Śaṅkarāchārya's construction, is considered. He has stated in his Devyaparādha Kṣ amāpana Stotram, second line in stanza five as "mayā pañchāsiteradhikamapanite tu vayasi" Which is interpreted by the modern Sanskrit scholars as "I have spent more than eighty five years of my age". It is fairly well known that Srimad Ādya Śaṅkarāchārya had taken a nirvana by the age of thirty four or so. Then one has to face a controversy. One of the many solutions is that this may be some different Śaṅkarāchārya. Some say, it is a reference to his psychological (adhyatmic) age and not the chronological age of his mortal body and so on and so forth. If however, the researched Vedic meaning is considered, the stanza will mean. "I have spent extensive period of full maturity of my age". With this interpretation the apparent controversy vanishes and there is no need for any explanations and the beauty of construction unveil without a blemish. How much heartening it is for the researchers to know that up to the period of Ādya Śaṅkarāchārya, the meaning of aśīti had been as "fully grown" and not eighty.

4. The fourth point to consider is to judge a technological level of interpretations. An example may be worth it. An excerpt from Kanada's Vaisesika Sastras @ 2.1.1 "Gandhavati Prithivi", may suffice. As per modern Sanskrit "gandha" is smell, and Prihivi is the Earth or soil. Thus, at the outset, the Sanskrit statement means the earth has a smell. May be, this is true at its own level. Let us not ask awkward questions and refrain from making such supplementary statement that dogs have much better sense of smell than men etc. and accept it as true. Now let us redefine, "gandha" as "ganam dhārayti iti gandhaḥ" and √Prath as the root of prithivi. Prath is to mean to increment. Then prithivi will mean an increment. Stated mathematically it may be written as δx or Δx. The whole statement may mean, summability (gandhavati) depends upon the incrementability (prathivi). Stated simply it is $y = x + \delta x$. The question is whether the traditional Sanskrit

scholars will agree with this effort of scientific conclusion. The answer is obvious negation. One may now appreciate that the two levels, differ very widely. The second level helps us to get an entry into the science of mathematics in Sanskrit, Where the first level of interpretation is by choice and is more a personality parameter rather than an attribute of a text.

5. Validity, reliability and authenticity :

Having pointed out, that VEDIC language may be a different language than modern Sanskrit the question will be asked as to who will be the "judge". It may be beneficial NOT TO EXAMINE the Vedic research EXTERNALLY. Built in INTERNAL CONSISTANCY shall be sufficient to accept the Vedic meaning. Another criterion of judging the validity is to watch the continuity. When the interpretation reveals that there is no continuity between consecutive Richas, the research study has to aim at providing the continuity.

6. To study the subject like ELECTRONICS in Vedas a great deal of ground work is required. At least elementary knowledge of electronics, is a must. More advanced knowledge may be a boon. An unbiased survey of Vedic literature may have to be taken because, there is no "subject inventory" or a between consecutive Richas. The research study has to aim at providing the catalogue to find out the location of electronics in Vedas. This is very crucial. This may take very long time. If one is fortunate he may find it earlier. Another difficulty is that, on its surface, the researcher may be drifted away by his preconcepts. The only method that is possible is to keep all technological senses open and look to the character of the text and try to match the character with the known character of the subject. If the match is satisfactory, then proceed to develop the directory of word characters. Then, test of the hypothetical meaning, for internal consistency. There is no short cut to this procedure. Technological word power so developed may be useful to other scholars. One may trust himself. This is a research work. The Sanskrit scholars who worked so far were not electronic engineers. Same is true for other branches of technology.

7. What are the opposition's points ?

Mostly, the historical calendars are based on Archaeological findings and are considered as final verdict, based on available MATERIAL EVIDENCE. The fact of life, remain that the archaeological artefacts are not interpreted by the broad based group of technologists. There are examples to demonstrate this fact. On the page 56 of India Today, dated 5 /4/2004 is presented a picture of perforated clay pottery found in excavation in India. On closer inspection, it is clear that the artefact is a broken piece of an earthen boiler as is evidenced by the ring marks surrounding the holes. This is an important historical epoch in itself, which is not highlighted. When boilers are considered the first name appears on the scene, is that of Hero of Alexandria, about 1st century AD. This was only a play thing or a toy. Then came James Watt's wagon in 1769 AD. The materials used for boilers in EUROPE during 18th century were lead, wood and brick work, which were inadequate to handle even low pressures like 5 pounds per sq. inch. The clay boiler pieces of Harappan age or Sindhu culture may date back to about 3000 B.C. The point of emphasis is that the Indian clay boiler is much more older than the Hero of Alexandria. But, the archaeological artefact has been labelled as perforated pottery. This shows the hollowness of todays archaeological evidence. Another archaeological evidence at Mansar near Nagpur in Maharashtra appears to be great thermal machine to the keen eye of an engineer. However, it is declared as the monument of saint of a particular religion. Thus evidence of "(महायन्त्रप्रवर्तनम्) Mahayantrapravartanam" of Manusmruti (11.63) which is lost for ever. As such, let the electronic engineer not be deterred by the pressure of archaeologists to confirm the Vedic research to their satisfaction. If Maitrayanisamhita mentions of systematic electronics, then it is an evidence, in itself.

8. DARK PERIODS

The Vedas and Vedic literature is not considered as a "document" in historical sense because it is already considered as Religious matter. This view needs a change. Basic handicap in doing this is the language problem. Every attempt to fathom the Vedic past, with modern Sanskrit, as a tool, is more likely to provide strange information, only to confirm, the historian's view, that the Vedas can not be accepted as a "documentary evidence". Every time commercial modernity has swept out, the history, under the banner of progress. The Language is not an exception. Burning of Nalanda and Taxshila Universities by foreign aggressors has further contributed to creation of dark period and destruction of technological Sanskrit. Every Vedic research scholar has to be aware of such problem and use corrective measures.

9. Another valid question is asked as to why the technology of Indian Origin has disappeared, if at all, it existed. This is not a unique question. There are many such questions. Why the "mighty" DINOSAURS have disappeared. The answer are only speculative. A possible explanation is outlined above. Commercial replacement is an important factor for extinction of past. This is applicable even to

so called modern Sanskrit. With the rise of Sanskrit grammar, by say, noble men like Bhattoji Dikshit, (1570 - 1635 A.D.), this is likely to happen. The attempt was re-grouping of the rules. Earlier "linguistic" rules had a different thread of sense but this was replaced by more "easy" and more "rational" system. The modern "Spirit" has "easiness" and "rationality". First statement of Panini is Vṛuddhirādaic (वृद्धिरादैच्). Generalised meaning of Vriddhi is prosperity.

Progress, growth, amplification, etc. The grammarians applied it to the Lengthening of Swaras, which was an occupational need. This was propagated so heavily that "Paninian Progress" has been "literally" washed off. Thus past has been overshadowed. Same fate is enjoyed by Maheshwar sutras. Considering only the first cluster, a few things can be unveiled. The first sutra reads, (aiu) ṇ. How the grammarians perceive it is well known. Seen from the angle of technologists, the sutra (aiu) ṇ = aṇ, iṇ, uṇ, where 'ṇ' is a subscript, defining the fields in which the variable `a', `I', or `u' is to operate. The subscript `ṇ' is understood to be an omnipresent field. The technologist considers that it is the subscript which defines the jurisdiction of a variable. It is represented by the etymology as Prati + hr. (Prati = nearby + hr = carry or convey). Prati + hr = to close, keep shut for pressure, MW673, I). This means that the a, i, u, are CLOSED for the force ṇ. Technologically, the subscript n of Maheshwar Sutras represent a CLOSED system of force field for subscripted variables. This technological grammar will help to under stand the motion of a particle (an electron) in the local electric field `n', and a negative reaction `u' (उ). This is the origin of electronics in the Maheshwar Sutras, which form the key to understanding of the Vedas. Vedas do contain these elements. Refer App- I diagram. An represent, simple harmonic motion, symmetrical or unsymmetrical in the force field represented by subscript `n'. This leads to the development of electron coupled Oscillators in the field of electronics. `Iṇ' (इण्) represents the velocity constants of the potential field `ṇ' and the Uṇ represents the potential drops in the potential field `ṇ'. All this technology, gets over shadowed if a, i, u, was to be restricted, to traditional Sanskrit.

Thus it will be observed that technology has not disappeared from Vedas. It is only obliterated which needs the clearing operations.

10. Basic purpose of the above chapter is not to blame the phonetic grammar and the learned grammarians. The purpose is to answer the question as to why the Vedic technology was lost.

It may be appreciated that the communication can be with 1) phonetic language and 2) the symbolic language. Modern Sanskrit with its grammar is of phonetic character. The Vedic language is symbolic language. Paninian system is a grammar of a symbolic language. There are a few peculiarities of both systems of languages. A phonetic language can transport only the known information known by both the transacting parties. The symbolic language transports EXTRAPOLATED information. Thus the information packaging quantity of symbolic language is very high. Examples of the symbolic languages are mathematics, technical graphics, road signs etc. Many times the symbols and words appear similar but the information is different. An example will make it clear. Let us consider a symbolic mathematical Statement Y = mX+C, This is not a phonetic statement. It may be seen, that, the symbolic statement is much more compact. Another point to be observed is that `y' of symbolic statement is not the `y' of the alphabets as y in yard or yarn, even though the forms are the same. The y of a symbol can take up various forms as y, y, y, but not the alphabet y. Now let us apply this, to phonetic and Vedic Sanskrit. Let us start from the word VEDA. In phonetics the word Veda means 1- knowledge, truth, sacred book, 2- acquisition, wealth, 3- tuff of bunch of grass, weave or bind together. And 4 - proper noun etc. (MW 1015to 1017) : The symbolic meaning of Veda is a `RELATION'. This Vedic relation is a `functional relation', and is expressed by various symbols in modern language viz. =, fx, Σx, +, AXB, A.B,ϕ, δ(and many others. This is Vedic language. Another `word' is say `Brahma' (ब्रह्म) In phonetic Sanskrit variously. In Vedic context Brahma is an `unknown' which is operated upon by an `operator' to make it known. Thus Brahma is an entity on which to operate, to produce results, or a Karma. The operator is Karta. The Karana is the characteristic of the operator. Represented graphically, the Karana is the graph, which may be linear or non linear. Sampradana and Apadana are the positive and negative slopes of the characteristic graph, depending upon, whether Y X (Sampradana) or Y 1/X (apadạna). `Genetive' provides the value on Y, when, the value of X is set to zero. Locative provides the location of a point on "X" axis (अक्ष). `Yajan' is the movement away from the origin and havan is a movement towards the origin. The `vocative' is the value invoked on the Y axis. Thus the information obtained from Vedic (symbolic) language is different than phonetic Sanskrit.

As far as the `Linga' is concerned in Vedic language the feminine gender (stri-linga) represents absence of testing. The masculine (pum-linga) represents tested and found correct and the neuter gender represents the rejection after testing. Thus the feminine gender = masculine + neuter. An example may be better. Let us take three statements in Vedic Sanskrit.

1. Kati māsā Gāyatrī
2. Kati māsā Gāyatraḥ

3. Kati māsā Gāyatram

Kati is how much or how many, the same as in phonetics, māsā = unit of power measurement, Gayatra = the maximum power developed. (Vedic meanings). In the first statement the SHABDA, Gayatri is in feminine gender. Therefore, the question relates to `how many units of unmeasured power (Indicated horse power, I.H.P.)' is there ? The second question asks, how many units of measured power.

(Brake horse power B.H.P., as the Gayatra is in masculine gender). The third question is how may units of power is lost (Gayatram, being in neuter, it is indicative of a loss). This, technological relevance of the gender (Linga), is available only with the Vedic Sanskrit.

For vachana, (number), in Vedic Sanskrit singular will represent one, uniform, homogeneous, The dual (dvivacana), the ambiguity, inadequacy, the plural (Bahuvacana) represents adequacy, unambigiuity. For Phonetic Sanskrit the singular number (Ekavacana) is almost the same except it also refers to critical and monostable state. The dual and plural are widely differing. The Vedic sense of dvi = ambiguous (MW 504 III) is reflected in the word `dvigata' as commented by Patanjali on Panini VIII. 2.3. on page 388 in Mahabhashya. For the use of plural, the phonetic language depends upon the number three or more, whereas, for Vedic language the criteria is `sufficiency'.

11. Samasa - Compound :

In forming the Samasa, in Vedic language, it may be remembered that, we are not compounding only two words or two padas, as in phonetic language. It is the compounding, of two symbols, in which there is Samasata or compatibility. As we can not connect a 6 volt Lamp to 230 V line, or a 5 A socket to 15 A pin. It is known that if A and B are symbol of force, velocity, (vectors), then AB = A.B = AXB= BXA etc. This is the peculiarity of a symbolic language. In Vedic language, every Samasa represents a motion in a force field. The `tatpurusha' represents a motion `against' the field, the avyayibhava represents a motion `along' the field, the bahuvrihi is along the equipotential i.e. at the same leval, without the gradient and the dvandva represents the motion under parallel forces. Thus, a samasa is considered as an ordered set, in mathematical language, and provides valuable information about the motion in a force field. The domain of Samasa is very wide. A statement of Patanjali in his Mahabhashya on Paninis sutra @ 8.2.3, page 388 vol III, states as "sarvavibhaktyantaḥ samāsaḥ", which means that all case-endings are samasa. Samasa itself being the field gradient every `pada', term of Vedic language is subject to a gradient. There is a difference between Samasa and

`Sandhis'. The nature in Samas, is `inductive coupling', through the common field and the nature of Sandhis is a `conductive' coupling. To summerise, an udatta mark, shifts the SVAR to right and the svarita shifts it to the left. Such Sandhis are the test points in the circuit.

Every rule has a DOMAIN of its operation. Rules of grammar are not exception. Paninian rules of grammar have wider domains, which includes symbolic languages of Vedas. Traditional phonetic grammar has much narrow domain. Thus an apparent connection is not an equivalence. Similarity is only incidental. Greater comprises the less but not the other way.

12. One of the strange outcomes of application of phonetic language is interpretation of Bhagavad Gita. To understand the context properly Let us imagine, that two medical practitioners are discussing something by the side of a patient and many more such situations. What is the probability of philosophy being discussed ? Is it fair to expect that Shrikrisna and Arjuna both war- technologists of top order, were discussing philsophoy on the back ground of IMMEDIATE war, on the battlefield ? If the context is immediate war then every word of the text has to be interpreted in that context without offering any reasons. The hundreds of commentaries on Bhagawad Gita, have ignored this point. If anybody tries to go on right track, then he is most likely to be blamed that the meaning is stretched. In fact it is the other way round. War technology interpretation has been STRETCHED to philosophy. India has been deprived of one more important WAR text. To provide a clue, reference is invited to 1.19. The great sound filled whole space between the earth and the sky. For sonography the projection of Sound is necessary. 1.20 Having clearly seen the enemy, Arjuna, prepared himself for the attack. The point to note is that at this time. Arjuna was not demoralized even after seeing everything clearly (atha vyaṣṭhitam dṛṣṭvā) and asked the driver to change the position of his war Vehicle. New position was nearer to the enemy. He wanted to Fix the target (Kairmayā saha Yoddhavyam). When war-craft went near to great warriors, like Bhisma and Drona (1.25) then only, he was disheartened and not earlier, for the same view. Thus it was a problem of change of location which took place between, Gita 1.20 and 1.25. This is important in war strategic control. Why imagine that Arjun was observing a vast army with naked eyes. He may have been observing on a radar screen actuated by sonar pulses (1.19). The image on screen is likely to be affected by the PROXIMITY of radiation from the opposite party. Pitri, Pitamahan and etc. can be the names of the flaws on the radar screen. At present also the names like `snow', `pincushion' etc are used, about faults in TV, when there is no `ice' in TV.

13. To summarise, Vedas and Vedic literature are full of technology including electronics. Vedic language is to be re-constructed before the interpretation is attempted. Success is sure, the Labour is immense, there is no short cut.

4

Few non-attended topics of physics in Sanskrit Sastric Traditions such as the proton, deuteron & neutron

(Saṃkṛtaśāstrīya paraṃparāyāṃ bhoutikaśāstrasya kāścana anālocitaḥ paksaḥ yathā: āpo-raso-reto-ghanatrayāṇi)

Dr.N.G. Dongre

1. Introduction:

The glorious past of India convinces that it must have earned an esteemed position in the world, in culture and civilization with its rich knowledge of science and development of literature high lighted by the fact that Vedic literature is the oldest is the oldest- known to mankind. But now a days even most of the easterners as well believe that the eastern knowledge is intuitive (mystic) and can be had only through meditation and cannot be experienced by a common man like the rational knowledge (science), which is earned through experience and experiments.

In our opinion Indian civilization had reached to its peak well before, around 2500 B.C. due to advent of Buddhism, Jainism and other 'tantric' systems, new conflicts between Vedic and non-Vedic concepts have aroused in India. It was the crisis that the ancient begin to advocate Vedic view through their various schools of philosophies (darsanas), a new trend emerged which sought the use of rational knowledge, which was already prevalent well before 2500 B.C. for explaining the human nature and uplift the moral and spiritual powers of human beings. Later on it seems, many different schools of philosophy came into being, which has no reference to the rational knowledge as such. As a consequence, the linking threads between the ancient science (rational knowledge) and the various schools of philosophy (now intuitive knowledge) have torn apart, due to rise and fall patters of major civilizations. After a civilization reaches to a peak of vitality, it tends to lose its cultural steam and declines.

However now in the ear both the ends of modern physics, that is (i) the world of subatomic particles and (ii) astrophysics and cosmology for the description of the universe at large have forced to search parallels in the eastern mystics. Several great physicists of late 20th century like Julius Oppenheimer, Niels Bohr, Werner, Heisenberg, Fritij of Capra, Donah Zohar and so on, have felt that the Veda, Upanishad, Dharshan, etc. in which the concepts of the universe, space and time show a keen parallel views with the rational knowledge of the physics, which as been earned through experience and experiments. But in general belief the eastern knowledge is considered to be very much intuitive (mystic) only to be experienced through meditation, a different source of knowledge other than experience and beyond the reach of a common man. The reason for this was that the majorities of manuscripts on various scientific topics are missing or have been destroyed by invaders. Few of many of remnant manuscripts on ancient scientific literature were taken away from India to elsewhere and rest are preserved in great control by the centres of various philosophical school of India, which have not been studies so for.

Even then the available traditional text such as Vaisesika Darsana of Kanada, Amsu Bodhini of Bharadvaja along with the text of Suryasiddhanta, Lilavati & Goladhyaya of Bhaskaracārya, Kiranavali of Udayanacharya, Nyaya Darsana of Gautama, Prasastapada Bhasyam, Nyayakandali of Sridhara Caraka Samhita, Sushruta Samhita, Astanga Hridaya of Vagbhata, Nidana of Madhavakara, Rasa Samuchchaya, Rasendra Chudamani of Somadeva, Rasa Chintamani of Madantadeva, Kautilya Arthasastra, Samaranga Sutradhara of Bhojadeva, Agastya Samhita are sufficient to describe the physical world more or less par with other similar scientific works. The text of Amśubodhinī not only reveals the deep scientific knowledge well known to India, but also provides a clue to that how the ancient scientific rational knowledge has formed the basis of the other so called intuitive (mystic) knowledge as expressed in the various schools of Indian metaphysics.

Upanishad and Darshana are considered as the various schools of metaphysics, where metaphysics is a philosophical term, the literal meaning of which is that the subject with which it deals in Metaphysics (behind the physics). The term has signified generally those parts of philosophy which deal with any thing beyond the natural objects and which are incapable of being dealt with by physical research. Thus in contrast to what we call natural sciences, this intuitive system deals with the ideas, the meaning, the cause and the principles of existing things. Its opposite, all systems of natural sciences, we call empiric systems, systems which are based on experience.

There are almost as many metaphysical systems as there are philosophers. Every philosopher who has intentionally experienced the 'metaphysical doubt', who has ever begun to consider whether things which appear real to us are actually real, will sooner or later develop his own system of metaphysics whether this system centres around the idea of motion, or of time, or of change, or of substance, or of identity, or of cause, or even of God, does not matter. As already told in the rational knowledge then of various stages of scientific deep know how about the Nature were chosen to build their own systems of metaphysics by their expanders. All various systems of metaphysics emerged as different schools of Darshan only to over come the non- Vedic concepts. Therefore every school of Darshan shows some sort of parallelism with concepts of modern physics to certain extent but least hope of advancement as the every fundamental axiom of the schools forbid them and so Darshanas have reached to the stage of stagnancy.

Science, on the other hand, a word from the Latin '*Scire*' meaning 'to known', and hence, any form of organized knowledge. In common usage, however, the word denotes only that aspect of learning which represent a systematic accumulation of knowledge pertaining to nature or the operation of general laws, in which sense the term is used as an abbreviation for 'natural science'. As all the systems of natural science e.g., physics mathematics, chemistry etc. being based on experience, the advances made in various system, where each generation knows more than the preceding generation and uses the accumulated empirical knowledge of the past in its attempts to develop the science further.

Besides all these in order that the human beings should not lose the treasures of the knowledge a comparative and scientific study of the texts of various manuscripts related to scientific knowledge as well as the text Indian metaphysics, which are probably a sort of relics of the rational knowledge gained by the ancients, with the modern science is necessary. A search for maiden manuscripts related, to rational knowledge should continue and an opportunity should be made available to the scholars to decipher them. All these attempts may certainly give a way to resolve the interlocked theoretical and experimental situations of the present millennium.

So far the traditional text of Vaisesika, Nyaya etc. mainly shed light only on the general principles of physics (classical) for example the laws of motion, gravitation, elasticity, etc.

2. **The manuscript of *Amsubodhini* of Maharshi Bhardvaja: A narration about the Big Bang to Formation of Galaxies** (A treasure of few non-attended topics of physics in Sanskrit sastraic Traditions):

The study of the manuscript of *Amsubodhini* of Maharshi Bhardvaja shows that the ancients were aware of the systematic steam lined process of investigation of 'modern times.' It will be interesting to note that the laws of mechanics and gravitation electrical charge were well known to them. There were sophisticated instruments like spectrometer and used to measure the solar radiation and applied for the spectral classification as it is done in astrophysics. Further in the light of the text of *Asmsu Bodhini* (अंशु बोधिनी) of Maharshi Bharadvaja one can imagine the evolution of the present universe has occurred in the following three convenient steps.

In the first step, the primeval atom Paramātmā (परमात्मा) due to some slight disturbance suffer a Big Bang and leads to the formation of galaxies (*Brahmāndas*). In this coarse of action quark gluon plasma (Cicchktiḥ lohita-capalā) separates first and gets surrounded by the lepton plasma (Maya/ moolaprakriti).

Then the interaction of them causes to the lepton plasma to be get separated as agni and soma, which is turn along with formation of positrons and electrons (dhvānta dvaya) give rise to background radiation (tama). Now the central core of the quark-gluon plasma breaks up into various pairs of quarks (lohita capala) and binding forces (vega's). Further the interactions with its surroundings give rise to proton plasma (vidyutodaka) which due to its internal repulsion, within a short interval of time of few minutes beings to stream along the regular tracks (सृष्टि-पक्ष-रेखा) in the form of galaxies, meanwhile due to gradual cooling the proton plasma, interacts with the electromagnetic radiations liberating neutrinos (rohini) and free electrons which are present and smeared all over the tracks on which the galaxies have to be formed and which results into the formation of the protons, neutrons and deuterons (*Ghanatraya: Āpoghana, Retoghana & Rasoghana)*

The second step starts after the emergence of the Sun and other stars, the subsequent compression and heating of the matter give way for nuclear reactions and the nuclei of heavier elements are formed.

In the third step, further cooling the nuclei beings to trap the electrons and the formation of atoms and molecules take place. Further cooling causes to exhibit the various states of matter and in the favourable situations just similar to our earth the life may come to existence.

These views are more or less just parallel to the modern thoughts. The European Laboratory for Particle Physics (CERNS- European Centre for Nuclear Research) claims first experimental creation of quark-gluon plasma and further it adds "it has compelling 'evidence' that its scientists have created the quark- gluon state of matter predicted to have existed shortly after the Big Bang. If confirmed, this would be the first time that that conditions within the first three minutes after the Big Bang the point at which the protons and the neutrons that make up atomic nuclei came into being– have been observed under the experimental conditions."

When CERN's Super Proton Synchrotron (SPS) fired very high beams of lead ions into gold or lead target, the collisions create a temperature 1000,000 times hotter than the centre of the Sun, which is sufficient to break down those confines quarks inside nucleons.

Inside the sun thermonuclear fusion is going on. For proton-proton and carbon cycles the temperature at the centre of the sun should be of the order of $2X10^6KX10^7$. Therefore the temperature at protons and neutrons came into being as a result of quark-gluon interactions may be estimated as of the order of $2X10^{12}K$.

Further in the same context we may refer. J.V. Narlikar In his book 'The Structure of Universe' he narrates the cosmological postulates, in which the way postulate states that that "The galaxies streaming along regular tracks drawn in a space-time diagram which do not intersects each other. The tracks of all galaxies in the way postulate meet at a point. The epoch 0=0 is taken as the event of 'creation'. Prior to this there existed no Universe."

Hanstephani in his book on 'General Relativity' has made "some brief remarks about the evolution of the Universe after the formation of the electromagnetic background radiation. During the gradual cooling of the primeval 'fire-ball' the chemical elements hydrogen and helium form in the preferred equilibrium ration of about 73:27, almost no heavier elements being synthesized, small disturbances to homogeneity of the universe then lead to galaxy formation, and there the subsequent compression and heating of matter in the stars leads to nuclear processes during which the heavier elements are produced. All these things are still the subject of research, however, and the details are not very well understood."

The search for such a non traditional text which might be describing the modern Particle Physics (from Big Bang to Formation of Galaxies) so far the present moment is concerned, is rested upon the printed manuscript of Amsubodhini, having a commentary by Bodhananda on Bharadvaja-sūtras, which seems to be describing. The Probable Structure of the Position & Electron (*Tamadhvānta* and *Kañcukadhvārita*), Photon (*Tama*), Proton, (*Āpoghana*), Neutron, (*Retoghana*) and Deuteron, (*Rasoghana*).

In the light of aforesaid keeping in view of the present non-traditional texts of Amsu-Bodhini (अंशुबोधिनी) Maharshi Bharadvaj one can imagine the evolution of the present universe has occurred in the following three convenient steps.

In the first the primeval atom/ Paramātmā (परमात्मा) due to some slight disturbance suffer a Big Bang and leads to the formation of galaxies (Brahmāndas).

In this coarse of action quark gluon plasma (Citcchktiḥ lohita- capalā) separates first and gets surrounded by the lepton plasma (Maya/moolprakriti).

यत्स्वमायोपाधिशक्त्या जगदेतच्चराचरम्।
सृष्ट्वा तदंतरे पश्चादन्तर्यामितया स्वयम्।।५७।।
स्थित्वा संधार्यते सर्वमसंगत्वेन केवलम्।
तदेवाधिष्ठानमिति प्रवदन्ति मनीषिण:।।५८।।

इतिप्रपञ्चलहर्याम्।। पृष्ठ २०

The attributes of the primeval atom/ Paramātmā (परमात्मा) are described as follows:

।। सृष्ट्याधिकार:।।

अध्याय: १ सूत्रम्।।

।। अथकल्पसृष्टिरात्मत्वात्सूर्येण।।

This very evolution in Milky way galaxy (श्वेतवाराह-कल्प) begins from sun star (ज्योति) similar to the creation of myriads of galaxies have begun by primal atom (ब्रह्मात्मा/ परम-आत्मा).

This very first aphorism is just introductory statement of the chapter the Universe (सृष्ट्याधिकार:). First he defines the aim of this chapter as to give the details of the evolution of the existing world that has started from Sun in our galaxy. The explanatory note given by Bodhananda as are given follows.

बोधानन्दवृत्तिः।।
पराशक्त्याद्यासदाशिवशक्तत्यान्तं यथाक्रमम्।
महाकल्पश्चतुष्टिरितिशस्त्रविनिर्णय:।।९।।
श्वेतवाराहकल्पाधिपतिरित्याहुरास्तिक:।
सूर्य आत्मेत्यत: प्राह श्रुतिस्साक्षात्सनातनी।।१९।।
पाद्यकल्पानन्तरितश्वेतवाराहकल्पके।
तस्मात्सृष्टिरभूत्सूर्येणात्मत्वाद्विवृतत्वत:।।२०।।

These clarify that one of the galaxy-forming regular track in space – time diagram (सृष्टि पथ सेवा)

१. अध्याय:, ९ अ सूत्रम्।।

।।आत्मसिद्धिरधिष्ठानात्तत्पञ्चदशन्यायेन।।

१५ स्वंतस्सिद्धन्यायसूत्रम्

।।शक्तित्रयविशिष्टत्वेसत्येकस्मिन्नात्मन्यधिष्ठानान्तराभावात्सर्वाधिष्ठानभूतब्रह्मात्मनास्याद्विवृताधिकारश्च स्वस्मिन्।।

एतासु विवृतशक्तिः प्रथमेति कीर्तिता। द्वितीयाव्यापका शक्तिशिवच्छक्तिस्स्यात्तृतीयका।।४३।।

The above-cited aphorisms describe three powers of primal atom (ब्रह्मात्मा/परम-आत्मा), which are the ability to expand, to pervade and to illuminate (विवृतशक्ति, व्यापकशक्ति चित् and शक्ति).

Then the interaction of them causes to the lepton plasma to be get separated as agni and soma, which is turn along with formation of positrons and electrons (*dhvānta dvaya*) give rise to background radiation (*tama*). Now the central core of the quark-gluon plasma breaks up into various pairs of quarks (lohita- capala) and binding forces (*vega's*). Further the interactions with its surrounding give rise to proton-plasma (*vidyutodaka*) which due to its internal repulsions, within a short interval of time of few minutes begins to steams along the regular tracks (सृष्टि-पथ-रेखा) in the form of galaxies, meanwhile due to gradual cooling, the proton- plasma, interacts with the electromagnetic radiations liberating neutrinos (*rohini*) and free electrons which are present and smeared all over the tracks on which the galaxies have to be formed and which results into the formation of the protons, neutrons and deuterons (Ghanatraya: Āpoghana, Retoghana & Rasoghana).

The second step starts after the emergence of the Sun and other stars, the subsequent compression and heating of the matter give way for nuclear reactions and the nuclei of heavier elements are formed.

In the third step, further cooling the nuclei beings to trap the electrons and the formation of atoms and molecules take place. Further cooling causes to exhibit the various states of matter and in the favourable situations just similar to our earth the life may come to existence. To incorporate the complete description is beyond the scope of this article. The elaborate description and the details of the *amsu-bodhini*, having the commentary of the fifty sutras of Maharshi Bharadvaja, along with the scientific notes of the present author are ready of publication.

This study is to bring out the salient information before the scholars for their perusal and critical suggestions.

References:

1 Nature/ Vol. 4031/pp 581

2 Nature/ Vol. 4031/pp 581

3 The Structure of the Universe, Oxford University Press, London (1990), pp 118.

4 General Relativity (An introduction of the theory of the gravitational Field), Hans Stephani, Cambridge University Press, Cambridge (1985)

References for general reading

1 Amsu Bodhini, Maharshi Bharadvaja, M.S. from the library of Oriental Institute Vadodara.

2 The Structure of the Universe, Jayant V. Narlikar, Oxford University Press. (1990) pp. 121-125.

3 The Structure of the Universe, op cit., pp 126-127.

4 The Structure of the Universe, op cit, pp. 230-232.

5 CERN, Nature, Volume 403/10 February 2000 www.Nautre. Com. pp 581.

6 Unified Theory of Elementary Particles, H. Heogry, Scientific- American (April 1981)

7 Dhvant Pramapaka Yantra of Maharshi Bharadvaja (Spectrometer/ Monochromator), N.G.Dongre, pub- Indian Journal of History of Science, 29 (4), 1994, pp. 611-626.

8 Spectroscopy in Ancient India, N.G.Dongre, Pub. Indian Journal of History of Science, 33 (3) 1998, pp. 229-238.

9 Prakasa Stambhanabida Lauha of Maharshi Bharadvaja (A novel IR transparent material of range 5000 to 1400 cm^{-1}), N.G. Dongre, S.K.Malaviya and P. Ramachandra Rao, Pub. Indiar Journal of History of Science, 33 (4), 1998, pp. 274-280.

10 CERN, Nature, Volume10 February 2000 www Nature. Com. pp. 581.

11 Prakasa Stambhanabida Lauha of Maharshi Bharadvaja (A novel IR transparent material of range 5000 to 1400 cm^{-1}), N.G. Dongre, S.K.Malaviya and P. Ramachandra Rao, Pub. Indian Journal of History of Science, 33 (4), 1998, pp. 277.

12 Metrology and Coinage in Ancient India and Contemporary World, N.G. Dongre, Pub. Indian Journal of History of Science, 29 (3), 1994, pp. 361-373.

13 The Structure of the Universe, op. cit. pp.62.

14 Physics in Ancient India (A Mathematical Analysis of Vaisheshika Philosophy), Narayana Gopal Dongre, pub. Wiley Eastern Limited (New age International Publishers Limited), New Delhi, 1995.

15 Vaisesika Siddhantanam Ganitiya Paddatya Vimarsah (Hindu Physics), Dr. Narayan Gopal, Pub. Director, Research Institute, Sampurnananda Sanskrit Vishvavidyalaya, Varanasi, (1979).

5

UNIVERSAL GRAVITATIONAL FORCE AS FOUND IN VAIDIKA LITERATURE

Prof. (Mrs.) Vijaya Rani

Introduction :

It is generally thought that modern science is originated and developed only recently about 5 or 6 centuries back and it has nothing to do with our Ancient Sanskrit Literature. But it is a wrong conception. A serious and deep investigating insight into the Sanskrit works including our Vaidika literature, reveals that many scientific facts and solutions to various problems related to science - physics, Chemistry, Botany, Zoology, Mathematics, Astronomy, Astrology etc. - are hidden in them. Sometimes they have been dealt with directly, sometimes indirectly and sometimes in a symbolic language. A learned Vaidika Scholar Pandit Satyavrata Samashrami, in the introduction of his work 'Trayī bhāṣa', writes about the vast nature of knowledge of the Vedas :

> "The four-fold Trayī is the repository of all Āryan knowledge, whether theological, practical, scientific, ethical and social. Our opinion is that in Vedic times our country had made extra-ordinary progress. In those days the sciences of Geology, Astronomy and Chemistry were known as 'Adhidaivika Vidyā' and those of Physiology Psychology and Theology as Adhyātma Vidyā."[1]

Further, a great thinker and yogi of modern India– Shri Arvindo, agreeing with the views of Swami Dayānanda, writes :

> "There is, then, nothing fantastic in Dayānanda's ideas that Veda contains truth of science as well as truth of religion. I

1. See Introduction to 'Trayī-Bhāṣa', p. 8-9.

will even add my own conviction that Veda
contains other truths of science, the
modern world does not at all possess.......".[1]

Similarly a great vaidika exponent of modern times - Pandit Damodar Satvalekar - opines about the Vedas :

"हमारे भारत के ऋषियों ने, परा और अपरा ये दो विद्याएँ हैं, ऐसा मानकर अपरा विद्या में सब विज्ञानों को माना और परा विद्या में आत्मा की विद्या मानी। इस वर्गीकरण से ही स्पष्ट प्रतीत होता है कि भौतिक विज्ञान भी प्राचीन ऋषिकाल में अच्छी प्रगति तक पहुँचा था...... इस समय पाश्चात्य विज्ञान जहाँ तक पहुँचा है, उससे भी आगे प्राचीन ऋषि पहुँचे थे।"[2]

Internal evidences from the Ṛgveda[3] suggest that the Vaidika sages placed themselves under the discipline of Satya (truth) Ṛta (eternal order) Tapas (spiritual austerity) and then they revealed the innermost reality through their Vāk. They put-forth the scientific facts before the world through the awakening of their purest intellect.

The present research paper proposes to explore some relevant references from the Vedas, which point out about the Law of Gravitation and the Universal Gravitational Force in the Universe.

2. What do we understand by Gravitation & the Universal Gravitational Force :

According to modern physics, there are four fundamental forces in nature, namely (i) Electric Force, (ii) Magnetic Force, (iii) Nuclear Force, (iv) Gravitational Force. Magnitude-wise Nuclear Force is considered to be the strongest and Gravitational Force is the weakest among the four, still Gravitational Force is supposed to be the supreme as it governs the motion of all the celestial bodies of the universe.

In the history of modern physics, it is only in 1665 A.D. that the credit for discovering the universal law of Gravitation goes to Sir Issac Newton.[4] He was first inspired by a simple incidence of falling of an apple from the tree to the earth, which made him contemplated. After deep contemplation, it striked to his mind that there is certainly some force responsible for the attraction of the apple to the

1. See Shri Arvindo, Bankim - Tilak-Dayanand, p. 67.
2. See Introduction to 'ऋषियों के विज्ञान की श्रेष्ठता' p. 9-10.
3. See Ṛgveda X.71. 1-2.
4. Resnick & Halliday, Physics, vol. I, p. 382.

earth and it was named by him as the 'Gravitational Force'. He further thought that the same Gravitational Force might attract the moon towards the centre of the earth, due to which moon continues to move along its orbital path around the earth (Fig. 1). Still further, he contemplated that the same force might be regarded as the cause of the motion of all the celestial bodies of the universe, due to which all celestial and terrestrial changes occur. As such, it was named as 'Universal

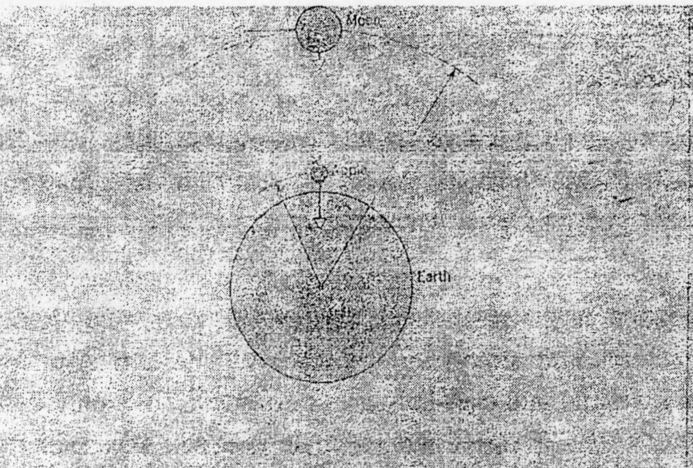

Fig. 1. Both the moon and the apple are attracted towards the center of the earth.

Fig. 2. The motion of the planets along different orbits around the sun

Gravitational Force.[1] Here lies the origin of the Law of Gravitation in the modern times.

This Gravitational Force is attractive in nature, due to which all material bodies remain attracted towards the earth. It is also responsible for the movement of earth and other planets of our solar system around the sun (Fig. 2). Particularly the rotation of earth about its own axis as well as around the sun is also due to this very force. it causes the cycles of day and night, months, years and other climatic and temporal changes on the earth.

The other special aspect of this force is that it causes stability as well as dynamism in the whole universe, because planets maintain almost a particular distance from the sun, while they keep on moving along the elliptical path around the sun.[2]

In fact, this force acts in the most secret manner and cannot be seen and realised unlike the solar rays and atmospheric winds etc. This universal force influences invisibly the terrestrial and oceanic waves movements. Not only this, it greatly impression upon various physical and psychological activities of human beings, animals, birds and plants. This reality has been explicitly observed in the recent destruction caused by Tsunami waves, during which the physiological and psychological behaviour of animals, birds and tribals were changed.

3. Universal Gravitational Force as found in Vaidika Literature :

Some indications about the Law of Gravity and the Universal Gravitational Force are traceable to some extent at different places of Vaidika Literature, such as:

(i) Nasadiya Sūkta (Ṛgveda)

(ii) Skambha & Jyeṣṭha Brahma Sūktas (Atharvaveda)

(iii) Īśopaniṣad

(iv) Praśanopaniṣad & its Śaṅkara Bhāṣya

In the present paper, I shall discuss the above subject only with regard to the first two references i.e. (i) Nāsadiya Sūkta and (ii) Skambha & Jyeṣṭha Brahma Sūktas.

---

1. *Op.cit.* p. 383.
2. R.M. Kai, the Vedic Experiences Mantrāñjalī, p. 66.

3. (i) Nāsadiya Sūkta and the Gravitational Force :

The seer of Nāsadiya Sūkta, describing the condition before the creation of this universe, states :

"There existed neither death nor immortality, neither day nor night. The only one that was there, breathed of his own with no breath & beyond this nothing existed."[1]

In this verse, the word "स्वधा" etymologically means, "स्वस्मिन् धीयते ध्रियत आश्रित्य वर्तत इति स्वधा।"[2] that which exists depending only on itself. The words 'अमृतम् and मृत्यु' are also to be analysed. These two are inter-dependent terms and are fundamentals for the cycle of creation. Here immortality (अमृतत्व) stands for stability and death (मृत्यु) denotes motion (गति). Pt. Madhu Sudan Ojha has also expressed the same view and accepts that these two (stability and motion) are the root-cause of the universe.[3]

Through the above analysis of the words, it can be deduced that whatever was existing at that time (at the time of creation), was 'svadhā' i.e. nothing but some vital force like gravitation. It was existent depending on itself and was bearing both the aspects of stability and motion simultaneously. Thus, in this verse and also in other verses of this sūkta, we can find only a reflection of an universal force responsible for the creation. It can be regarded as the Universal Gravitational Force.

3. (ii) Skambha and Jyeṣṭha Brahma Sūktas & the Gravitational Force :

Now we have to discuss some verses of Atharvaveda related to the Skambha Sūkla and also of Jyeṣṭha Brahma (A.V.X.7 & X.8).

(i) According to Pāṇini, the word 'स्कम्भु' is derived from the roots 'स्कम्भु' or 'स्कुम्भ'[4] the meaning of which is 'रोधन' i.e. arresting or holding or stopping[5], (ii) According to Apte, the word 'Skambha' is derived from the root 'स्कम्भु' which means support, stay fulcrum and supreme-being.[6] Hence the word 'Skambha' in general, refers to some Supreme-power which supports or beholds the universe.

---

1. न मृत्युरासीदमृतं न तर्हि न रात्र्याः अह्नः आसीत्प्रकेतः।
   आनीदवातं स्वधया तदेकं तस्माद्धन्यन्न परः किंचनास।। Ṛgveda X.129.02
2. See Sāyaṇa-bhāṣya on Ṛgveda X.129.02.
3. See Pt. Madhu Sudan Ojha, Daśavāda-Rahasya, p. 15.
4. See Pāṇini -Aṣṭādhyayī 3.1.82.
5. See M.M. William's Sanskrit - English Dictionary, p. 1257.
6. See Apte's Skt. Eng. Dict., p. 618.

While studying the Skambha Sūkla, it appears in the beginning that the Vaidika seer has a strong feeling of astonishment as to what is that vital force, which governs the universe and puts it under a definite law and order, what power beholds the sun, the earth, the moon and the other planets of our Solar System and the stellar world of galaxies on their respective places.[1]

However, despite a strong feeling of astonishment, he himself accepts that there is certainly a power, which acts as the supreme force of the creator Lord and it is named as 'Skaṁbha'.[2] The seer has a deep sense of reverence for the Skaṁbha and treats it as par excellence. He says : Skaṁbha is the supreme power, in which whole of the universe resides, Austerity (तपस्) and Holy Laws (ऋत) repose in it.[3] He further acknowledges that the Skaṁbha sustains both heaven and earth, the antariks a and the six wide directions, rather this whole existence (the universe) enters into it.[4] The same supreme power has been named as the 'Jyeṣṭha Brahma' also.[5]

From all this description of the Skambha-Sūkta, one can assume that our seer, about whatever supreme-power, (with the name of 'Skambha') has described, is nothing but the Universal Gravitational Force. It is named as Skambha, because this force beholds the whole universe.

In another verse, the Vaidika seer says : A great god (yakṣa) resides at the centre of the universe and all other deities are set into it, just as all the branches of a tree are set around its trunk.[6] Similar ideas are also stated by the seer in the verse 6, 15 & 16 of the Jyeṣṭha Brahma Sūkta (A.V. X.8).

All these verses from Skambha and Jyeṣṭha Brahma Sūktas must be referring to the Universal gravitational force of the planets in the whole solar system, under which all planets move around the sun, the centre of the solar system.[7]

---

1. See A.V.X.7.1-4, 12 & so on.
2. स्कम्भे लोका: स्कम्भे तप: स्कम्भेऽध्यृतमाहितम्।
   स्कम्भ त्वा वेद प्रत्यक्षमिन्द्रे सर्वं समाहितम्।। A.V.X.7.29.
3. See AV.X7.29-34.
4. (i) स्कम्भो दाधार द्यावापृथिवी उमे इमे स्कम्भो दाधारोऽन्तरिक्षम्।
   स्कम्भो दाधार प्रदिश: षडुर्वी स्कम्भ इदं विश्वं भुवनमाविवेश।। A.V. X.7.35.
   (ii) See A.V.X.8.02.
5. See A.V.X. 7.36.
6. महद्यक्षं भुवनस्य मध्ये तपसिं क्रान्तं सलिलस्य पृष्ठे।
   तस्मिन्छ्यन्ते य उ के च देवा वृक्षस्य स्कन्ध: परित इव शाखा।। A.V.X.7.38.
7. See Resnick and Halliday, Physics Vol. I, p. 385-6.

4. Conclusion :

From the above discussion, we can say that the Universal Gravitational Force which is the supreme power, is traceable at various places in the Vaidika Literature. In Nāsadiya Sūkta, there is only an indirect indication to this force, responsible for the creation of the universe. But in Skambha and Jyeṣṭha Brahma Sūktas of Atharvaveda, it is very clearly stated by the seer that there is some supreme god (force), who beholds the sun, the earth, the moon and other planets of our solar system and the stellar world of galaxies at their respective places. it has been named as 'Skambha' and Jyeṣṭha Brahma by the seer. If we analyse it scientifically we can deduce that this supreme god 'Skambha' or Jyeṣṭha Brahma, is nothing but the Universal Gravitational Force which beholds and governs the whole universe.

* * *

6

Weather Forecast in Ancient India

Dr. Ravi Prakash Arya

In the Vedic times, the Vedic seer had a total control over the weather. He was able to modify weather to the extent of his needs and requirements. That is why, he dared to proclaim as, 'Let the cloud precipitate as and when we desire.[1]' In the earliest Vedic phase, the science of Mantra was all the more practised by the high spirited Vedic Ṛṣis at the level of their Asaṁprajñāta Samādhi. At this stage, the Yoga was the main tool or instrument for modifying weather. He could make it to rain, or ward off the same instantly at his will (Mantra). Later on, owing to the decline in his spiritual powers, he could practise yoga only up to the level of saṁprajñata samādhi and at this stage, his means of weather modification also changed. Instead of *Mantra*, he could induce rain with the help of Tantra or the divine power achieved for rainmaking through *saṁprajñata yoga*. At the same time, keeping in view the decline of the tradition of Brahmarṣis (the seers practising *asaṁprajñata yoga* or Brahma yajña) and Ṛṣis (the seers practising asaṁprajñata yoga or ātmayajña), a more gross form of the means of the weather modification was also developed by the Vedic Ṛṣis which was christened as Devayajña or Yajña. The Devayajña form of Yajña was applied by the lay Ṛṣis in the Vedic times. Thus all the three forms of Yajña were the main tools or instruments of the Vedic Ṛṣis for weather modification. This fact was alluded to by the Vedic seer as:

'Make it to rain for me with the help of Yajña.[2]

Thus weather was totally under the control of Vedic seers. They could modify it to such as extent as it could suit their requirements and needs. As such, there was no need to think about the weather forecasting or weather predictions and so the Vedic seers at the earliest stages didn't develop any means of weather forecasting. But by and by the people forgot all the three means of weather modification and thus they had to remain dependent upon the day to day weather for their needs and requirements. Instead of reviving the old Vedic techniques of weather modification, they carried out the indepth studies to understand the

1. nikāme nikāme naḥ parjanyo varṣatu
2. vṛṣṭiśca me yajñena kalpatām , VS. 18.9

behaviour of the atmosphere and change in the weather conditions, so that a reliable and efficient technique of weather forecast may be developed.

During the course, the change in weather conditions and the behaviour of the atmosphere was studied (1) in the context of the movements of various stars and planets; (2) in the context of physical process involved in the materialization of meteorological phenomena on the earth; (3) the astronomical impressions left by the resultant meteorological phenomena on the Earth; (4) in the context of premonitions had by the various sensitive creatures on the planet; (5) and in the context of the representation of the entire season by one single day.

Based on the results of these studies five means/ methods of weather forecast were evolved in the Vedic age which may be summarized as a under:

Forecast of the basis of Physical Process

The Vedic Ṛṣis had studied deeply the physical process involved in the materialization of various meteorological phenomena. Later, when the science of artificial weather modification was forgotten, the knowledge of the same process was applied for making weather predictions.

The Vedic seers discovered that the cloud that seeded in a particular period marked by the Moon's conjunction with a particular constellation would precipitate after six and half month i.e. 195 days when the Moon comes in conjunction with the same constellation.[1]

For instance, if the cloud begins to seed in the dark fortnight of Pauṣa, it will precipitate in the bright fortnight of Śrāvaṇa.[2] If the cloud begins to seed in the bright half of Māgha, the precipitation will take place in the dark half of *Śrāvaṇa*.[3] If, however, the seeding starts in the dark fortnight of Māgha, the precipitation will occur in the bright fortnight of *Bhādrapada*.[4] Similarly, if the seeding offsets in the bright fortnight of *Phālguna*, the precipitation will occur in the dark fortnight of Bhādrapada.[5] The clouds those begin to seed in the dark fortnight of *Phālguna*, will precipitate in the bright fortnight of Āśvina.[6] Likewise, the clouds those

1. pauṣasya kṛṣṇapakṣeṇa nirdiśecchrāvaṇasya sitam. BS. 21.9
2. māghasitotithā garbhāḥ srāvaṇ prasūtimāyānti. BS. 21.10
3. māghasya kṛṣṇapakṣeṇa nirdiśedabhādrapadaśutkam. BS. 21.10
4. phālgunaśuklasamuthā bhādrapadasyāsite vinirdeśyāḥ. BS. 21.11
5. tasyaiva kṛṣṇapakṣodbhavāstu ye teśvayukśukle BS. 21.11
6. caitrāsitapakṣjātāḥ kṛṣṇeśvayujasya vāridā garbhāḥ
caitrāsita sambhūtāḥ kārika śuklebhivarṣanti. BS. 21.12

conceive their foetuses in the bright and dark fortnight of Caitra will precipitate in the bright fortnight of Āśvina and the bright fortnight of Kārtika respectively.[1]

It is peculiar about the Pakṣaja clouds that should they start seeding in the dark fortnight, they will precipitate in the bright fortnight and vice-versa and should they start seeding in the day-time, they will precipitate at night and vice-versa, that which at dawn, in twilight and vice-versa, that which in the forenoon in late night and vice-versa, and that which in midnight in the mid-day and vice-versa.[2]

One more eccentric feature about the precipitation behaviour of these clouds is that they precipitate in the direction opposite to that of seeding. For instance, if they seed in the east, they will precipitate in the west and vice-versa. The same rule holds goods in case of other pairs of directions and also in case of the position of winds at two periods.[3]

Forecast on the basis of Extra-planetary Factors

The investigation of ancient meteorologists regarding the weather control on the earth led them to conclude that the movement of the heavenly bodies also to some extent affect the various meteorological phenomena on the Earth. Hence keeping in new the same fact, a new technique of weather-forecast was developed on the basis of the extra-planetary factors.

The planetary weather is controlled to some extent by the movement of star, planets, constellations and comets, etc. Since, the star, planets, constellations and comets in a particular solar-system owe their birth to one particular solar nebula (*hiraṇyagarbha*) and they are sustained by the gravitational pull[4] of one another. In other words, they are gravitated towards each other. Thus by way o the gravitational pull, i.e., either by being gravitated to others or by gravitating others they affect at large each other in many aspects.

The following effects of the heavenly bodies on the Earth were studied which were utilised for weather forecasting.

Sun (Ravi)

The sun is a star composed mainly of ionized hydrogen and helium. It has been identified as a high temperature zone. Its temperature at photo-sphere is about 6000^0c, at chromo-sphere about 32400^0c at transition region about 324000^0c and

1. sitapakṣabhavāḥ kṛṣṇe śukle kṛṣṇā dyusambhavā rātrau
naktaṁ prabhavāścāhani sandhyājātāśca sandhyāyām. BS. 21.8
2. purvodbhūtāḥ paścādaprotthāḥ prāgbhavanti jīmūtāḥ
śeṣāsvapi dikṣvevaṁ viparyayoḥ bhavati vāyośca. BS. 21.13
3. Saṁ no iṣiro vātu vātaḥ RV. 1.35.4
4. ākārṣaṇena rajasā vartamānaḥ VS. 33.43

Corona about 2,700,000^{0}c hot enough to emit X-rays. It radiates intense light and heat on the earth. So, if the year, month or day is presided over by the Sun, the weather of that particular year, month or day will remain warm (even in winter).[1]

Astronomically also Sun dominates the fire elements.

Hence, the Sun's ruling thought causes lot of *vaporization* and hotness in the atmosphere, yet the air remains unsaturated. This leads to scanty clouding and scarce rain fall.[2]

Moon (Soma)

The Moon is the only satellite of the Earth. The Moon is deficient in vasus, i.e. it has no atmosphere as its gravitational power is too weak to hold down *vasus*.[3] The Moon along with the Sun is responsible for the tides, as the Moon is held by the Earth and the Earth is held by the Sun. But Moon being nearer to Earth then Sun exerts greater effect on the Earth. Astronomically, the Moon dominates the water element. This is why, under the ruler ship or dominance of the Moon, the evaporation takes place enough to give rise to thundering clouds often yielding heavy rain.[4]

In addition, Moons' diverse movements have been observed to affect the weather diversely. For instance, if the Moon moves to the north-east of *Rohiṇi*, there will be good and timely rains.[5] When the Moon touches *Rohiṇi* while moving to the north, there will also be copious rain giving rise to lot of disaster. On the contrary, when it moves north-ward without touching *Rohiṇi*, there will be abundant rain, but no disaster whatsoever.[6]

The Moon's being aspected by benefices, when it occupies the 7th house from Venus, or 5th, 7th or 9th from Saturn during rainy season, adds to its power in the positive direction and hastens the materialization of the resultant meteorological phenomenon, i.e, rain.[7]

---

1. tīkṣaṇaṁ tapatyadittijaḥ śiśire'pi kāla BS. 19.2
2. natyambudājalamuco'calasannikāśāḥ BS. 19.2
3. See Origin of star and planets and Origin of Midsphere in authour's book: Origin of Universe.
4. vyāpataṁ nabhaḥ pracalitācala sannikāśaiḥ
   vyālāñjanāliga valacchavibhiḥ payodaiḥ
   gāṁpūrayabhirakhilāmamalābhiradbhi
   sutkaṇṭhitena guruṇādhvanitena cāśāḥ. BS. 19.4
5. yāte sthāṇudiśaṁ guṇāḥ subahavaḥ sasyārdhavṛṣṭyā dayaḥ. BS. 24.33
6. spṛsannudagyātiyadā śaśāṅka stadā suvṛṣṭirbahulopasargā asaṅspṛśan yogamudak sametaḥ karoti vṛṣṭiṁ vipulāṁ śivam. BS. 24.29
7. prāvṛṣi śitakaro bhṛguputrātsaptamarāśigatataḥ

Though the Moon, by nature, is rain-promoting factor, but when it is eclipsed by the strong planets like Mercury or Saturn or the Ketu (a comet) that promote anti-rain phenomenon on the Earth by their influence, the rainlessness, or famine/drought will take place due to the dominance of anti-rain causing factors. This is why; ancient meteorologists studies the Moon's eclipse by Saturn, Mercury or Ketu as consequent upon famine/drought or rainlessness.[1]

Mars (Maṅgala)

Mars is a close neighbour of the Earth. It is smaller in size than the Earth. It has a thin atmosphere. It is barren place covered in the pink soil and boulders, as per modern scientific investigation. Long ago it was more active. Its surface is marked with dormant volcanoes. Its volcanic activity rendered it a dry land. Astronomically, it dominates the five elements. Hence its effect on the Earth also causes the conditions of evaporation and hotness. This is why, the ancient meteorologists investigated that if the year is ruled by Mars, the clouds, due to heavy evaporation from oceans, though appear thick and piled up in the sky, don't yield sufficient rain.[2]

Its movement via different constellations generally accounts for the phenomenon of rainlessness. For instance, should the Mars pass through the southern side of *Rohiṇi*, the rain would be marred.[3] If it makes a move through *Rohiṇi Srāvaṇa, Mūla, Uttaraphālguni, Uttarāṣāḍhā, Uttarabhādrapadā, Jyeṣṭhā* it will even deseed the clouds.[4] If it is retrogressed from the 10th, 11th or 12th asterism counted from the one wherein it rises, it will give effect to the conditions of rainlessness.[5] If the Mars commences its retrograde motion from 17th or 18th constellations, it will give rise to the condition of drought.[6] Should the Mars not

---

śubhadṛṣṭaḥ sūryesutannavapañcamago vā saptamagaśca jalāgamanāya. BS. 28.19

1. śastra kṣudbhayakṛdyamena śaśijenā vṛṣṭidurbhikṣakṛt kṣemārogya subhikṣavināśī sītamśu yadi bhinnaḥ. BS. 4.27
2. abhyunnatā viyati saṁhatmūrtayo'pi
muñcaniti na kvacidapaḥ pracuraṁ payadāḥ BS. 19.9
3. dakṣiṇato rohiṇyāścaram mahijo'rgha vṛṣṭye nigrahakṛt. BS. 6.10
4. prāpāpae śravaṇe mūle triṣu cottareṣu śakre ca
vicaran ghamnivahānāmupaghāta karaḥ kṣamātanayaḥ BS. 6.11
5. dvādaśa daśamaikādaśanakṣādvakrite kuje'śrumukham
dūṣayati rasānudaye karoti rogānavṛṣṭiṁ ca. BS. 6.2
6. asimuśalaṁ saptadaśadaṣṭādaśato'pi vā tadanuvakre dasyugaṇebhyaḥ pīḍāṁ karotyavṛṣṭiṁ saśastrabhayam. BS. 6.5
Parāśara also records similar effects
saptadaśe'ṣṭādaśe vā dasyugaṇaiḥ prajānāmupadrava mvrṣṭiṁ śastrabhayaṇ ca. (Bhaṭṭ pala on BS. 6.5)

commence its retrograde motion from the middle of the *Maghā* asterism, there would be no rain.[1] If it passes through *Viśākhā* after having passed via *Maghā,* the famine will take place.

In spite of its anti-rain effect on the Earth, it is also noticed that some of its retrograde motions heralded food or immediate rain. For instance, the retrogression of Mars from the 15th and 16th asterism counted from excluding the one wherein it rises after eclipse, gives effect to immediate rain.[2] If the Mars is retrogressed from the 13th or 14th asterism, there would be good rains.[3]

Mercury (Buddha)

It's a tiny planet, just larger the size of Earth's Moon. Like that of Moon, it, due to its smaller size, has no atmosphere to hold down gases and water. This is why, the cratered planets, days are scorching hot and nights are frigid like that of Moon. Though it has all characteristics (qualities) like that of Moon. astronomically, it differs from Moon. Moon dominates the water elements, so it is responsible for creating the phenomenon of rain. The Mercury, whereas, rules the elements of air. Due to this reason, the planet is also responsible for giving effect mainly to the meteorological phenomenon of rainlessness or drought except for few cases of rain, for instance, the Mercury while passing through the zones of *Śrāvṇa, Dhaniṭhā, Rohiṇi, Mṛgaśirā, Uttarāṣāḍhā* cuts anyone of them across will give effect to the meteorological phenomenon of anti-rain.[4] when the Mercury enters the zones of the constellations of Ārdrā, Punarvasu, Tiṣya, Āṣleṣā and *Maghā*, it will create the conditions of drought.[5] On the other hand, while passing through the zone of Bharaṇī, *Kṛttikā*, Rohiṇi or Svāti, it causes good rains.[6] In its saṁkṣipta and miśra coures, i.e., while passing through *Puṣya Punarvasu*, two

---

1. madhye na yadi maghānāṁ gatāgataṁ lohitaḥ karoti
   pāṇḍyo nṛpo vinaśyati śastrodyogādbhyayamavṛṣṭiḥ. BS. 6.8
2. rudhirānanamiti vaktraṁ pañcadaśāt ṣoḍaśācca vinivṛtte
   tatkālaṁ mukharogaṁ sabhayaṁ ca subhikṣamāvahati.
3. vyālaṁ tryodaśarkṣāccaturdaśādva vipaccyate' stamaye
   daṅṣtri vyālamṛgebhyaḥ karoti pīḍāṁ subhikṣaṁ ca. BS. 6.3.
4. vicaran śravaṇa dhaniṣṭāprajāpatyenduvaiśvadevāni
   mṛdnan himakara tanayaḥ krotyavṛṣṭiṁ sarogabhayāṁ. BS. 7.2
5. raudrādīni yadā pañca nakṣatrāṇindunandanaḥ
   bhinatti śastra durbhikṣa-vyādhibhiḥ piḍyate jagat
   Kaśyapa quoted by Bhaṭṭpala on BS. 7.3
   Cf. also
   raudrādīni maghāntāni upāśrite candraje prajāpiḍa śastranipata
   kṣudbhayarogānā vṛṣṭisantāpaiḥ BS. 7.3
6. prākṛta gatāmārogyavṛṣṭī sasyapravṛddhayaḥ kṣemam BS. 7.14.

Phālgunis, Mṛgaśirā, Ārdrā, *Maghā*, Āśleṣā etc., it causes effects of mixed nature, i.e. sometime rain and othertime drought.[1]

Jupiter (Bṛhaspati)

This is the largest planet in the solar-system with the turbulent cloud cover marked with enormous eddy in them. Astronomically, it dominates fire and water elements. On account of its nature of always beings covered with turbulent clouds and its domination of fire and water elements, it gives effect to the numerous towering clouds yielding sufficient rain whenever it comes closer to the Earth.[2]

Its movement through various constellations creates various types of meteorological phenomena on Earth, on account of collusion of its basic character with those of various asterisms coming into its contact. For instance, if the Jupiter is eclipsed while moving to the zone of the 1st quarter of Dhaniṣṭhā i.e., the tenth houses and rises after eclipse in the month of Māgha (which is represented by fire element), it will instead give rise to drought, outbreak of fire's and storms.[3]

Should the Jupiter rise after eclipse in the zone of Mṛgaśirā asterism (1st two quarters of which lies with the second house, i.e the house of fire element and the last two quarters lies with the 3rd house, i.e the house of air), it would likely to cause the rainlessness.

It is because the ratio of air and fire get over water. Keeping in view the fact behind this statistics, the later meteorologists declared that should the Jupiter rise with Mṛgaśirā asterism, there would be drought.[4]

If the Jupiter rises in the zone of *Maghā* constellation (forming fifth house represented by fire element), it adds to good rains. The same fact been observed by Varāhamihira.[5]

Should it rise in the zone of *Phālguins* (constituted of 5 portions of fire and three portions of air), its new constituent ration of fire and air being 6:4, it will give effect to sporadic rain-fall.[6]

---

1. saṁkṣipta miśrayormiśram. BS. 7.14.
2. vividhairviyadunnataiḥ payodairvṛtanurvīṁ payasābhitarpa yadabhiḥ. Surarāguroḥśubhevarṣe bahusasyākṣitiruttamaraddhiryukto. BS. 19.15.
3. ādyaṁ dhaniṣṭhāśamabhiprapanno māghe yadā yātyudayaṁ surejyaḥ. ṣaṣṭyābdapūravaḥ prabhavaḥ sanāmnā prapadyate bhūtahitastadābdaḥkvacittvvṛṣṭiḥpavanāgnikopaḥsantitayaḥśleṣ makṛtāścarogāḥ. BS. 8.28.28.
4. saumye'bde'nā vṛṣṭiḥ. BS. 8.4.
5. pitṛpiyāparivṛddhirmāghe hāridiśca sarvebhūtānāam ārogya vṛṣṭidhānyārdhasampado mitralābhaśca. BS. 8.6
6. phalgunavarṣe vindyāt kvacitkvacitkṣemavṛṣṭī sasyāni. BS. 8.7

Its rise in the zone of Citrā constellation change its elementary composition as air 4, fire 9, water 9, and so it will be able to effect only scanty rain-falls. This is why, later meteorologists maintained that Caitra month have scanty rain-fall.[1]

With its rise in the zone of Viṣākhā, the proportion of new composition works out as air 3, water 10, and fire 9 which is sufficient to bring good rains for the growth of crops and so was also observed by ancient meteorologist Garga.

'In the month of Viṣākhā, rain sufficient for the growth of crops takes place.'[2]

In the zone of Aṣāḍhā constellation, the proportion of fire, earth and water being 14:3:9, it effects the phenomenon of rain at some place and rainlessness elsewhere.[3]

In the Bhādrapadas the revised statistics being air 3, water 13 and fire 9, it will give rise to rain sometimes and rainlessness othertimes.[4]

In the Aśvinī, the elementary proportion of fire and water being 13:9, it cause constant rain-fall because its power is boosted with the addition of four portions of fire.[5]

The Jupiter's movement to the north of the constellation in the zone of which it rises, was also observed by the ancient meteorologists as effecting the phenomenon of rain of Earth.[6]

Venus (Śukra)

Venus is Earth's twin so far as its size and mass is concerned. It is also Earth's next door neighbour like that of Mars. It is seemingly hot and perpetually veiled behind the reflective sulphuric acid clouds. It has a carbon-dioxide environment over flat, rocky plains. The sings of volcanic activity have also been detected by the modern probe. Astronomically it is said that it rules the 2nd and 7th houses i.e., it dominate the elements of water. Mythologically, it is known as the Purohita (precursor) of *asuras*, i.e., clouds. Perhaps, this is why, during the

---

1. caitre mandā vṛṣṭiḥ. BS. 8.8
2. vaisākhe tu sasyajanmā vṛṣṭayaḥ sambhavanti hi. BS. 8.9
3. kvacidavṛṣṭirnyatra. BS. 8.11
4. bhādrapade- kvacitsubhikṣaṁ kvacicca bhayam. BS. 8.13.
5. āśvayuje'bdi jasraṁ patati jalam. BS. 8.14.
   See also Vṛddha Garga quoted by Bhaṭṭpala on BS. 8.14
   paryātasasyāṇnajaalakṣeśāśvayujaḥśivaḥ
6. udgāroary subhikṣemakaro vākpatiścaṭṭran bhānām. BS. 8.15.

governance of Venus, the year witnesses heavy rains poured out of mountain like clouds. The ancient meteorologists also noted the same phenomenon.[1]

Its various movements give effect to various types of weather conditions of the Earth.

It has often been observed that its movement through anti-rain constellations brings about rain on the Earth and through rain-producing constellations prevents rain on the Earth.[2]

For instance, if the Venus passes through *Kṛttikā*,[3] Ārdrā,[4] Puṣya,[5] *Maghā*,[6] Pūrvaphalgunī,[7] *Uttaraphālgunī*,[8] *Swāti*,[9] *Viśākhā*[10] and *Pūrvabhādrapada*,[11] it will cause heavy rain. On the contrary, its movement through Anurādhā, Jyeṣṭhā, Mūla and Hasta causes drought on the Earth.[12]

Similarly, if the Venus is housed in the eastern horizon during rainy season (monsoons) there will be scanty rain-fall. On the contrary, its presence in the western horizon causes copious rain.

Venu's rising or setting in different lunar days or tithis gives effect to different type of weather conditions on the Earth. For instance, if it rises after eclipse or gets eclipsed on the 8th, 14th or 15th (i.e. new moon day) or the dark fortnight, there will be heavy rain storms causing floods.[13]

---

1. śalīkṣumatyapi dharā dharaṇī dharābha
   dhāsādharojjhitapayaḥ paripūrṇavaprā — BS. 19.16
2. Cf. Parāśara on BS 9.36. quoted by Bhaṭṭpala
   avārśake bhe vicaran yadi varṣati bhārgavaḥ
   vārśakarkṣagato vyaktaṁ sodaśārcina varṣati
3. bhindaṅgato'nalarkṣam kūlātikrāntva vārivā hābhiḥ — BS. 4.24
4. ārdrāgatastu salilanikarakaraḥ — BS. 4.26
5. puṣye puṣtā vṛṣṭiḥ — BS. 9.27
6. bhindan maghāṁmahāmātrā doṣakṛd bhūrivṛṣṭiḥ. — BS. 9.28
7. bhāgye ambunivaha mokṣāya — BS. 9.29
8. āryamṇe saliladāyi — BS. 9.29
9. citrāsthe śobhanavṛṣṭiḥ. — BS. 9.30
10. Svātau prabhūta vṛṣṭiḥ aindrāgn'pi suvṛṣṭiḥ — BS. 19.31.
11. karoti cosmin sitaḥ salilam — BS. 19.34
12. triṣvapi caiteṣvanāvṛṣṭiḥ — BS. 19.32.
    haste piḍā jalasya ca nirodhaḥ — BS. 19.30
13. caturdaśīṁ pañcadaśīṁ tathāṣṭamīṁ
    tamisrapakṣasya tithim bhṛgoḥ sutaḥ
    yadā varjed darśanamastameti vā
    tadā mahī vārimayīva lakṣyati — BS. 9.36
    Cf. also Kāśyapa quoted by Bhaṭṭpala on — BS. 9.36
    kṛṣnapakṣe himāvasyā caturdaśyaṣṭamīṣu ca
    udayaṁ bhārgavaḥ kuryāttadā vṛṣṭi pramuñcati.

If it is eclipsed or rises after eclipse in the month of Kārtika, there will be no rain at all for the 90 days.[1]

Should the Venus appear or disappear at the time when the Sun or the Moon are completing their circuit, the rains, would be effected.[2]

Saturn (Śanaiścara)

Saturn is the second largest planet in our solar-system. Its closeness to Earth gives effect to the rainlessness.[3]

When it moves through the 'Front gate', viz. the seven asterisms commencing with *Kṛttikā* and becomes retrograde, there will be a dreadful and long-drawn famine and drought.[4]

Thus, we have studies above the weather-control on the Earth by the individual planets passing through different constellations housed around the Earth. Now we shall see the combined effect of the two or more planets moving through a particular house or houses.

Combined effect of two or more planets

If the combination of four or five planets takes place in the house of *Śvāti, Bharaṇī* and *Kṛttikā* constellations (astronomically known as *Nāgavīthi*), there will be heavy rain.[5]

If this combination occurs in the constellations of *Pūrvāṣāḍhā* and Uttarāṣ āḍhā (popularly known as *Amalavīthi*), there will be no rain at all.[6]

---

1. kārttike tu yadā māsi kurute'stamayodayau
   tadāhnāṁ navatiṁ pūrṇām devo bhuvi na varṣati.
   Parāśara on BS. 9.36
2. astodaye tu śukrasya yadi candradivākarau
   āvṛttimārgaṁ kurvāte tadā varṣati bhārgavaḥ.
   Parāśara on BS. 9.36
3. sasyāni mandamabhi varṣati vṛtra śatrur
   varṣe divākarasutasya sadāpravṛtte. BS. 19.21.
4. prāgdvāreṣu caran raviputro nakṣtreṣu karoti ca vakram
   durbhikṣaṁ kurutemahadugaṁ mitrāṇāṁ ca virodhamavṛṣṭim. BS. 47.13.
   atijala mokṣaṁ kuryunāgākhyāyāṁ ca sarve tu.
   Samāsa saṁhita quoted by by Bhaṭṭpala on BS. 20.9
5. durbhikṣa roga taskara śastrā vṛṣṭī kṣudhaṁ grahāḥ kuryuḥ
   Samāsa saṁhita quoted by by Bhaṭṭpala on BS. 20.9
6. dehe krūra nipīḍite'gnyanilajaṁ (bhayam) BS. 8.19
   See also Kāśyapa quoted by Bhaṭṭpala on BS. 8.19]
   krūragrabhāhatedehedurbhikṣānalamārutāḥ

When the Sun, Mars and Saturn (known in astrology as malefic or Pāpagrahas) are in conjunction with *Kṛttikā* and *Rohiṇī* stars, there will be danger from fire and winds.[1]

Similarly, if this combination is in conjunction with the constellation of Pūrvāṣāḍhā and Uttarāṣāḍhā, there will be a threat of famine.[2]

If the Jupiter and Venus are situated in the Western and Eastern horizon respectively or vice versa and are housed in the 7th zodiac sign from each other i.e. 180^0 apart, the Earth will not experience rain-fall.[3]

If at the time of dawn, Venus and Jupiter sojourn the eastern and western horizons respectively or vice-versa opposing each other, there will be rainlessness.[4]

Should the Jupiter, Mercury, Mars, and Saturn precede Venus in their transit, there would originate hurricanes dashing trees, etc. to the ground thunder-bold smashing the tops of mountains and not even a single drop of water would be there in the name of rain.[5]

When the Mars should precede Venus in its transit, there would be complete rainlessness, and lightning, fire and duststorm would be active.[6]

---

1. nābhyāṁ bhayaṁ kṣutkṛtam — BS. 8.19
   See also Kāśyapa quoted by Bhaṭṭpala on — BS. 8.19
   kṣutbhayaṁ tu bhavennābhyām.
2. gururbhṛguścāparapūrvakāṭhayoḥ
   parasparaṁ saptamarāśigau yadā
   tadāprajārugbhayśokapiḍitā
   navāripaśyantipurandarojjhitam — BS. 9.37.
3. pratyuṣe prāksthitaḥ śukraḥ pṛṣṭhataśca bṛhaspatiḥ yadā'nyo'nyaṁ nirīkṣete----- avṛṣṭiśca tadā bhavet.
   yadā tu pṛṣṭhataḥ śukraḥ puracaśca bṛhaspatiḥ
   yadā valokapetāṁ tau tāvadeva phalaṁ bhavet
   Bhadrabāhu quoted by Bhaṭṭpala on — BS. 9.37
4. yadā sthitā jīvabudhārasūryajāḥ
   sitasya sarve'grapathānuvarttinaḥ
   ------- bhavanti vātāścasamucchriāntakāḥ — BS 9.38.
   nacālpamapyambu dadāti vāsavo
   bhinatti vajreṇa śirāṁsi bhūbhṛtāṁ — BS. 9.39
5. nihanti śukraḥ kṣitije'grataḥ prajām
   hutāśaśastra kṣudvṛṣṭitaskaraiḥ
   carācaraṁ vyaktamathottarapathaṁ
   diśo'gnividyadrajasā ca pīḍayet. — BS. 9.41.
6. bṛhaspatau hanti puraḥsthite sitaḥ
   site samastaṁ dvijagosurālayān
   diśaṁ a pūrvāṁ karkāsṛjo'mbudā

If the Jupiter takes precedence over Venus, hailstones will precipitate instead of rain.[1] But the precedence of Mercury over Venus gives effect to rain.[2]

Comets

In addition to the planets, comets, which originated at the time of origin of the planets and satellites, too play a vital role in affecting the atmospheric conditions on our planet. According to the investigation carried out by the ancient sages, there are 22 comets originated from the Earth. They are round in shape like a mirror devoid of crest possessed of rays and are luminous like water or oil. They are housed in the north-east horizon. There appearance portends famine.[3] According to another investigation, there are three comets originated from Moon, which are as white as the Moon herself. They tenet northern horizon. Their visibility brings about rain-fall.[4] Some other comets were also discovered that affected the atmosphere of Earth considerably. For instance, there is a big comet named *Vasātketu* which has elongated glossy body stretched out northward. it sojourns in the western horizon. Its rise brings about rain.[5]

There is another comet named *Asthiketu*, endowed with all the characteristics of *Vasātketu*, but is rough instead of glossy. Its appearance portends famine.[6]

There is yet another comet christened as *Jalaketu*, tenanting the western horizon and possessed of a crest also protruding towards west. Its rise gives effect

---

mlegadā bhūri bhavecca śārdam. BS. 9.42.

1. saumyo'stodayayoḥ purobhṛgusutasyāvasthitastoyakṛd BS. 9.43.
2. darpaṇavṛttākārā viśikhāḥ kiraṇanvitā dharātanayā
kṣudbhayadā dvāviṁśatiraiśānyā mamabutailanibhāḥ BS. 11.13.
Cf. also Garga quoted by Bhaṭṭpala on BS. 11.13
samastavṛttā viśikhā raśmibhiḥ parivāritāḥ
ambutaila pratikāśa dvāviṁsad bhūsutāḥ smṛtāḥ
aiśānyāṁ diśī dṛśyante durbhikṣabhyadāstute.
3. śasikiraṇa rajatahimakumudakunda
kusumopamāḥsutāḥ śaśinaḥ
uttarto dṛśyante tryaḥ subhikṣāvahāḥ śikhinaḥ BS. 11.14.
Cf. also Garga quoted by Bhaṭṭpala on BS. 11.14
candraraśmisavarṇābhāhimakundendusaprabhāḥ
trayaste śaśinaḥ putrāḥ saumyāśāsthāḥ śubhāvahāḥ
4. udagāyato mahān snigdhamūrtiraparodayī vasāketuḥ
sadya karoti marakaṁ subhikṣamopyuttamaṁ kurute. BS. 11.29.
5. tallakṣaṇasthi ketuḥ sa tu rukṣaḥ kṣudbhayāvahaḥ proktaḥ. BS. 11.30.
6. Jalaketurapi ca paścātsnigdhaḥ śikharyāpareṇa cannatayā nava māsān sa subhikṣam karoti śāntiṁ ca lokasya. BS. 11.46.

to good rain.[1] One more comet called *Bhavaketu*, which is glossy, tiny and appears only for one night in the eastern horizon. It is endowed with a crest similar to that of a lion's tail curling to the right. it amounts to good rain for as many months as many *muhurtas* it remains visible.[2] The comet named Kapālketu which is hosed in the east and makes its appearance only on the Amāvāśyā new moon day. Its rays and crest are of smoky colour. It traverses half of the horizon. It causes famine and drought.[3] The comet known as Raudra will also causes famine and drought like that of Kapālaketu, if it rises in the east near Purvāṣāḍhā and Uttarāṣāḍhā. It has a tridental crest. Its flame is grey, rough and red. It traverses the third of the firmament.[4] The comet called Calaketu tenets the western direction. It has a crest with its tip protruding south-ward. Its length increases with its advancement to north. it makes its retreat after touching the seven sages, i.e. the Great Bear, the Pole star and the asterism Abhijit. It sets in the south traversing half of the firmament. Its appearance portends famine.[5]

---

1. bhavaketurekarātraṁ dṛśyaḥ prāk sūkṣmatārakaḥ snigadhaḥ
   harilāṅgulopamayā pradakṣiṇāvartayā śikhavā
   yāvat muhūrtān darśanamāyāti nirdiśenmāsān
   tāvadatulaṁ subhikṣaṁ subhikṣaṁ rūkṣe prāṇāntikāṁ rogā. BS. 11.47,48.
2. dṛśyo māvasyāyāṁ kapālaketuḥ sadhūmaraśmiśikhuḥ
   prāṅnabhasorddha vicāri kṣunmarakā vṛṣṭirogakaraḥ. BS. 11.311
3. prāgvaiśvānaramārgeśulāgraḥ syāvarūkṣatāmrārciḥ
   nabhasastribhā gagāmī raudra iti kapālatulyaphalaḥ BS. 11.32.
   See also Paraśara quoted by Bhaṭṭpala on BS. 11.32
   nabhastribhāgacārī sa śastrabhaya rogadurbhikṣa nā
   ṣṭimarakairyāvānmāsān dṛṣyatetāvadvarṣāṇitribhāgaśeṣāṇI
   prajāṁ kṛtvārdhaṁ ca śāradadhānyamāḍhakamastaṁ vrajati.
   Also Vṛddha Garga quoted by Bhaṭṭpala on BS. 11.32
   tribhāgaṁ nabhaso gatvā tato gacchatyadarśanam.
   yāvato divasāṁstiṭettāvadvarṣāṇī tad bhayam
   śastrāgnibhayarogaiścadurbhikṣamaraṇairhatāḥ.
4. aparasyāṁ calaketuḥ śikhayāyāmyāgrapaṅgulocchritayā
   gacchedyathā yathodakt, tathā tathā dairghyamāyāti saptamuīna saspṛśya dhruvamabhijitam eva ca prati nivṛttaḥ. BS. 11.33.34.
   nabhaso' rddhamātramitvā yāmyenāsaṁ smupayāti
   anyāmapi ca sa deśān kvacit kvaciddhanti rogadurbhikṣaiḥ BS. 11.36
   Cf. Paraśara quoted by Bhaṭṭpala on BS. 11.33-36.
   teṣvapi kvacit kvacicchastra durbhikṣavyādhimarakabhayaiḥ kliśnātityaṭdaśa māsāniti. Also Garga on BS. 11.33-36.
   kṣucchastra marakayādhi bhayaiḥ sampīḍyet prahāḥ
   māsān daśa tathāṣṭau ca calaketuḥ rudāruṇaḥ.
5. snigdhau subhikṣaśivadāvadhikaṁ dṛśyate ka nāmāyaḥ BS. 11.38.
   Cf. Paraśara quoted by Bhaṭṭpala on BS. 11.38.
   Kaḥ prajapatiputra yadyadhikaṁ dṛśyate tadā dāraṇataraṁ
   prajāñaṁ śastrapaṁ kuryāttathaiva snehavṇayaktān

There is a comet named Śvetaketu, whose crest is inclined south-ward and which sojourns in the east at midnight. There is another comet named Ka which has a yoke like form and is housed in the west. Both can be seen at the same time for a period of 7 days. They are glossy and bring about rain leading to prosperity.[1]

The comet called Kumuda has a white lustre and a crest inclined towards east. It appears for a single night in the west. It bestows upon the countries it is sighted a unique prosperity caused by rains for a period of 10 years to come.[2]

The comet named Maṇiketu is tiny comet housed in the west and is visible only once for a period of three hours. Its crest is straight and white like a streak of milk ejected from the breast, as soon as it appears, it causes rain.[3]

Forecast on the basis of the Astronomical Impressions

The ancient meteorologists found that the various meteorological phenomena occurring in the atmosphere of the Earth either due to the geocentric formative factors or the extra-planetary controlling factors leave various astronomical impression in terms of giving various shapes, sizes, appearances, colours, images, halos, etc. to various heavenly bodies, rainbows, twilights and clouds. As such, the ancient meteorologists made all these resultant astronomical impressions as their tools to forecast/predict weather in the days to come. Hereunder a humble attempt will be made to assess the astronomical impression left by the various meteorological phenomena.

Appearance

If humidity reaches the saturation point, it will make the astronomical phenomena appear glossy, lustrous, brilliant, dense or compact. For instance, glossy features of both the twilights signal the immediate rainfall.[4] If the Sun-rays appears to by glossy, there will be rain.[5] Lustrous and glossy Sun in the sky portends sufficient rainfall.[6]During rainy season, the Sun having a dazzling shining

---

kṣemārogyasubhikṣadau bhavataḥ.

1. kumuda iti kumudakāntivārunyāṁ prākṣīkhoniśāmekām
 dṛṣṭaḥ subhikṣamatulaṁ daśa kila varṣāṇī sa karoti. BS. 11.43.
2. sakṛdekayāma dṛśyaḥ susūkṣam tā ropareṇa maṇiketuḥ
 ṛjvī sikhāsya śukla stanodgatā kṣīradhāreva
 uḍayanneva subhikṣaṁ caturo māsān karoṭyasau sarddhān. BS. 11.45
3. sarvairetaiḥ snigdhaiḥ sadyo varṣam BS. 47.22.
4. snigdhā dīdhitayaī----- vṛṣṭī syād BS. 47.23.
5. ayane suprabhaḥ snigdhaḥ sevate yadi bhāskarah BS. 3.5.
6. prāvṛṭa kāle sadyaḥ karoti vimaladyutir vṛṣṭim. BS 3.28

will effect rain within a short span of time.[1] Also the Sun bearing a shining like *śriīsa-puṣpa*, i.e. yellowish blue will effect rain shortly.[2]

The dense or compact Moon also point to rain-fall.[3] The Moon with glossy and whitish rays in an omen of rain-fall.

If the Saturn, only when it looks glossy, takes its course through *Śrāvaṇa*, *Svāti, Hasta, Ārdrā, Bharaṇī*, and *Pūrvaphālgunī*, the Earth will be inundated with waters.[4]

If the mountains wear the look like the heaps of collyrium or the cave wrapt in vapour, the rain will take place immediately.[5]

Colour

It has also been that meteorological phenomena also leave its astronomical impression in the form of colour. For instance, saturated air of the atmosphere gives a blue colour to the star Canopus (Agastya).[6] It marks the white colour in the black *Rāhu.*[7] Sometimes humidity also provides the grey colour or yellowish shining to the *Rāhu.*[8]

The phenomena of rain is also indicated by the yellow colour of lightning.[9] It makes the Sun shine white, yellow or in golden shade.[10] It attributes the sky with the white and blue colour or with colour of cow's eyes or crow's eggs.[11] It attributes the Moon with red hue like the eyes of a parrot or a pigeon or of honey.[12]

On the contrary the phenomena of anti-rain in marked by the multi-colours or the smoky colour of the Sun-rays, or a black coloured Sun in the rainy-season.[13]

---

1. varṣākāle vṛṣṭiṁ karoti sadyaḥ śirīṣapuśpābhaḥ. BS. 3.27
2. sthūlaḥ subhikṣakāri BS. 4.20
3. singdhaḥ prasanno raśmivānśvetaḥ kṣemasubhikṣavṛṣṭikaraḥ BS. 4.30
4. sravaṇānilahastārdvā bharaṇī bhāgyopamaḥ suto'rkasya
   pracurasalilopagūdāṁ kaoti dhātriṁ yadi snigdhaḥ BS. 10.1
5. girayoñjana curṇasannibhā yadi vā vāṣpaniruddha kandarāḥ----vṛṣṭidāḥ BS. 28.6
6. nilo tivarṣāya See Paraśara quoted by Bhaṭṭpala on BS. 12.22
7. śvete kṣema subhikṣam. BS. 5.53.
8. ādhūmre kṣema subhikṣamādiśenmanda vṛṣṭiṁ caBS. 5.55
   vimalakamaṇī pītābho vaiśyadhvṁso bhavet subhikṣāya BS. 5.57
9. pitābhavati sasyāya. PM. 2.3.13.
10. varṣāsu śuklaśca BS. 3.23
11. gonetrābhyāṁviyadvimalādiśo---
    kākāṇḍābhaṁ yadā ca bhavennabhaiḥ BS. 28.4
12. śukakapota vilocanasannibho madhunibhaśca yadā
    himdī dhitiḥ----patati vāri tadā na cireṇa. BS. 28.11
13. -----salilaṁ nāśu pātayati BS. 3.22
    varṣāsvasitaḥ karotyanāvṛṣṭim BS. 3.26

Similarly, white hue of lightning indicates famine conditions.[1] Red colour in *Rāhu* is also portentous of famine.[2] The meteorological phenomenon of anti-rain makes the Venus appear grey like ashes.[3] It attributes the star Canopus with red or russet colour.[4]

Shape

The changing meteorological phenomena also give rise to the various astronomical shapes in the sky. For instance, the phenomenon of rain is marked by the Moon with the northern peak looking up.[5] It also attributes the Moon with the boat like shape with its northern peak having attained more shining.[6] The phenomenon of rain-fall also provides the Moon with a shape of a tambourine.[7]

On the other hand, the meteorological phenomenon of anti-rain is marked with the sign of the form of a Kīlaka in the Sun.[8] Its impact on Moon reflects in the Vajra Moon. i.e. the Moon with a slender waist.[9] It also makes the Moon sometime appear with the southern peak looking up.[10] Sometimes, it attributes the Moon with a shape called *āvarjita*. i.e. the Moon with the upward front.[11]

Size

Sometimes, the meteorological phenomena also tell upon the size of the various heavenly bodies. For instance, the phenomenon of rain-fall is indicated by the big size of the Moon. On the other hand, extremely small size of Moon fortells about the possible occurrence of famine.

---

1. durbhikṣāya sitā bhavet. PM. 2.3.13.
2. aruṇakiraṇānurupe durbhikṣāvṛṣṭayo
   CF. also pāṁśuvilohitarūpaḥ-bhavativṛṣṭeśca BS 5.55
3. patati na salilaṁ khāt bhasmarukṣā sitābhe BS 9.44
4. kapilastva vṛṣṭim māñjiṣṭha rāgaśadṛśaḥ kṣudham BS. 12.19
5. *proktasthānābhāvādudagaccaḥkṣemavṛddhivṛṣṭikaraḥ* BS 4.16
6. arddhonate ca lāñgulamiti--- subhikṣaṁ ca BS. 4.9
   Cf. also Vṛddha Garga quoted by Bhaṭṭpala on BS 4.9
   yadā samaḥ pratipadi nausthāyi sampradṛśyate
   kṣema subhikṣamārogyaṁ sarvabhuteṣu nirdiśet.
7. candro mṛdaṇgarūpaḥ kṣemasubhikṣāvaho bhavati BS. 4.19.
8. durbhikṣaṁ kīakerkasthe BS. 3.17
9. eko durbhikṣakaro BS. 3.14
10. madhya tanurvajrākhyaḥ kṣubhayadaḥ BS. 4.19
11. dakṣīṇatuṅgaścandro durbhikṣabhayāya nirdiṣṭaḥ
    See also
    yagameva yāmyakoṭyām kiñcituṅgaṁ
    sa pārśvaśāyīti------ vṛṣṭeśca vinigrahaṁ kuryāt. BS. 4.13

Image

The phenomenon of rain-fall causes the development of the image of the Sun to the north of it and also of the Moon reflected in the sky.[1]

The phenomenon of anti-rain is reflected by the image of the Sun developed to the south of it. Also the image to the both sides, i.e., north and south foretells the water crisis.[2]

Halo

The meteorological phenomena also leave their impression in the form of halos formed around various heavenly bodies. For instance, the halos of Sun and Moon that resemble the peacock's neck in colour are impressed by the phenomenon of rain.[3] Similarly, a thick and glossy halo that possesses the single colour assigned for the season and is covered with little razor like clouds, or a yellow halo accompanied by the fierce Sun is also formed to signal rain-fall. Moreover, all the haloes around Saturn and Mercury are the heralder of rain.

A deep red coloured halo like cock's eyes around the Moon is also impressed by the saturated air.[4]

Rainbow

Rainbow is also one of the forms of astronomical impressions left by meteorological phenomena of rain. Rainbows are formed of various sizes, shapes and colours according to the various rainy conditions. For distance, if a rainbow is unbroken, bright, glossy, thick multi-coloured and touching the Earth at both extremities, and if it appears double, it is a sign of good rain.[5]

A rainbow seen in the east when there is no rain will produce rain and vice-versa; one seen in the west always indicates rain.[6]

Forecast on the basis of Prognostics

During the course of studying the effect of meteorological phenomena in the context of astronomical as well as human environment, it was found that living

---

1. āvarjitamitya subhikṣakāri tad yodhanasyāpi BS. 4.12
2. divaskṛtaḥ pratisūryyo jalakṛdudag
   pratiśaśī ca yadā divi rājate patati vāri tadā na cireṇa BS. 28.11
3. dakṣiṇe sthitenilakṛt ubhayasthaḥ salilabhayam BS. 3.37.
4. śikhigalasame tivarṣaṁ BS. 34.6
5. kṛkavākuvilocanopamāḥ pariveṣāḥ śaśinaś ca vṛṣṭidaḥ BS 28.6
6. acchinnamavanigādhaṁ dyutimat
   snigdhaṁ ghanaṁ vividhavarṇam.
   Dviruditamanulomaṁ ca praśastamabhbhaḥ prayacchati caBS 35.3

creatures were also very sensitive and alive to the natural happenings or climatic conditions. They had the power to premonition the future weather conditions. Hence the behaviour of atmosphere/weather was also studies in the context of the special behaviour observed by the living creatures. A few investigations made in this direction by the ancient meteorologists may be cited as under.[1]

For instance, much tumbling of fishes ashore and repeated croaking of frogs are prognosticative of rain-fall.[2] Immediate rain is also heralded by cats scratching the ground vehemently with their nails, and the construction of bridges on roads by children.[3] If the dogs weep standing on the roofs of houses or looking up towards the sky, the Earth will be unlevelled with the water.[4] If the birds bath in water or sand and reptiles are seated on the tips of grass, their will fall rain immediately.[5] Should the ants shift their eggs without any trouble, snakes mat or climb down the trees, cows leap, the chameleons perch on the tips of trees and fix their gaze on firmament, and cows or (bulls) look up towards the Sun, there would be rain ere long.[6] If the domestic animals like cows are reluctant to go out of the house and shake their ears and hoofs, or if the dogs behave in the same manner, it should be known that there will be rain soon.[7]

Forecast on the basis of one single day

Apart from the aforementioned inductive methods of forecast, ancient sages also discovered a deductive method of weather forecast. In course of investigating the various laws of nature, they found that the whole rainy season beginning from Śrāvaṇa to Āśvin is represented by the one single day, i.e. the day of Moon's conjunction with the star *Rohiṇi*, The conjunction of Moon with the *Rohiṇī* takes place in the dark fortnight of the lunar month of Āṣāḍha. The seers like *Garga*,

1. vṛṣtiṁ karotyavṛṣtyāma vṛṣṭiṁ vṛṣṭyāṁ nivārayatyaindryāṁ
   paścātsadaiva vṛṣṭiṁ kuliśabhṛtaścāpamācaṭe BS. 35.6
2. Jhaṣāḥ sthalagāmino rasanamaskṛnmaṇḍukānāṁ jalāgama ketavaḥ. BS. 28.4
3. marjārā bhṛṣam avaniṁ nakhair likhanto
   rathyāyāṁ śisuracitāśca setubandhāḥ
   samprāptaṁ jalamacirannirvedayanti. BS. 28.5
4. yadā sthitā gṛhapaṭaleṣu kukkurā rudanti vā
   yadi viyanmukhāḥ ------kṣamā bhavati samavia vāriṇā. BS. 28.10
5. snāyante yadi jalapānsubhirvihaṁgāḥ sevante yadi ca sarīsṛpāstṛnāgrāṇya sanno bhavati tadā jalasya. BS 28.13.
6. vinopaghātena pipīlikānāmaṇḍopasaṁkrānti rahivyavāyaḥ
   drumāvarohaśca bhujaṅgamānāṁ vṛṣṭernimittāni gavāṁ plutaṁ ca. BS. 28.7
   tarusikaropagatāḥ kṛkalāsā gaganatalasthiadṛṣṭinipātaḥ yadi
   ca gavāṁ ravivīkṣaṇamūrdhvaṁ nipatati vāri tadā na cireṇa. BS. 28.8
7. nechanti vinirgamaṁ gṛhādhunvanti śravaṇān khurānapi
   pa śavaḥ paśuvacca kukurā yadambhaḥ patatīti nirdiśet. BS. 28.9

Parāśara, Kāśyapa and *Maya* investigated that the weather conditions to prevail in the entire rainy season may be predicted on the basis of weather conditions prevailed in one single day. Now the question arises as to how to get the prospects of weather of the entire season consisting of four months on the basis of one single day? The answer is very simple. The rainy season generally consists of four months commencing with the lunar month of *Śrāvaṇa*. These four months have eight fortnights and on the other hand, a day consists of eight watches of three hours each. Hence, the eight watches of a day can be equated with the corresponding eight fortnights of a season, and its sub-divisions as days, i.e. twelve minutes of a day can be taken for one day of the rainy season and one hour for four days.

One should observe the behaviour of the wind throughout the day when the Moon is tenanting the star *Rohiṇī* and predict the weather for the rest of the season accordingly.[1] Varāhamihira gives here some guidelines as to how to predict weather on the basis of the wind's behaviour. According to him, if the wind blows round from east to the South and so on, i.e. in a clock-wise manner. it is always indicative of good rain. When two contrary winds blow, the prediction should be made from the point of that wind which one is stronger.[2]So, it can be maintained at the end that the position and behaviour of wind in each hour or each prahara (watch) of the day foretells the prospects of weather to be prevailed in each corresponding group of days or fortnights of the rainy season.

Thus from the foregoing discussion it can be inferred that the ancient Indians, who were not able to modify the weather as per their needs and requirements, developed techniques (mentioned above) of exact forecasting of all short, medium and long-ranges. They were able to predict weather for more than six months through their efficient means of effective forecasting. They didn't develop any satellite system to get the synoptic view and computer facility to compute the statistical and numerical data taken for the purpose of weather forecasting. In other words, they were not merely puppets of these instruments; rather they were seers- the masters of vision. They observed the whole natural phenomena through Yoga and framed such natural laws as were capable of defining the various secrets of nature beyond time and space.

---

1. tatrārddhamāsāḥ praharair vikalpyā
   varṣānimittaṁ divasāstadaṁśaiḥ BS. 24.10
2. savyena gacchañchubhadaḥ sadaiva
   yasminpratiṭhā balavān sa vāyuḥ BS. 24.10.

On the other hand, modern meteorologists have evolved three methods to understand and forecast the behaviour of atmosphere. The important one being the synoptic technique wherein dominant weather producing systems are traced and evolution and movements are monitored. The advent of satellites has added new dimensions to this old method. The second and third methods being statistical and numerical respectively. The invention of super-compute has also revolutionized these methods. With all these sophisticated instruments though the old techniques have been revolutionized to the extent of analyzing data hastily and accurately and getting the synoptic view without any flaw of subjective prejudice, but the drawbacks that lies in the very nature of these techniques cannot be removed. This is why, they modern meteorologists, even with the help of all the sophisticated devices, sometimes fail even in short range forecasting, i.e. weather forecast upto 48 hours in advance. The medium (3 to 5 days) to long range (i.e. a month or more than a month) predictions are still largely a gamble.

Select Bibliography

Aeolus	Meteorology, Eng. U. Press. Ltd. 1952.
Avadha Behari Tripathi	Bṛhat Saṁhitā of Varāhamihira, Varanasi, 1968
Bellikoth Rama	Jaiminīya Upaniṣads Brāhmana, K. Chandra Sharma Sanskrit Vidyāpiṭha Tirupati, 1967
Bhagavaddatta	Ṛgveda par Vyākhyā, Dayānanda College, Lahore, 1920
Brahmadatta Jijñāsu	Dayananda's Yājuṣa Bhāṣya Vivaraṇa
Brahma Muni Parivrājaka	Nirukta sammarśah, Sanskrit Tr. of Nirukta, Ajmer, 1996
Dayananda Saraswati	Yajurveda Bhāṣya, Ajmer.
Dayananda Saraswati	Ṛgbhāṣya, Ajmar.
Eggeling, J	Śatapatha Brāhmaṇa Oxford, Clarenden Press.
Griffith R.T.H.	Text of white Yajurveda, Varanasi, E.J.Lazarus, 1957
Herbert Riehl	Introduction to the Atmosphere, 3rd ed, Mac-Graw-Hill Kgakusha Ltd. 1965
John Daires	Saṁkhya Kaṁka, Indological Book House, Delhi, 1957
Jwālā Prasad Miśra	Vājasaneyī Saṁhitā, Mumbai, 1969

Kaegi, A — Ṛgveda Eng. Tr. Boston,Gim. 1886

Kashinath Shastri — Aitreya Brāhmaṇa with Sāyaṇa- Bhaṣya, Poona, 1977.

Keith, A.B. — Ṛgveda Brāhmaṇas, Aitareya and Kauṣitaki Brāhmaṇas of Ṛgveda, Delhi Motillal, 971.

Kṣemakaraṇa Trivedi — Gopatha Brāhmaṇa, Prayāga, 1920.

Macdonell, A.A — Bṛhaddevatā, Harward Oriental Series, 1904

Macdonell, A.A — Vedic Mythology, Varanasi Indological Book House. 1963.

Mahavir Sastrin — Taittirīya Brāhmaṇa with the commentory of Bhaṭ abhāskara Miśra, Mysore, 1908-21.

Mason, B.J. — Clouds, Rain and Rain making, 2nd ed. Cambridge U. Press 1975

Max Muller, F. — Vedas, Kolktta, Sushil Gupta, 1956

Max Muller, F. — Vedic Hymns, Motilal, Delhi

Motilal Sharma — Śatpatha Brāhmaṇa,Vijñāna Bhāṣya, Jaipur.

Oldenberg, H — Sāṅkhāyana Gṛhya Sūtra,Oxford, 1886.

Pandit, M.P — Gems from the Vedas,Ganesh, 973.

Raghunandan Sharma — Vedic Sampatti, Bombay, 206

Raghvir — Atharva Veda, S.V. Granthamala, Lahore, 1936.

Raghvir, Lokesh Chandra — Jaiminīya Brāhmaṇa, SVS. Nagpur, 1954.

Ram Kumar Rai. — Śaunakīya Bṛhaddevatā, Varanasi.

Ravi, Prakash Arya — Researches into Vedic and Linguistic studies, Delhi. 1991.

Ravi, Prakash Arya — Vedic Science of weather, 1993.

Ravi, Prakash Arya — Contrastive Morphological Analysis of Vedic and Classical Sanskrit, Parimal Publications. Delhi. 1995.

Ravi, Prakash Arya — Ṛgveda Saṁhitā

Ravi, Prakash Arya — Yajurveda Saṁhitā

Ravi, Prakash Arya — Sāmaveda Saṁhitā

Samarpaṇānada — Śatapatha Brāhmaṇa Sāmarpaṇa Bhāṣya.

Swami Shridhara Shastrin	Śāṅkhāyana Āraṇyaka, Anandashram Poona, 1922.
Shridhara Sharma	Kaṭha Saṁhitā, Aundh, 1942.
Shridhara Sharma	Kāṭhaka Saṁhitā, Aundh, 1943.
Shridhara Sharma	Maitrāyaṇī Saṁhitā, Aundh, 1943.
Shridhara Sharma	Maitrāyaṇī Saṁhitā (Brāhmaṇa (porion), Aundh, 1943.
Maitrāyaṇī Saṁhitā,	Taittirīya Saṁhitā, Aundh, 1945.
Uvaṭa & Mahidhara	Vājasaneyī Saṁhitā, Varanasi,
Vedapala Sunitha	Śatapatha ke Daśapatha, Part I & II, Tilaura, Rajasthan, 1991-92.
Vedapala Sunitha	Darśapūrna māsesṭirahasya Prakāsa, Tilaura, Rajasthan, 1991.
Vishva Bandhu	Ṛgveda Skandha Udgitha, Veṅkaṭamādhava Kṛtā Vyākhyā Sahitā, Hoshiarpur.
Vishva Bandhu	Vedic Word Concordance, Lahore.
Yudhiṣṭira	Ṛgvedādibhāṣyabhūmikā Bahalgarh, Sonepat, Mīmāṁsaka. 1984.

Propagation of Sound in the Past and the Present

Dr. V.K.Goyal

Acoustics is the science of Sound that deals with properties of sound waves, the laws of their origination, propagation and action on the obstacles. Here, in the present paper, we propose to study mainly about the origination and propagation of sound on the basis of philosophical Sanskrit literature in the past and the modern science.

1. View of the Mīmāṁsakas and the Grammarians:

The Mīmāṁsakas and the Grammarians postulate that the word (Śabha) is eternal and ubiquitous (Vibhu) which requires no substrate. According to Bhartṛhari (C. 450-510 A.D) the whole world is the manifestation of one eternal entity, having neither a beginning nor an end (anādinidhanam) which is called Śabda-brahman.[1] He has re-affirmed his view with the view of the Vedic Ṛṣis who already had accepted that the world is the result of word, because at the beginning of the Kalpa, the first creation was done by the Vedas which are Śabdātmaka.[2]

Mīmāṁsakas have also declared the eternality of world[3] which has been named by the Grammarians as 'Sphoṭa' and can be revealed by sound (dhvani) i.e. sound is the revealer (abhivyañjaka) of sphoṭa.[4]

Now the question arises: how does the sound originate and propagate? Regarding the origination of sound (word), Bhartṛhari has given the view of a Śikṣ ākāra like this: When the speaker wants to speak, he makes some effort, by which a motion originates in his internal air (Prāṇa-vāyu) and the same motion is result into sound (word) after having impact (abhighāta) with the speech-organs like Kaṇṭha, tālu etc.[5]

---

1. अनादिनिधनं ब्रह्म शब्दतत्त्वं यद्क्षरम्।
   विवर्ततेऽर्थभावेन प्रक्रिया जगतो यतः।। वाक्यपदीयम् I.१.
2. शब्दस्य परिणामोऽयमित्याम्नायविदो विदुः।
   छन्दोभ्य एव प्रथमेतद्विश्वं व्यवर्तत।। वाक्यपदीयम् II.१२०.
3. See (i) Mīmāṁsā-Sūtra. 1.1.18.
   (ii) Mīmāṁsā-Śloka-Vārttika, 1.1.5.
4. See K.A.S. Aiyer (trans. by R.C. Dwivide), Vākhyapadīya of Bhartṛhari. p. 195.
5. लब्धक्रियः प्रयत्नेन वक्तुरिच्छानुवर्तिना।

For the propagation of sound the Mīmāṁsaka, Śabara Svāmin (C. 200 A.D. or earlier) holds.

The airy waves (nāda) produced by the continuos conjunction and disjunctions of air particles are induced by some impact (abhighāta) of our internal air (after having contact with out speech-organs). These airy waves make their way through the neighbouring stagnant airs producing conjunctions and disjunctions in all possible directions which go on reproducing the other ones till the velocity (Vega) is fully diminished. They manifest the sound in the sense of hearing (of a Śrotā) till they are in existence, although we are unable to see them due to the imperceptibility of air.[1]

The method of propagation of sound has been discussed in detail in the works of almost all the later Mīmāṁsakas like Kumārila Bhaṭṭa[2], Pārthasārathi Miśra[3] and Nārāyaṇa Bhaṭṭa.[4]

2 View of the Nāya-Vaiśeṣikas:

The Naiyāyikas and the Vaiśeṣikas are the upholders of non-eternality of word. According to them ; the word is produced by the speech-organs and vanishes away within a moment, giving rise to the other ones. Nyāya-Sūtrakāra Gautama (C. 200 A.D.) has pointed-out about the vanishing nature of the word in two Sūtras.[5] Although Bhāṣyakāra Vātsyāyana (about 200A.D. or later) has also given some explanation of these Sūtras[6] but Vārttikarkāra Udyotakara's (About 600 A.D.) explanation is more elaborate. For the mode of propagation of sound, he has given an example of Kadamba-golaka. He says:

It is inferred through Pariśeṣānumāna that the word propagates through series (Santa). The first word is cased by the conjunctions and disjunctions (of the internal-air with the speech-organs), but the other words are reproduced by it (the original one) in all the directions in the mode of a Kadamba-golaka. Each and

---

स्थानेष्वभिहतो वायुः शब्दत्वं प्रतिपद्यते।। वाक्यपदीयम्, I. १०८.

1. संयोगविभागनैरन्तर्येण क्रियमाणाः शब्दभव्यञ्जन्तो नादशब्दवाचाः। **शाबर भाष्य: 1.1.17.**

अभिघातेन हि प्रेरिता वायवः स्तिमितानि वाय्यन्तराणि प्रतिबाधमानाः सर्वतोदिक्कान्संयोगविभागानुत्पादयन्ति। यावद्वेगमभिप्रतिष्ठन्ते। ते च वायोरप्रत्यक्षत्वात्संयोगविभागा नोपलभ्यन्ते। अनुपरतेष्वेव तेषु शब्द उपलभ्यते नोपरतेषु। **शाबर भाष्य: 1.1.13.**

2. See Mīmāṁsā-Śloka-Vārttika, Śabda-nityatādhikaraṇa, Kārikās 121. ff.
3. See Nyāya-ratnākara on the above Kārikās.
4. See Mānameyodaya, (Varanasi), pp. 208-215.
5. See Nyāya-Sūtra 2.2.12 and 2.2.16.
6. See Nyāya-bhāṣya (Cal. Skt. Series) pp. 594-96, 609.

every new word is reproduced with the help of a medium (*āśraya*) but the velocity of world goes on diminishing as it proceeds further and at last, it decays completely. Among all these words, reproduced through series, only that one is apprehended by sense-of hearing (of the *Śrotā*) which is confronted with the sky of his ear.[1]

Here it is necessary to clarify the meaning of '*kadamba-golaka-nyāya.*' Just as in kadamba flower, one kadamba-bud, which first shoots up at the centre, causes several other kadamab-buds to shoot up simultaneously in all possible directions, these several buds again cause new several buds around themselves; in a similar way the first word, produced at some point, causes other words to spring up simultaneously in all the ten directions and these words again cause some other new words around themselves. The series goes on and thus the word can be heard in all possible directions.

Representing the Naiyāyika's veiw Bhartṛhari has used the term *dīpa-prabhā*[2] instead of kadamba-golak. In fact both the usages convey one and the same method, because the way of propagation is similar in both the examples i.e. in kadamab-golak and in *dīpa-prabhā*.

Now according to the Vaiśeṣikas the word in an attribute of ākāśa, apprehended by the ear, non-eternal, and produced by the conjunctions and disjunctions.[3] It is inferred by pariśeṣānumāna that the word goes on spreading in series like the wave of water (vīci-santānavat)[4]. This view has also been accepted by the later Vaiśeṣika teachers like Śankara Miśra,[5] Śridhara etc. The explanation given by Śrīdhara. (C 991. A.D.) is explicitly clear.

He says:

Just as one water wave gives rise to the other one in its close proximity and the new wave again gives rise to the next one and thus, series of wave goes on, similarly the word produced (by the impact of internal-air and the speech-organs)

---

1. परिशेषात् तु सन्तानसिद्धिः। तत्र आद्यशब्दः संयोगविभागहेतुकः तस्माच्छब्दान्तराणि कदम्बगोलकन्यायेन सर्वादिक्कानि, तेभ्यः प्रत्येकमेकैकः शब्दो मन्दरतमारिन्यायेन आश्रयाप्रतिबन्धमनुविधीयमानः प्रादुरस्ति, ततोऽन्त्यस्य अतिमान्द्याच्छब्दान्तरोत्पत्ति शक्तिविघातो येन केनचित् प्रतिबन्धात् भवतीति। अतः शब्दसन्तानोच्छेदः। तत्र यः कर्णशष्कुलीमन्तमाकाशदेशं प्राप्नोति स उपलभ्यते।। Nyāya-vārttika, (Cal. Skt. Series) p. 596.
2. See Vākyapadiyam, I. 04.
3. शब्दोऽम्बरगुणः श्रोत्रग्राह्यः क्षणिकः,– संयोगविभागशब्दजः, Praśastapāda-bhāsyam, (Indo-vision, Ghaziabad) p. 213.
4. ibid., p. 215.
5. See Vaiśeṣikasūtropaskara, 2.2.31.

creates the other one in its close proximity and this new word again creates, the other ones and so the series of words goes on. The word apprehended by the ear is the last one which reaches the sense-organ through series.[1]

3. View of the later Naiyayika-Jayanata Bhaṭṭa:

From the above description we find that the Naiyāyikas and Vaiśeṣikas in their respective treatises set two independent models to explain the mode of propagation of sound–1. *Kadamba golaka* 2. *Vīci taraṅga.* It seems that either they both were ignorant of each other's hypothesis or they have difference of opinion about the propagation of sound. But the later Naiyāyika, Jayanta Bhaṭṭa (C 900 A.D.)[2] seems to be fully conversant with both the models. It is he only who has presented a combined view and says:

> A Sound is produced either by conjunction or by disjunction. It, after being produced, thus spreads in all directions – upwards, downwards and obliquely –just in the way of a round kadamba flower giving rise to new buds in all directions. With the analogy o kadamba flower, a sound also produces other similar sounds at its very near sounds in their close proximity. This propagation of sound, thus goes on in the way of water-wave-series (vīci-santāna-vṛttyā) and the last word is produced in the *Śrotrākāśa* (of the listener) which is apprehended by him them and there.[3]

From the above, it appears that according to Jayanta Bhaṭa, the words are produced in their close proximity spherically in all the directions in the mode of a kadamba-golak, while they propagate from one place to the other in the mode of a Vīci-taraṅga (water wave).

4. View of Navya-Naiyāyika, Gaṅgeśa Upādhyāya:

The Navya-Naiyāyika, Gaṅgeśa Upādhyāya also has presented his views about sound in his *magnum-opus* Tattvacintāmaṇi. About the origination and propagation of sound, he opines:

---

1. यथा जलवीच्या तदव्यवहिते देशे वीच्यन्तरमुपजायते, ततोऽप्यन्यत् इत्यनेन क्रमेण वीचीसन्तानो भवति,, तथा शब्दादुत्पन्नात् तदव्यवहिते देशे शब्दान्तरम्,........ ततोप्यन्यत् ततोऽप्यन्यत्, इत्यनेन क्रमेम शब्दसन्तानो भवति। एवं सन्तानेन श्रोत्रदेशे समागतस्य अन्त्यशब्दस्य ग्रहणम्।– Nyāya Kandali, p. 694. (pub by Vārānaseya Skt. Viśva-vidyālaya, Varanasi, 963.)
2. See, Shastri, D.N. Critique of Indian Realism, P. 102.
3. संयोगाद्विभागाद्वा शब्द उपयाजते जातश्चासो सर्वतोदिक्कानि कदम्बगोलकाकारेण सजातीयनिकटदेशानि शब्दान्तराण्यारभते तान्यपि तथेत्येव वीचिसन्तानवृत्त्यारम्भप्रबन्धप्राप्तो अन्त्य: श्रोत्राकाशजन्मा शब्दस्तत्समवेतस्तेनैव गृह्यते। —Nyāya-Mañjari. Pt. I, p. 197. (Ch. Vidya Bhavan, 1971)

> Production of a sound (word) from the previous sound (word) is not in the serial of one molecule, but it is caused by the conjunction (Saṁyoga) with the air, capable of moving everywhere, and thus, a sound (word) is produced by the original one (*ādi śabda*) in the *mahat deśa*. This process goes on successively covering more and more space just like the spreading a water-wave (vīci-taraṅga). Just as eyes are very quick in apprehending their object, air also is very quick and fast-moving; therefore, the sound (śabda), produced by the previous one, is apprehended by the listener immediately i.e. without any delay (due to the fast-moving air-waves).[1]

Explaining how the sound (word) propagates in all the ten direction simultaneously, Gaṅgeśa further says:

> The sound (word) spreads in all the ten directions, having been produced by the original one in the way of a kadamba-golaka and its intensity goes on feebler and feebler successively in the air.[2]

From the above, three points can be posited as Gaṅgeśa's view– (1) Sound (word) requires a medium for its propagation which is air, (ii) Velocity of sound is very swift due to the high velocity of air-wave, (iii) Sound (word) originates in all directions like kadamba-golaka and propagates further in the way of vīci-taraṅga.

5. View of the Modern Science:

Now we should turn towards the other side of the topic i.e. the view of the modern scientists about the propagation of sound.

If we go to the depth of the history of science, we find in 600 B.C. the name of a Greek philosopher and mathematician, Pythagoras who talked about the quality (tone) of sound, produce by the vibrating strings.[3] But it seems that no enhancement in the knowledge about sound was made in science from Pythagoras till 1600 A.D., as no reference is traceable during this period.[4] It is only in 1638

---

1. शब्देन चाव्यवहितपरमाणुदेशोत्पादक्रमेण न शब्दारम्भो। किन्तु मेघाद्यभिहतं सर्वतोगामि-महावायोः महति देशे संयोग निमित्तमासाद्य आद्यशब्देन सर्वदिग्वत् शब्दः एक एव जन्यते, उत्तरोत्तरेणापि अधिकाधिकदेशतः सर्वत्र एकैक एव शब्दः वीचितरङ्गवदुपपद्यते, वायोश्च नयनवत् महावेगतया शीघ्रगामित्वेन अग्रिमशब्दारम्भात् आद्यशब्दानन्तरं अचिरेणैव शब्दोपलम्भः। — Tattva-Cintāmaṇ
2. वायोरेव च मन्दतरतमादिक्रमेण मन्दादिशब्दोत्पत्तिः यथा आद्यशब्देन कदम्बगोलकवत् दशदिशि दशशब्दा आरभ्यन्ते, तैश्च दशशब्दसन्ताना इति। —Tattva-Cintāmaṇ
3. See, New Standard Encyclopedia, Vol. 12. Standard Eductional Corp. Chicago.
4. Max Von Laue, History of Physics, Academic Press, New York, P. 24.
   (ii) New Standard Encyclopedia, Vol. 12.

A.D. that Galilei Galile, an Italian, scientist and astronomer in his famous book '*Discorsi*' discussed the detailed relations of the frequencies of sound produced by vibrating strings.[1] But no light was thrown by him on the mechanism by which sound travels from its source to the listener. However, Robert Boyle, a British physicist and chemist proved n 1660 A.D. that for the propagation of sound, a medium is a must.[2] After him, Christian Huygen, a Dutch physicist, in 1678 A.D., clearly mentioned about the propagation of sound and light on the basis of wave-theory. Regarding sound he says:

> "We know that by means of the air, which is an invisible and impalpable body, sound spread around the spot where it has been produced, by a movement, which is passed on successively from one part of the air to the another; and that the spreading of this movement taking place equally rapidly on all sides, ought to form spherical surfaces ever-enlarging and which strike our ears."[3]

A similar statement has been given by him about the propagation of light in which he explains that light moves from its source to our eyes in succession and consequently " it spreads as sound does, by spherical surfaces and waves."[4]

He has also accepted that the example of water wave is taken only due to the similarity in the spreading of water wave and sound wave, although their causes and conditions are different.[5]

From the above description it is evident that sound is originated from the vibratory motion of a body, although its motion may not be visible. Further the vibratory motion of the source of a sound will originate condensation or rarefactions (conjunctions and disjunctions) in the air in the form of spherical wave fronts (in all directions). And points on a wave-front can be considered as point-sources for the production of spherical secondary wavelets.

6. Conclusion:

A comparative study of sound on the basis of ancient India thought and modern physics, sheds light on the following points:

(i) Necessity of a medium (air) for the propagation of sound was realised by the thinkers of both the times (the old and the new)

---

1. See, History of Physics, p. 24.
2. See, New Standard Encyclopedia, Vol. 12.
3. Sambursky, Samuel, Physics Thought, Hutchinson of London, 1974, p. 287.
4. See Supra, pp. 287-288.
5. See. Supra.

ii) The sound is originated either by conjunctions or by disjunctions (called as condensation and rarefaction in physics) in the air spherically for which the ancient thinkers have given the example of a kadamba-golaka to indicate its spreading in all possible directions.

iii) The sound propagates from one place to the other in the mode of the water-wave (Vīci-taraṅga). This example is acceptable to the ancient thinkers as well as to the modern ones, but it is only due to the analogy of their way of propagation of water-wave and sound-wave, not because their causes and conditions are same.

iv) Both the thinkers accept that sound, having been produced by a motion, goes further and further to reach the ear of the listener through series (santāna) successively.

(v) And the last and the most important point is that the sound is produced spherically in all directions like a kadamba flower and propagates very swiftly like a vīci-taraṅga or a water-wave.

Section IV

Medical Science and Genetics

1

Some Aspects of Medical Science in the Indian Shastras

Dr. Padmakar Vishnu Vartak

Shastra means sacred treatise or scripture. All the Shastras of India have their origin in the Vedas. Later on in the epic Mahabharata and the Puranas these Shastras are explained nicely and elaborately. Let us see here the original Medical science present in India during the Vedic era, the Mahabharata and the Puranas. Medical science is presented in the books of Ayurveda as a separate Shastra; however I find that the original medical science present in the Vedas, the Mahabharata and the Puranas is far more superior to the Ayurveda. The Vedas are composed in the Vedic era extending from 23960 years B.C to 5560 BC. The Mahabharata is composed during 5560 BC and the Puranas still later.[1,3,7]

We shall proceed chronologically from the Vedas to the Puranas. In the Rigveda lot of medical advances are seen recorded if observed critically. During 1976, I found mentioned in the Rigveda, that a horse was developed from another horse and a cow was developed from skin of a cow. This miracle was done by three brothers called as Rubhus. I told this finding to an augustus audience of scientists and doctors during May 1976; when all of them laughed and said this is impossible even for today's science then how can it be in Vedic era? At that time the word 'clone' was not coined at all.

And in 1997 one Irish scientist from Edinburgh, U.K., prepared a lamb from udder of a sheep, the technique being called as 'Cloning'. Was that not possible for our scientists? If the Indian scientists believed in the Vedic Truths they could have done this miracle, but they did not believe and did not try. The Vedas have given the principle that another animal can be produced from an animal's skin. The principle came true in 1997, when Dolly was cloned from udder; that means from the skin of the lamb. Some scholars critisize that when foreign scientists discover or invent something, you say that it was in the Vedas. For those scholars I would like to tell that the same Rubhus, who were the human beings as told in the Rigveda 1-110-2,4, made their old parents youthful again (Rigveda 1-20-4, 1-110-8, 4-33-3, 4-35-5, 4-36-4, 4-39-9, 1-161-3,7, 1-111-1, 3-60-3). Giving youthfulness to an old animal is not yet done by the modern science upto 2005 A.D. The miracle of Rubhus was similar to that of Ashvinau, who rejuvenated old sage Chyavana (Rigveda 1-116-10). How can one deny a fact having so many references? We have to accept the superiority of Vedic sages in the physical sciences, and our scientists have to experiment on such Vedic concepts.

The first step in their work on cows was to produce a cow which yielded copious milk. (RV 1-20-3). This fact shows that Rubhus were working on some living cows. Probably the cows then, were not giving as much milk as present days' cows. This is quite possible because female elephants, camels, deers, donkeys, sheeps and mares do not give a good quantity of milk and therefore human beings cannot depend on them for milk. Similarly cows also would not have been giving profuse milk at that ancient Vedic time. Therefore, Rubhus might have taken that project of cattle rearing as a 'Vrata'(व्रत), might have worked on the project scientifically and got success in breeding a cow, which produced copious milk. This might be a root of genetic engineering.

The Rigveda has recorded how systemically the Rubhus worked on the project 'cow'. Rig.

1-111-1 tells that the Rubhus gave new mothers to orphaned calves. This fact definitely shows that Rubhus were working on cows. RV.1-161-3 states that Rubhus were fully engrossed in the work of creating a cow etc. and therefore they refused to attend a Yajña. However, they promised to attend the Yajña after completing the work. This statement suggests that Rubhus were working on the project for a long time and were not interested in minor works like Yajñas. This throws light on the importance of their work. For that project they protected cows like good sons, organized mothers and supported cow shed.(Bhasa भास) (RV. 4-33-4). For their great research they had to maintain a cow shed and protect their cows, just like a good son protects his mother. For their experiments one cow was not sufficient. They needed many cows. Therefore, they organized cows in great numbers, which were kept safe and protected in the cow sheds. During their experiments many cows might have died, leaving behind their calves orphaned. For those orphans they had to give new mothers and for that reason they organized mothers. (Apinshat अपिंशत् RV. 4-33-4).

To produce plenty of milk a cow must have her calf nearby. This psychology of cows must have been noted by them so that they worked on mothers and calves. They must have examined human mothers and children also while doing that research. (RV. 1-111-1)

While doing this type of research on cows and calves, Rubhus attained a great knowledge and were inspired to produce **a living cow from a cow's skin.** This was held by the sages as a divine work of great talents. Hence it cannot be neglected as a toy making. A hidebound man without thinking seriously may say that Rubhus prepared a toy, having a shape of a cow, from some hide. But then that work would not have been held with high esteem as is done by the Rigveda.

Moreover, toy making needs no preliminary pains of working continuously on cows and calves for a prolonged period, neglecting Yajñas, as was done by Rubhus. Again, the Rigveda says that Rubhus made a living cow which was made a mother by producing a calf from it. (RV. 1-110-8). That newly prepared cow was named as 'Vishvaroopa' विश्वरूपा and it was 'Vishvajuvam' विश्वजुवम् (RV. 4-33-8), which means matching with the world. It was similar to the other cows in the world. Bruhaspati accepted this cow Vishvaroopa happily, as a gift, which was created by Rubhus.(RV. 1-161-6).

The Rubhus created a cow from skin of a cow. (RV. 1-161-7, 3-60-2, 4-36-4). It was the first ever production of a living clone, from a part of a body namely skin of a cow. The modern science has now proved it possible by producing a lamb Dolly from udder of a sheep. Udder is nothing else but a part of skin. The modern science is not clear whether the lamb, produced in a womb of a sheep, was from the skin of the same sheep or the skin was from some other sheep. Therefore we cannot blame the Rigveda for not giving the full details. Moreover, the Rigveda has preserved the knowledge for at least 25000 years.[1]

Then how can we expect all the fine details about an experiment done 25000 years ago. The seed of the knowledge is preserved by the Vedic Seers, we have to work hard on that seed so as to get its fruits.

The Rubhus produced another clone by their supreme intelligence.(RV. 3-60-2). They created a horse from another. (RV. 1-161-7). From which part of the horse they produced another horse is not recorded by the Rigveda. One may say that a horse may be produced by copulation of a horse with a mare. But in that case no sensible person will ever say that Rubhus created a horse. It is a nature's creation. In the Rigveda the verb used is 'Atakshat' अतक्षत्. This verb may suggest carpentry; but that horse was not created from wood or timber, it was created from another horse. (RV. 1-161-7). It took involvement of Rubhus for a prolonged period. (RV.1-161-3). Therefore carpentry is out of question. Then the only thing remains that the Rubhus cloned a horse from another horse.

To achieve that knowledge Rubhus worked with horses too. (RV. 1-111-1). To begin with they produced two horses, which were well trained to be yoked to a chariot. (RV. 1-20-2). These two horses were presented to Indra. They were very fast and powerful. (RV. 4-35-5). After this work they produced a horse from another horse with their intellectual skill. (RV. 1-161-7, 4-33-10). That horse was called as 'Hari' and was presented to Indra, who accepted it. Hari was yoked to a chariot. (RV. 1-161-7). This fact proves that Hari was a real living horse, and not a toy.

Considering all these facts we have to admit that the creation of a horse 'Hari' from another horse and creating a cow 'Vishvaroopa' from a skin of a cow were experiments of cloning, just like that of producing a lamb Dolly from a sheep. During 1997 many newspapers compared the scientists, who prepared Dolly by cloning, with the Gods. In the same way the Rubhus were compared with the Gods by the Vedic sages. (RV. 1-161-6,7).

The same Rubhus had divided one **'Chamas' (चमस),** which was prepared by the God Twashta, into four parts.(RV.1-20-6, 1-110-5, 4-33-5, 4-35-3,4, 4-36-4) I think that 'Chamas' means an urn containing life. Because the word 'Chamas' is formed of two components Cham + Asa. The verb 'Cham' चम् means to drink, to eat. The verb 'Asa' अस् means to live, to exist. Therefore, Chamas means a thing which eats and drinks to live. It means in the modern language a primary cell, full of life or a fertilized ovum. Modern science has done this miracle of producing four animals from a single fertilized ovum, not in India, but in foreign countries, though the basic concept is Vedic. RV 1-110-3 calls it 'Chit-Chamasa' चित् चमस Chit means energy or rather life energy. From this we can imagine a pot containing life energy. It is nothing else but a living cell of a living animal or a plant. The basic unit of life is a living cell. Therefore Chit-Chamasa means a cell, full of life.

Soorya, the Sun gave 'Amruta' (अमृत) in a Chamasa. Amruta means an immortal thing, Brahman or Atman. Amruta also means life, because life is a part and parcel of Amruta Brahman or Atman. Therefore a Chamasa full of Amruta means a cell full of life. Any cell is a pot or a vessel, minute in form, but it contains life. Therefore, Chamasa means a cell full of life, a living cell. Chamasa cannot be an ordinary pot as supposed by other scholars, because any ordinary pot cannot be divided into four parts. The famous Ishavasya Upanishad tells about Hiranmaya Patra, which refers to body of a living animal or a living cell.[2]

A living cell has four walls in which is stored a fluid called as Protoplasm. This cell may be called as Chamasa, because it contains Amruta i.e. life in it. A living cell or an embryo is prepared by God, Twashta. Twashta had produced such a Chamasa or a living embryo. Rubhus tried to divide that embryo.

While doing that experiment the eldest brother said, "we can make two cells from the original one." (RV. 4-33-5). The modern science has proved that an embryo can be divided into two, so that two animals can be produced. However, the younger brother said, "We can make three." He said so because he observed that out of the two cells, which the eldest brother saw, one cell had divided into two. Therefore there were three cells present. The youngest brother observed more, thought more and exclaimed, "We can make four. ''

Let me explain here the facts. An embryo is a fertilized egg. It is a living cell, which has a capacity to divide and redivide to form billions of cells, which arrange themselves in a particular way to mould into an animal at its final stage. The first cell divides into two cells. Each of these two cells has a capacity to divide into two cells each. Each of these four cells has a capacity to produce a fully developed animal. This is proved by experiments by the modern science. The experiments were carried out on a sea animal called as Salamandar. A fertilized egg of a Salamandar was taken and observed when it gave rise to two cells. When these two cells were separated apart by a fine needle and kept separately in separate petridishes, they grew more and divided and redivided to become two salamandars. A similar experiment was carried out again. This time the prime embryo divided to produce two cells. These two again divided to produce four cells. These four cells were separated and grown separately, when four Salamandars were formed. A similar experiment might have been done by the Rubhus.

They observed one embryo which divided into two cells. Therefore, the eldest brother said that they can produce two animals or two Chamasas. On hearing him the younger one went eagerly to see the state. Till then, one of the two cells had divided into two, so that he could see three cells. Hence he said that three Chamasas can be made. After him the youngest brother peeped. By that time the second cell was also divided and he could see four cells. Therefore, he proposed to prepare four cells or animals from the original one Chamasa.(RV. 4-33-5). These four cells were separated by a superfine and sharp instrument (RV. 1-110-5) and grown in four separate vessels so that four animals came into existence. Twashta wanted to produce one animal, but Rubhus produced four animals from a single zygote or a fertilized ovum.

Working on the Vedic principles the sages in the Puranic period prepared a human Clone named as 'Pruthu', from one dead king Vena. (Vishnu Purana 1-13-8,9,33 to 39) In the modern era a human clone is not yet developed either from a living person or a dead. Modern science dreams that it can be done and the foreign scientists are after it; but the Indian scientists are not even thinking of it. Modern science says that an exact copy of Einstein could be produced by cloning. But I say that it is not possible. Bodily that clone may resemble Einstein; but may not be as intelligent, because there is a precedence in India, which shows that from a villain king Vena, a good-natured Pruthu was cloned. Mind and intelligence depend upon the previous deeds or Kārmas as told in the rebirth theory, which is discovered by the ancient India, in Vedic era and has no parallel in the modern science. Here I want to point out that the name 'Pruthu' is a very scientific term, because it means an extension. The root 'Pruth' means to extend, so 'Pruthu' means an extension. The extension was from only one person Vena. The same is the process in cloning, which is an extension of only one animal. Normal offsprings are not extensions of only one, they come after union of a male and a female. The word 'Clone' has no such root meaning. The Puranas say that the sages did churning of a thigh or a hand. Churning means 'Manthana'. It is not only mechanical churning using a churner, but includes 'Vichara-Manthana', strenuous mental contemplation, along with dissection of dead body. By the strenuous intellectual efforts along with skill of hand, the seers produced a clone from a dead king Vena. From which tissue the clone was produced is not reported by the Puranas or the Vedas. But please note that in the present experiments it is not reported from which tissue of the udder the clone Doily was produced. Recent reports state that the Stem cells are used for cloning an animal. The stem cells can be supposed to be the Vedic Chamas.

It is important to note that in the ancient experiment an abnormal clone 'Nishada' was produced, who had small head and large body. In the modern experiments, too, abnormal clones are produced. Many modern clones have died immediately. The survival rate of clones is minimal.(London news 11-4-2002. Yuwa Sakal 12-4-02).(www.eSakal.com) It is reported in the Vishnu Purana that the Nishada, though an abnormal clone, could live long and could produce children.

So far upto July 2005, no clone from a dead animal is produced by the modern science, though it is a dream of science; but Pruthu was a clone produced from a dead king 'Vena'. We have ample scope to try in this field.

A male clone is developed from a male in the past history. It is not yet done by the modern science. Two different clones were produced from two different parts of one body in the past. This is not yet done in the modern era.

Shrimad Bhagawata (4/15) states that the sages, later, also had cloned a girl from the left hand of king Vena. This girl later married Pruthu. Such type of experiment of producing a male and a female from a dead male is not yet done in the modern science.

Similar other clones are also reported in the Puranas. Lord Ganapati was produced from the 'Mala' (मल), which means the superficial layers of skin of Parvati. This is possible according to the modern science. However, here a male is cloned from a female. Whether it is possible or not is not conceived by the modern science. But I think it is possible. Science knows that a single dose of X chromosome produces a male, while a double dose of X chromosome produces a female. Parvati might have used her skin-cell with two X chromosomes in its nucleus. She might have removed that nucleus having XX and replaced it by a nucleus of her genetic cell or an ovum, which contained a single X chromosome. This might have produced a male clone Ganapati. Ahi-Ravana and Mahi-Ravana used to produce Rakshasas from blood drops. It might be cloning from the white blood corpuscles, which contain nuclei. On 1st May 1999, Saturday, there appeared a news that Japanese scientists have prepared two clones from cells of cow's milk. If the cells from milk can be cloned, it is certainly more possible to clone an animal from the blood cells.

Other achievements like Parthenogenesis, Test-tube-baby etc. are also reported in ancient literature of India. The technique of division of Chamasa might have been used by sage Vyasa in producing 100 Kaurawas and their sister Dusshala. Gandhari became pregnant naturally from her husband Dhrutarashtra. However she did not deliver a child even after nine months were completed. By that time Kunti gave birth to a child named as Yudhishthira. Therefore Gandhari became angry because her son's right to the throne was lost. That right went to Yushishthira, the first born. Total two years passed but Gandhari did not deliver a child. Therefore, burning with anger, Gandhari aborted her foetus. In fact it was not a foetus, not a developed child; but a mole, a mass of living cells having no shape like a human child. As soon as the sage Vyasa heard about her abortion, he came immediately, took the aborted cell mass in possession and carried on further work over it. He divided that mole, it is said. This means that Vyasa dissected the mole carefully and found out normal living cells, which he kept separately in 'Ghruta Kumbha' (घृत कुंभ.). Ghruta Kumbha does not mean an ordinary pot of clay, filled with Ghee. 'Ghruta' means a nutrient medium supplying life energy, which was kept in a special Kumbha, a highly sophisticated urn. The normal living cells were implanted one each in a separate Kumbha. There were 101 cells. Those cells were grown separately in vitro, i.e. out of any living body, in Ghruta Kumbhas or nutrient urns. In due course children were born, 100 Kaurawas and their sister Dusshala. (Mahabharata, Adi, 115)

Normally, a fertilized ovum starts dividing. One cell divides in to two. Two cells divide to produce four cells and so on, till billions of cells are produced. In the initial stage, every new born cell has a capacity to divide and redivide till a full grown foetus forms. Rubhus took four cell stage and produced four animals. However, Vyasa was not doing an experiment of cloning on one fertilized ovum. He had collected one aborted mass of cells, which contained some normal living cells. Out of these primary cells, Vyasa found 101 cells which had a capacity to begin multiplication from the original stage. Therefore all the 101 cells developed in to 101 children. The modern science does not know upto what stage of multiplication the newly born cells have a capacity to multiply from the beginning so as to form a child. But from this experiment of Vyasa it appears that upto 128 celled stage, each cell may have a capacity to begin multiplication anew to produce a child. It seems that Vyas found 101 living cells out of 128 cells. These cells may be termed as Stem cells of the modern science.

Thus sage Vyasa relied on the principle of 'Chamasa division', laid down by Rubhus in the Rigveda and worked further to give birth to 100 Kaurawas. The Rigvedic science was traditionally used then, by the expert scientists, till at least, upto the time of the Mahabharata, 5561 years BC.3 Sage Vyasa was a scientist himself, he composed the epic Mahabharata to narrate the true history. Following Vyasa the modern scientists must work on the principles stored in the Rigveda and other ancient literature, so as to produce new miracles.

Modern scientists boast and dream that they can produce a clone of Einstein who will be exactly like the famous scientist Einstein. However, looking towards the Vedic precedence I think that the clone of Einstein may look exactly like Einstein, bodily, but may not be as brilliant in intelligence as Einstein. I say so because Vena was a tyrant, but Pruthu was a nice king. The mental state was far different in the clone Pruthu than king Vena. This happened because the mind cannot be cloned. Body may be cloned because, as the Rigveda (1-161-9) says, the Pruthvi, Teja and Apa Tattwas of the body cell can be divided and extended to form a clone. Mind is above these three Tattwas and so mind cannot be cloned. After a body is formed, mind comes to stay in it. This is explained by the theory of Panch-Koshas of Taittiriya Upanishad.[2]

The Anna-Rasa-maya Kosha i.e. the physical body can be cloned. After the formation of the Annamaya Kosha, the Pranamaya Kosha comes in it. Within this Pranamaya Kosha is present the Manomaya Kosha. Inside it is the Vijnanamaya Kosha and still inside is the Anandamaya Kosha. The Pranamaya Kosha with all the internal three Koshas comes from some other body after its death, and enters into a new body for rebirth.(Aitareya Upanishad[2]) He has to pay for his past Karmas. Therefore even though Einstein is cloned bodily, his Pranamaya, Manomaya, Vijnanamaya and Anandamaya Koshas may not come in that cloned outer body; some other Pranamaya may enter in it, with his past Karmas and their fruits. Therefore there may be difference in the mind and intellect.

Test tube babies were prepared then. Rigveda (7-33-13) mentions that Agasti and Vasistha were born out of an urn called as 'Pushkara'(पुष्कर), from the semen of Mitra-Varuna. The word 'Pushkar' is formed of two componants. 'Push' means to nourish and 'Kara' means to do. The urn, which nourishes or a vessel which has nourishing nutrient matter, is Pushkara. From such a Kumbha Agastya and Vasistha were born. They had no mother. They were developed only from a male. Modern science accepts theoretically that a sperm is fully genetic matter and so it can reproduce. An ovum contains nutrient matter.

Draupadi and Dhrushtadyumna got birth from an utensil only from the semen of king Drupada. They had no mother. Drona was born from the semen of Bharadvaja in an urn known as Drona, without any help of any mother. Such an experiment is not yet done in the modern science. All these experiments were done in vitro, i.e. out of any living body of any animal. On the contrary in the modern experiments only fertilization of an ovum is done in a test tube, while further growth takes place in the uterus of a living woman after implantation of the fertilized ovum in it, and therefore, the baby is born out of a living uterus, not out of a test tube. In the ancient experiments no mother was used. Therefore the babies were termed as 'Ayonija'. Further observations were also done in the Mahabharata period that one test tube baby Drona was married to another test tube baby Krupi and the behaviour of that couple was observed. They lead normal married life producing one son Ashvatthama. As a control test Krupa, the test-tube baby, was wedded to a normal girl. This couple too behaved normally. All these records of the Mahabharata Adi Parva, show that there were real test tube babies and the experiments were carried further till they reproduced child. Such an experiment has not yet been done in the modern science; because the eldest test tube baby, which was born in 1979, is not yet married.

An embryo of Devaki was taken out and implanted in the uterus of Rohini. Rohini delivered a child, which was termed as Samkarshana (संकर्षण), who later became famous as Balarama, Krishna's elder brother. The word Samkarshana is a scientific term, which explains the precess behind his birth. [Hariwansha, Vishnu 4, Vishnu Purana 5-1-75, Bhagawata 10-2-13]

Kunti used a technique of **Parthenogenesis**. She stimulated her ovum with the help of solar energy and produced Karna. (Mahabharata Adi 111.) Other energies were used to produce other children. Such experiments are done in the modern science to produce young ones only in the lower animals and not yet in the human beings, but Mahabharateeya scientists used the technique in the humans. The technique is termed as Parthenogenesis by the present day European scientists. The term Parthenogenesis contains a word 'Partha' which means 'of Prutha'. Prutha is Kunti's second name. Was this a chance happening or did they coin the term from Kunti's experience? In the modern times the technique of Parthenogenesis produced all the females from a female. But the Mahabharata technique produced sons from Kunti as well as Madri. Does this not show supremacy over the modern technique? Durwasa, the original scientist might have told Kunti what to eat during pregnancy period because sex is determined on the rate of metabolism and the quality of food. Insects like Vasps, bees, ants do produce males when the eggs are not fertilized by sperms from a male. This is a puzzle before the modern science, but it might have been solved by Durvasa and used in the experiment on Kunti. We must learn from these true stories.

Life science was fully studied, so the sages knew about the chromosomes which were termed as **'Gunavidhi'**. (Mahabharata Shanti 308) This is a perfectly scientific term; because these principles control the characters i.e. Guna and also Vidhi, the functions, of a person. Even the chromosome number 23, specific for human beings is recorded, in the Mahabharata and the Bhagawata.[4] It is also stated that these principles give rise to genetic diseases. A list of genetic diseases is given in the Mahabharata, which is similar to the list in the modern science. Bhagawata (3/6/1-7) has recorded that these 23 Gunavidhi of a male gamete enter the female gamete and unite with her 23 Gunavidhi to stimulate them so that a zygote is formed. Then the zygote, which is termed as 'Kalala', goes on dividing once, ten times, three times. It means that the cell divides once to give rise to two cells. This happens ten times so that [2^{10}] two raised to the power of ten cells form. This happens thrice, at three layers namely Endoderm, Mesoderm and Ectoderm. This multiplication produces two raised to the power of thirty cells, [2^{30}] which are really present in a new born baby, according to the modern science. How could the sages achieve such a vast and microscopic knowledge? All the stages of growth of an embryo are given in the Bhagawata commencing from the first day upto the full term. These stages concur with those proved by modern science, and some times supersede the modern science. Bhagawata records that heart of a foetus begins its work in the second month of pregnancy. This fact was unknown to the modern science till December 1972. Medical science presumed till then, that the foetal heart began in the fifth month of pregnancy, but then with the use of Disonar apparatus it was confirmed that the foetal heart begins in the second month of pregnancy.[5] This is now well known due to ultrasonography. Aitareya Upanishad, which is composed in about 6000 B.C., much earlier than the Bhagawata [of 1652 B.C.] also states the same fact.[6]

The Bhagawata (2/1022, 3/26/55) states that ears are responsible for recognising the directions. This is proved by the modern science after 1935, when the Labyrinth or the Vestibular apparatus was discovered in the internal ear.[4]

Surgery was far advanced in the Rigveda era, therefore it is recorded that Queen Vishpala was fitted with an artificial lower limb when it was severed in a battle.(Rg.1-116-15), and was made fit, again, to fight in a battle. Modern surgeons do fit an artificial limb; but a patient takes about a year to walk with it, and he cannot take part in a battle again.

There is a surgical concept of transplant operations in India since the Rigveda. One sage Daddhyanga got some disease in his head. To cure it Ashvinau took his head away and in its place implanted a head of a horse for a while.(Rg.1-116-12). After repairing his human head they replaced it again removing the horse's head. The sage became normal. One may laugh at this statement because at present no such operation on head or brain is possible. But is it really impossible ? At present heart can be transplanted. During an operation on heart, it is totally stopped by freezing and with the help of Heart-lung machine circulation is maintained and after correcting the defect in the heart it is connected to the circulation, removing the heart-lung machine. In the same way, was not it possible to implant a horse's head in place of human head to maintain the give and take of sensory and motor impulses and to continue the vital functions, which are similar in man and horse? Serum of horse is tolerated by humans, then nervous tissue may also be tolerated. We have to experiment on this technique.

Testis of a goat were implanted on Indra. Modern surgeons must try this technique, but being unknown to the western world our surgeons will not even think on it.

Because of this Vedic tradition of surgery, Sushruta could develop plastic surgery to repair severed or deformed nose - Rhinoplasty. He used a flap of skin from forehead to repair nose. I had an occasion to perform an operation of Rhinoplasty with my senior surgeon Dr. Sulakhe P. B. during 1957. We used a skin-flap from abdomen and transferred it to face and repaired nose, according to the then surgical concept. Fortunately, I was then working as a surgeon in S.T.R.Hospital and Tilak Ayurvedic college, where I read Sushruta Saṁhitā. I suggested to use that old technique but it was neglected. Twelve years later one German surgeon read Sushruta and operated accordingly with good results. Since then all the surgeons are using the same technique thinking that it is a German technique. Thus our own seeds of knowledge are neglected.

Extension of Life by Space Travel

Shrimad Bhagawata 9-3-27 to 36 gives a story that king Kakudmi went with his daughter Revati to the Brahmaloka to find her a husband. Brahmadewa told him that in a second of his presence in that Loka 27 Tetra-Yugas on the Earth had passed and that there was nobody from his family alive. Brahmadewa advised him to return and wed Rewati to Balarama, who had then born. Accordingly Kakudmi returned with Rewati and gave her to Balarama in marriage. The important scientific points in this story are :

(1) The two went on another solar system leaving the Earth. They spent one second there, but during that second 27 Ttreta-Yugas passed on the Earth. It shows that Vyasa knew that the time scales are different on different planets or solar systems. This is agreed to by the science now.

2) They returned after 27 Ttreta-Yugas. Yuga means minimum 2. TETRA-YUGA means four times the Yuga. Therefore 27 x 2 x 4 = 216 years had passed; but still Rewati was in puberty and her father was alive.

(3) Ageing factor had stopped working on them.

(4) After their return, they could not see their descendants, so at least 216 years had passed. They had defeated the death. Their life had extended. The only cause attributed is the space travel to another solar system with tremendous speed.

One may not agree that the two actually travelled to some other Loka; but one has to accept the theoretical knowledge present in the story, which is the same as that propounded by Einstein in the 20th century, that if some body travels with a great speed to another place in the space, its or his life extends. Modern science states that some Subatomic particles, whose life is usually a fraction of a second, do come from the space, travelling for many days to the Earth. Even in the present world nobody has travelled to another solar system, even then the theory is accepted by science.

This story was altogether neglected, nobody ventured to find out the scientific truth in the concept. Now I have put it before the scholars.

There is one more important point for consideration. The Prashnopanishad states at 3/8 that Deity present in the Earth attracts Apana of a living person and helps its action. Adi Shankaracharya comments on this verse that if the famous Goddess of the Earth did not help Apana and did not hold down by its attraction force or Gurutwa i.e. Gravitation, then this physical body would have floated in the space or fallen anywhere. Thus Prashnopanishad of 6000 years B.C. and

Shankaracharya of, at least, 800 A.D. knew the **gravitation force of the Earth** much before Newton of 18th century A.D.[6]

Apana forces down anything through the living body. It forces the food and drinks from the mouth downwards.(Aitareya Upanishad 1-3-10). Apana forces down urine, faeces, flatus, foetus with the help of gravitation. It is proved beyond doubt by science. According to Aitareya 1-3-10, death is also a function of Apana.6 If gravitation is absent there is difficulty in the functions of Apana. This is also proved by the modern space research. Naturally, if gravitation is absent, Apana would be unable to cause death by forcing down Prana from the physical body. This is the reason why death was defeated in the space travel of Rewati and Kakudmi.

We have this scientific concept in the Puranas and the Upanishads and we have to work on it.

In the story of Rewati, Brahmaloka is mentioned. Where was it ? I think that it may be in the constellation of Abhijit because its Deity is Brahma. Hence a planet moving around Abhijit i.e. the star Vega might have been named as Brahmaloka. Vega is 26 light years distant from us. It is surprising to see the news in the Indian Express dated 5th May 1984 that the modern Soviet scientists have found the presence of a solar system around the star Vega. They found that Vega is girdled by swarm of particles, grains of dust and even stones. These observations are based on the measurements taken by an international satellite 'Iras'. The slow rotation of Vega similar to our Sun confirms that it may have a planetary system. It might have been called as Brahmaloka. Therefore it is our duty to work hard on the ancient scientific concepts.

The Aitareya Upanishad (1-1-4) states how the organs develop in the foetus. It says that first the mouth comes into existence, then the nostrils, then eyes, then ears, then heart, Nabhi i.e. navel, then the penis. The same order of development is given in the Bhagawata. It is surprising that the same order is discovered by the modern science, recently.[4]

The Bhagawata (3/31) narrates how the embryo develops. It states that after 5 days a Budbuda appears. Modern embryology calls this as 'Bubble stage' Bubble means Budbuda. On 10th day 'Karkandhu' is formed, says Bhagawata. 'Karkandhu' means a fruit 'jujube' or 'zizyphus', called as 'Bora' in Marathi, or 'Bera' in Hindi. Embryology calls it a spherical mass of cells. On the 15th day an egg is formed, says Bhagawata. Embryology accepts it by describing that stage as having oval shape.

The Bhagawata (3/31/3) states that the head of the foetus is formed at the end of the first month. This appears true in the light of the modern investigations in embryology. The Bhagawata gives further information that as soon as the nostrils develop there appears Prana and vocal chords also develop. Modern science has the same opinion.

The Bhagawata (2/10, 3/6,26) clearly tells that foetal heart begins working in the second month of pregnancy. I was shocked to learn that, because I had studied in the modern books that the foetal heart begins in the fifth month of pregnancy. I taught the same to my students from 1957 to 1972. However, during December 1972, the news flashed in the Medical times that Dr. Robertson, Queen Mother's hospital, Glasgow, England, used disonar apparatus to tap the heart sounds of the foetus and he discovered that the foetal heart begins working during 7th and 8th weeks of pregnancy. 8th week means the second month.

Bhagawata (3/31) tells that first the penis is developed, then the anus. This is also according to the modern science.

The Bhagawata (2/10/22, 3/26/55) as well as the Aitareya Upanishad (1/1/4; 1/2/4) tell that the ears developed, in them the hearing sense appeared and the directions are born by the ears. Apparantly the statement appears ridiculous if you think that the directions are east, west, north, south etc. Such directions have no relation with the ears. However, here the directions are personal, e.g. in front, at back, to the right or left., up or down. These six directions are original and they are sensed by the ears. The Bhagawata is composed around 1650 years BC and the Aitareya is composed around 6000 years BC. But the statements recorded in them are found true by the modern science during 1935 AD when Ross and Tait wrote an essay showing that the Labyrinth or the vestibular apparatus situated in the internal ear recognise the directions. He discovered that there are three semi-circular canals in the internal ear which are at right angles to each other and which are filled with lymph. As our head tilts towards left, the pressure in the canal in the left side increases giving a signal to the brain, which concludes that the head is tilted to the left side. Is it not miraculous that the discovery of 1935 is recorded in scriptures, thousands of years old? It proves that the medical science was well advanced during the ancient era in India.

While describing when the organs and their functions come into being, the Aitareya Upanishad (1/1/4; 1/2/4) states that when Nabhi came into being, there appeared Apana Vayu and from Apana the Death appeared.[6] If we think deeply we have to accept this fact. See how.

Nabhi or navel is a part of abdomen. It comes into existence during the second month of pregnancy. Thus it seems that the death also comes into existence during second month of pregnancy. Upto the second month all the cells are multiplying rapidly. When one cell divides into two, we cannot say that the first cell died. Upto the second month each and every cell exists independently. It takes nourishment, oxygen and all other requirements for living by itself. For that purpose it does not depend on anything else. It divides and redivides to form many cells. There is no death to these cells. But in the second month, the Nabhi appears and begins functioning. The function of Nabhi i.e. Umbilical chord is to supply nourishment, oxygen etc. to all the parts of the developing body. This Nabhi means the Connecting Stalk of the modern embryology. The connecting stalk takes all the requirements from the mother and supply to the growing parts. From this time onwards all the cells become dependent on the connecting stalk for their requirements to live. If the connecting stalk does not supply properly, they are bound to die. Thus the death begins working since the second month of pregnancy. When the connecting stalk begins its work, there appears differentiation of the cells and particular cells develop a particular tissue. Such cells can not live independently. Due to this dependence, death appears. So it is true that the death appears from second month of pregnancy onwards.

The Functions of Apana Vayu

It is usually said that Apana digests everything eaten up. It is quite true, but digestion is not the only function of Apana. To gulp or swallow is a function of Apana. The word 'Apana' has two components. 'Apa' means downwards. 'Ana' means to carry. Actually, Apana takes food and liquids downwards. While digesting the food Apana moves it downwards through the gut. The remnants i.e. faeces are thrown downwards out of anus, the Guda. Urine also is carried downwards from Kidneys through Ureters to Urinary Bladder and then still more down through Urethra and passed out of body, downwards to the ground. Thus ingestion, digestion, micturition, defecation are the functions of Apana and all are downward movements. In addition parturition or delivery of child is also attributed to Apana. Apana takes the baby downwards from uterus via maternal passages. Thus delivery is a function of Apana. All these functions are helped by the gravitation force of the Earth. This fact is stated in the Prashnopanishad 3/8, which says,' The Deity of the Goddess Earth attracts and helps Apana of a person.'

Prashnopanishad is composed around 5761 years Before Christ. I say so because one Hiranyanabha is shown to be present then, who was the 15th descendent of Dasharathi Rama. (Bhagawata 9-12-3,4). I have proved, on astronomical evidences, the birth date of Rama to be 4th December 7323 BC, Rama won the war against Rawana on 15th November 7292 BC. Rama died at the age of 64 years. 15 generations means about 1500 years, because all the descendents are not recorded in the royal genealogies. Only the enthroned kings are enlisted.[7]

This proves that the gravitational force of the Earth was known to the Indians 7000 years earlier than Newton, who is credited with the discovery of the gravitational force. The famous Nasadeeya Sukta of the Rigveda, also, indicates knowledge of the gravitational force. Adi Shankaracharya, who was at least 1000 years prior to Newton also has recorded his knowledge about gravitation in his commentary on the Prashnopanishad. Shankaracharya says,' If the famous Goddess of the mother Earth had not helped Apana by attraction downwards, this body would have floated anywhere in the space.' Both these records show that the ancient Indians had knowledge of the gravitational force of the Earth on the body.

It is true that the gravitational force does help Apana to defecate, micturate, digest, deglutate. We can experience this fact ourselves. Take a sip of water and drink in sitting posture. We notice that it easily goes down the oesophagus. Now take another sip of water, lie down and try to drink it. We will find difficulty in drinking, because gravitational force is not helping Apana. Now take one more sip of water, take an upside down posture and drink. Water will not go down the oesophageal tract. On the contrary it may come out of nostrils. This happens because the gravitation of the Earth does not help Apana in upside down posture, where Gravity works against Apana. Every person finds difficulty in passing out urine and stools in lying down position, because Apana does not help.

In the absence of the gravitational force in the space the astronauts face a great difficulty to pass urine and feaces. It is not yet possible to develop toilets in the space-ships to fascilitate the astronauts.

Thus the sages were correct to state that Apana is helped by the Earth's gravitation. If this appears true, their further statement that Apana causes the death must also be true. At the time of the death, Apana forces down the Pranamaya Kosha and pushes it out of body. If we observe that at the time of death, many times, urine and stools are passed out, we have to believe that death is the result of the function of Apana. During death even the foetus also is expelled from the uterus, because Parturition is a function of Apana, which takes a fierce form to force out the Prana for effecting death.

From the fact that Apana is responsible for death, we can imagine that if Apana is absent, death can be avoided or delayed. Therefore in the space death can be defeated for want of gravitational force which is essential for Apana to work. It is for this reason that Rewati remained in Puberty, though she returned from Brahma Loka after 216 years of the Earth, and her father lived even then, to see her marriage with Balarama.

Thus it seems that medical and biological sciences were very well developed along with other physical sciences in ancient India and all that knowledge is stored in our Shastras. We have to decifer their meanings on the basis of modern science. Scientists should do this work because they are well versed with the present scientific knowledge, which is helpful to expose the knowledge concealed in the Shastras. Our Rishis were, in fact, scientists; therefore only the science scholars can reveal what the sages have said. If our scientists do so, we can progress much in the scientific field.

Here, I have presented only a few points from the ancient Indian medical science, though I have discovered a vast ocean of scientific knowledge in the ancient Shastras of India.

Bibliography

1 (A) The Date of the Rigveda - by Dr. P.V.Vartak, - Vishveshwaranand Indological Journal, Vols xxxii, Pts i-iv, June-Dec. 1993-94, published by Vishveshwaranand Vishva Bandhu Institute of Sanskrit and Indological Studies, Punjab University, Sadhu Ashram, Hoshiarpur.

(B) Vedic Astronomy, page 121, authored by Dr. Vartak P.V., Facets of Vedic Studies, edited by Bidyut Lata Ray, 2000, Kaveri Books, New Delhi.

(C) Astronomy to fix the dates of Sanskrit Texts or Historic events, authored by Dr. P.V.Vartak, pages 71 to 78 - Research in Indology, a new perspective, edited by Dr. Rabindra Kumar Panda, Bharatiya Kala Prakashan, Delhi. - 1998

2 Scientific Interpretation of the Upanishads (in Marathi) by Dr. P.V.Vartak.

3 (A) Scientific Dating of the Mahabharata War, authored by Dr. P.V.Vartak. Published by the Veda Vidnyana Mandal, 497 Shaniwar Peth, Pune, 411030

(B) Astronomy to fix the dates of Sanskrit Texts or Historic events, authored by Dr. P.V. Vartak, pages 71 to 78 - Research in Indology, a new perspective, edited by Dr. Rabindra Kumar Panda, Bharatiya Kala Prakashan, Delhi. - 1998

(C) Swayambhu, a Marathi and a Hindi book, authored by Dr. P.V.Vartak.

4 (A) Ancient Indian Embryology- authored by Dr. P.V.Vartak, printed in the Maharashtra Medical Journal, March 1973, on the pages 379-391

(B) Chromosomes in Ancient India, authored by Dr. P.V.Vartak, printed in the Maharashtra Medical Journal January 1972.

5 The Medical Times, December 1972 - Dr. Robinson, Queen Mother's Hospital, Glasgow, England

6 Scientific Interpretations of the Upanishads, volume 1, Aitareya, in Marathi, authored by Dr. P. V. Vartak

7 Scientific Dating of the Ramayana and the Vedas - authored by Dr. P. V. Vartak - Published by the Veda Vidnyana Mandal, 497 Shaniwar Peth, Pune 411030.

2

Identification of *Tridoṣa* :

Vāta, *Pitta* and *Kapha* of Caraka-Suśruta-Vāgbhata Saṁhitātrayī

as

the Bloods of Arterial, Venous Capillaries and Lymph of Lymphatic Capillary.

(A Scientific Bridge between Ayurveda and Allopathy)

N.G.Dongre

Abstract

Bṛhattrayī or Caraka-Suśruta-Aṣṭāṅghṛdaya *Saṁhitā*s are works on Indian therapeutics based on the principles of *tridoṣa*, which are comprised of *vāta*, *pitta* and *kapha*. The aim of this paper is to thrash out and decipher the meaning of *vāta*, *pitta* and *kapha* and identify them in the allopathic terms. They may be 'the lymph enriched purified bright red blood of arteries', 'the dark blood of veins loaded with the discarded wastes from the cells' and 'the white lymph drained from the tissue spaces to lymphatic vessels' respectively. Further modern techniques to test the bloods of artery and vein origin and lymph from the lymphatic vessel may be useful to determine the states of *vāta*, *pitta* and *kapha* and may also provide an elegant method for the diagnosis of a disease on the *tridoṣa* principle of Ayurveda.

Keywords: Ayurveda, *tridoṣa*, *vāta*, *pitta*, *kapha*, artery, vein, lymphatic vessel, cell, tissue.

Bṛhattrayī or Caraka-Suśruta-Aṣṭāṅghṛdaya *Saṁhitā*s[1] are the works on Indian therapeutics. According to these the causes for a disease are essentially four in number, viz., *āgantu*, *vāta*, *pitta*, and *kapha*. (In the Ayurvedic literature *śleṣma* term is also frequently used as a synonym for *kapha*.). Amongst them *agantu* is an external cause, where as *tridoṣa* i.e., *vāta*, *pitta*, and *kapha* are internal causes (*nija* or *śarīra*) as expressed in Caraka-Saṁhitā as below:

Catvaro rogā bhavanti āgantu-vāta-pitta-śleṣma nimittaḥ.

Teṣām caturṇāmapi rogāṇām rogatvamekavidhām bhavati ruksāmānyāt. Dvividhāḥ punaḥ prakṛtireṣām, āgantunijavibhāgāt.

चत्वारो रोगा भवन्ति- आगन्तुवातपित्तश्लेष्मनिमित्तः। तेषां चतुर्णामपि रोगाणां रोगत्वमेकविधां भवति, रुक्सामान्यात्। द्विविधाः पुनः प्रकृतिरेषां आगन्तुनिजविभागात्। च.सू.20

Similarly Suśruta says

Vātapittaśleṣmāṇaḥ eva dehasaṁbhavahetavaḥ. (3)

Visargadānavikṣepaiḥ somasūryanila yathā.

Dhārayanti jagaddeham kaphapittānilastathā. (8) Sūtrasthāna 21

वातपित्तश्लेष्मण: एव देहसम्भवहेतव:।। 3।।

विसर्गादानविक्षेपै: सोमसूर्यनिलायथा।

धारयन्ति जगद्देहं कफपित्तानिलस्तथा।। 8।। सु. सू. 21

Similar statement may be had from *Aṣṭāṅgahṛdaya* of Vāgbhaṭa, which is as follows:

Vāyuḥ kaphaśceti trayodoṣāḥ samāsataḥ. 6.

Vikṛtāvikṛtadeham ghnanti te vartayanti ca...7. A.S. 1

वायु: पित्तं कफश्चेति त्रयो दोषा: समासत:।।6।।

विकृताविकृतादेहं घ्नन्ति ते वर्तयन्ति च।। 7।। अ. सू. 1

Thus what so ever may be the reason for a disease, ultimately for the cure, *vāta*, *pitta*, and *kapha* have to be considered. The whole architecture of this therapy, Ayurveda is based on the knowledge of *tridoṣa*. Therefore their understanding in the modern scientific terms has a vital importance and only then the system of Ayurveda will have dialogue facility with the modern Allopathic system and then both the therapeutic systems may get enriched with each other.

Before starting the main discussions regarding *vāta*, *pitta*, and *kapha*, a few words about the modern anatomy and physiology are essential.

1. The salient features of modern anatomy and physiology[2]:

The general health should depend upon the proper functioning of the various organs of the human body. Human body is made of organs, which are nothing but an assembly of tissues. Therefore the structure and the function of an organ depend on the tissues, which are classified as connective tissues, muscles, nerves and all other forms of tissues. At the same time a tissue is a group of cells with varying amounts of inter-cellular cement, fibers, ground substances and fluid, where cell is a bit of protoplasm capable of (i) assimilating food and oxygen, (ii) discarding wastes, responding to stimuli, reproducing itself and (iii) drains the residual food in the form of lymph into the tissue space. Thus for the healthy body, the proper functioning of the cell (śarīra-aṇu) and tissue will be responsible. This may be pictured as follows:

(i) The purified blood i.e., full of oxygen enters into artery through the left auricle of the heart. Then the lymphatic of right thorax, right arm, right side of the body and neck transports lymph to the right lymphatic duct, while the lymph of from those of rest body and from the lacteals, which absorb the products of digestion of fat from the stomach and small intestine transport to thoracic duct. Finally the right lymphatic and thoracic ducts pour the lymph into the blood streams via the right and left subclavian veins. Thus the bright red blood received from arterial capillary by a cell will contain oxygen and food as well. (ii) Now in the cell repair is made and growth take place. In the process of repair oxygen carried tears down worn out part and burns them as fuel and is discarded as waste in the form of dark blood and unused food is thrown into the tissue space, which are their after carried out through venous capillary to vein and (iii) the unused food (lymph) of white colour is carried out through lymphatic capillary, originating through tissue space to lymphatic vessel.

Thus for an healthy body one can always assume that within a cell (*śarīra-aṇu*) of a tissue of any organ there must be a dynamic (*cara*) equilibrium (*sāmya*) between the incoming bright blood full with lymph (food) through arterial capillary (*śrota*) and outgoing fluids namely dark red blood through venous and white lymph through lymphatic capillaries, i.e. if any abnormality (*doṣa*) of these fluids carried by artery, vein and lymphatic vessels are detected then it may direct the cause of illness and helpful to diagnose a disease. These fluids may be considered as *vāta*, *pitta*, and *kapha* respectively as the conclusion drawn from the forth-coming discussion.

2. Vata, Pitta, and Kapha:

The vessel through which the blood is conveyed from the heart to different parts of the system is termed as arteries. 'These arteries[3] so named because the ancients found that the arteries of dead bodies contained air and supposed then to be air tubes leading through the body.' Ancient Indians knew this vessel system and named as *vāta-nāḍī*, which is conveying '*prāṇavāyu*', the oxygen along with the blood. Like wise the other fluids *Pitta*, and *Kapha* were having their separate systems as *Pitta-nāḍī* and *Kapha-nāḍī*. Suśruta, the father of śalya cikitsā (surgery) describes them as red vessel to convey vāta, hot blue to convey pitta and cold and whitish to convey less mobile kapha as given below:

Tatrāruṇāḥ vātavahāḥ pūryante vāyunāsirāḥ.

Pittāduṣṇaśca nīlāśca śītagouryāḥ sthirakaphat. 18 S. S.7

तत्रारुणा: वातवहा: पूर्यन्ते वायुनासिरा:।

पित्तादुष्णाश्च नीलश्च शीतगौर्या: स्थिरकफात्।। 18।। सु. शा. अ. 7

They knew that through these three *nāḍīs* the fluids of *vata*, *pitta*, and *kapha* respectively are mobile throughout the body.

Sarvaśarīracarastu vāta-pitta-śleṣmāṇaḥ. 9

सर्वशरीरचरस्तु वातपित्तश्लेष्माण: ।।9।। च. सू. 20

In order to diagnose a disease apart from the symptom there was a tradition to follow by the expert practitioners to cross check the disorder by critically examining the variations of motion of these nāḍīs (vessels)[4], which may be more prominently felt in the wrist just below by one angula (about 1 cm) from the root of the thumb as given below:

Ekāṅgulam parityajyādhastādaṅguṣṭhamūlataḥ.

Parīkṣet yatnavān vai sāhyābhyāsādeva lakṣyate.

--*Rāvaṇakṛtanāḍīpariksā*

एकांगुलं परित्यज्याधस्तादंगुष्ठमूलत:।

परीक्षेत् यत्नवान् वै साह्याभ्यासादेव लक्ष्यते।। रावणकृतनाडीपरीक्षा।

In this course keeping index, middle and ring fingers over the wrist, vat, pitta, and kapha nāḍīs (vessels) may be examined as given below:

Ạdau ca vahate vāto madhye pittam tathaiva ca.

Ante ca vahate śleṣma nāḍīkatrayalakṣaṇam.--Nāḍīvijñāna

आदौ च वहते वातो मध्ये पित्तं तथैव च।

अन्ते च वहते श्लेष्मा नाडिकात्रयलक्षणम्।। नाडीविज्ञान

Naturally experts might have felt these vessels in the background of radial and ulna bones. In case when there is serious disorder, these nāḍīs (vessels) may not be functioning the selective role in that case every nāḍī may give a mingled effect, then they may be considered to allowing partially all to flow (*Sarvavaha*) as given below:

Nahi vātam sirā kāścin pittam kevalam tathā.
śleṣmāṇam vā vahantyetā ataḥ sarvavahāḥ smṛtāḥ. 16
Praduṣṭānām hi doṣāṇam mūrcchitānām pradhāvatām.
Dhruvamunmārgagamanataḥ sarvavahāḥ smṛtāḥ. 17,S.Sa.7

नहि वातं सिरा: काश्चिन् पित्तं केवलं तथा।

श्लेष्माणं वा वहन्त्येता अत: सर्ववहा: स्मृता:।। 16।।

प्रदुष्टानां हि दोषाणां मुच्छितानां प्रधावताम्।

ध्रुवमुन्मार्गगमनत: सर्ववहा: स्मृता:।।17।। सु. शा. 7

Apart from these there are other systems of nāḍīs (vessels) for various other functions. *Sonita nāḍī* system is also one of them, which gains importance while the surgery problems of abscess; tumor, ulcer (व्रण) etc. come under consideration. These systems transmit the blood generated from the bone marrow; the blood after filtration through kidney, etc. to the main streams and in general such vessels are situated well in depth of the body. They have been also described as given below:

Gūḍhāḥ samasthitāḥ snigdhārohiṇyāḥ suddhaśoṇitam.

गूढा: समस्थिता: स्निग्धारोहिण्या: शुद्धशोणितम्।।

Since these do not have primary importance towards the general health at present we are interested about the identification of *vata*, *pitta*, and *kapha* in the terms of modern anatomy and physiology.

2.1 *Vāta*:

The attributes of *Vata* as enumerated in Caraka-samhita are rough, soothing, thin, pervading, mobile, bright and corpuscular.

'*Rukṣaḥ śīto laghuḥ sūkṣmaścalo(a)tha viśadaḥ kharaḥ*' caraka sūtra. 59

रुक्ष: शीतो लघु: सूक्ष्मश्चलोथविशद: खर:।।59।। च.सू.1

Vāgbhaṭa also enumerates the same list as given bellow:
Tatra rukṣo laghuḥ śītaḥ kharaḥ sūkṣmaścalo(a)nilaḥ.11 -A.S.1

तत्र रुक्षो लघु: शीत: खर: सूक्ष्मश्चलो निल:।।11।। अ. सू. 1

Vāta exhibits two functions burning (*dahakrt*) due to *teja* (*pranavayu* i.e. oxygen) and soothing (*sitakrt*) due to *soma* (nutrition in the form of *kapha*)

Yogavāhaḥ param vāyuḥ samyogādubhayakṛtaḥ
Dāhakṛt tejasā yuktaḥ śītakṛt somasaṁśrayāt //

योगवाह: परं वायु: संयोगादुभयकृत:।

दाहकृत् तेजसा युक्त: शीतकृत् सोमसंश्रयात्।। च.सू. 3

Thus we may identify *vata* as the arterial blood, which is issued to cells since the properties and functions resemble with each other.

2.2 *Pitta*:

The attributes of *pitta* as described in Caraka-samhita are as nominal sticky, hot, sharp, fluid, acid, mobile and tastily hot.

Sasnehamuṣṇaṁ tīkṣṇam ca dravyamamlam saram kaṭu -60

सस्नेहमुष्णं तीक्ष्णं च द्रव्यमम्लं सरं कटु।।60।। च.सू. 1

At the other place in Caraka-samhitā the properties of *pitta* listed as hot, sharp, fluidity, nominally sticky, the colour neither white nor bright red, stinky, sour as well hot and mobile.

Auṣṇyam taikṣṇyam dravatvamanatisneho varṇaśca śuklāruṇavarjyo gandhāśca visro rasouca kaṭukāmlau saratvam ca pittasyātmarupāṇi
Caraka Sūtra. 20

औष्ण्यं तैक्ष्ण्यं द्रवत्वमनतिस्नेहो वर्णश्च शुक्लारुणवर्जो गन्धाश्च विस्रो रसौ च कटुकाम्लौ सरत्वं च पित्तस्यात्मरूपाणि।।15।। च.सू. 20

Similarly Sushruta also describes the attributes as sharp, fluidity, stinky, dark (bluish yellow), hot, tastily hot, burned and sour.

Pittam tīkṣṇam dravam pūti nīlam pītam tathaiva ca
Uṣṇam kaṭurasaścaiva vidagdhām cāmlameva ca S.S.2

पित्तं तीक्ष्णं द्रवं पूति नीलं पीतं तथैव च।

उष्णं कटुरसश्चैव विदग्धां चाम्लमेव च।। सु. सू. 2

Similarly Vāgbhaṭa describes as:

Pittam sasnehatīkṣṇoṣṇam laghuvisram saram dravam.A.S.1

पित्तं सस्नेहतीक्ष्णोष्णं लघुविस्रं सरं द्रवम्।। 11।। अ. सू. 1

Thus *pitta* being acidic having odd smell etc. can always be identified as venous blood since these properties are always expected in the venous blood as it contains carbonic acid and discarded waste from the cells.

2.3 *Kapha*:

The properties of *kapha* as described in Caraka-samhita are thick, nutritious, soft, fatty, sweet, steady and sticky.

Guru- śīta- mṛdu-snigdha-madhura-sthira-picchilaḥ Caraka Sūtra.61

गुरुशीतमृदुस्निग्धमधुरस्थिरपिच्छिल:।।61।। च. सू. 1

According to the second list provided in Caraka-samhita the attributes of *kapha* are as fatty, soothing, white, dens, sweet, steady, slippery and sticky.

Sneha-śaitya-sauklya-gaurava-mādhurya-sthairya-picchilya-mātsrnyāni 18

स्नेहशैत्यशौक्ल्यगौरवमाधुर्यस्थैर्यपिच्छिल्यमात्स्र्यानि।।18।। च. सू. 20

Suśruta describes them as:

Śleṣmasvetoguruḥ snigdhaḥ picchilaḥ śīta eva ca.
Madhurastāvidagdhāḥ syādvidagdho lavaṇaḥ smṛtah. S.S.21

श्लेष्मा श्वेतो गुरुः स्निग्धः पिच्छिलः शीत एव च।

मधुरस्ताविदग्धाः स्याद्विदग्धो लवणः स्मृतः।। 15।। सु. सू. 21

And according to Vāgbhaṭa they are:

Snigdhāḥ śītam gururmandaślaṣṇomṛtsnaḥ sthiraḥ kaphaḥ.12.
-A.S.1
स्निग्धाः शीतं गुरुर्मन्दश्लक्ष्णोमृत्स्नः स्थिरः कफः।। 12।। अ. सू. 1

Thus we find that kapha resembles to the properties of lymph which is also a white fluid and carrier of all the nutritious substances for the growth of the cells, therefore one may identify kapha or *slesma* as lymph of lymphatic vessels.

3. Inference:

Thus from these discussions *vata*, *pitta* and *kapha* of Caraka-samhita may be identified as the blood of the artery the blood of the vein and lymph of lymphatic vessel.

References:

1. *Bṛhattrayī* or Caraka-Suśruta-Aṣṭāṅghṛdaya *Saṁhitā*s :

(i) Caraka-saṁhitā, Chaukhamba Surbharati Prakashan, Varanasi (1991)

(ii) Suśruta Saṁhitā, Chaukhamba Surbharati Prakashan, Varanasi (1994).

(iii) Aṣṭāṅgahṛdaya, Vāgbhaṭa, Chaukhamba Orientalia, Varanasi (1998).

2. Devid-Christopher. Textbook of Surgery (The Biological Basis of Modern Surgical Practice). Vol. II, 12th Edition. Pp1787-1872. Igaku-Shoin/Sounders Tokyo.1981.

3. The World University Encyclopedia. Publishing Company, Inc., Washington, D. C. (1965).

4. Śarīra-Kriyā-Vijñana, Ranajit Desai, Shri Vaidyanath Ayurveda Bhavan Samvat 2002(1945)

3

Genetics in the metaphors

Dr. C.P. Trivedi

All the living creatures exhibits same functions at the cellular level, under which energy transformation, growth and reproduction are their natural qualities. In all the living beings RNS-Ribose Nucleic Acid or DNA-Deoxy Ribose Nucleic Acid in the main genetic material, which decides the course of life.

The credit for discovering the Deoxy Ribose Nucleic Acid goes to the Waston and Crick, it has brought a revolution in the filed of biotechnology.

In the biology it is a kwon fact that in the human beings there are 23 pairs of chromosomes, in which one pair, xx, or xy decides the gender and one pair is having the information for the life and death. Which regulates the life through its Genetic hereditary characters. The chromosomes consists of DNA a characteristic feature of the DNA is the fixed content an sequence of nucleotide pairs along the entire length of its double helix. In this respect any one species of DNA differ from the other species. Accordingly the species diversity is there is the nature with different hereditary characters. A definite number of nucleotides constitute a gene, which have the information about the synthesis of specific kind of the proteins and the proteinacious enzymes regulates the metabolic reactions.

The DNA is having only four types of bases Adenine, Guanine, Thymine and Cytosine, they are arranged of the DNA spiral strips in different sequences. The two base Adenine and Guanine are known as purine and two base Thymine and Cytosine are known as pyrimidine, purine and pyramiding are complimentary with each other, such as Adenine = Thymine and Guanine =Cytosine and vice versa. These complimentary bases are bind with each other by hydrogen double bond and triple bond. The sequence of three bases call a triplet known as codon, with four kinds of bases in the DNA. There are 64 codes, which is known as genetic code, which is a code for aminoacid and constitute a language code for the protein synthesis. These 64 codes may constitute a language code for recording the thoughts and deeds of an individual, which is imperishable and get stimulated by the enzymes as indicated the Vedas. The DNA through transcription synthesise mRNA and mRNA through translation synthesise proteins.

The same has been expressed symbolically in the Veda, through association and interrelationships of the different forms with one another. Their relationship has been expressed through the family relationship as parents, spouse-wife, husband, son and daughter, union of two have been indicated through Asvin.

The living cell has been termed as Ṛbhu, nucleus as Brihaspati, chromosomes as Yama, genetic material as Tvastra and Vivasvata.

The DNA has been termed as Tvastr, who is shaper of all creatures, hence it has been termed as omniform (Visvarupa), it means the DNA is present in all the living beings.

Tvastr the lord hath made all forms and all the cattle of the field: Cause them to multiply for us Rg. 1.-188-9

Tvastr begot Brihaspati (nucleus), The Brihaspati represent nucleus, which has been expressed as regulator an guide. Rg. 2-23-7.

The Tvastr- DNA synthesis new DNA from the DNA template, the same has been termed as the bridal car, the single strand of the DNA template has been termed as the daughter of Tvastr and the new DNA has been termed as Vivasvata. He is the father of Asvin (nucleotide pair) it means Asvin represent nucleotide pairs, which get synthesized from the DNA template. He is the father of Yama-Yami (Twinned pair of chromosome). It means chromosomes get synthesised from the DNA. He is the father of Manu-human race, which signifies that DNA is responsible for the evolution of the human species. His wife is Saranyu daughter of Tvastr, who get disappear. It means the daughter represents the DNA template and after the synthesis of the mRNA, the DNA template lost its identity, which has been expressed metaphorically.

Tvastr makes a wedding-car for his daughter; by reason of this, all this creation comes together, the mother of Yama, wife of great Vivasvata, being drawn about, disappeared. Ath. 18-1-53.

The synthesis of the complimentary base pairs of the new DNA has been expressed in this fashion "having made one of like colour, they gave her to Vivasvat."

They hid away the immortal (biosoul) one from mortals; having made one of like colour, they gave her to Vivasvata; what that has carried also the two Aśvins (Nucleotide); and Saranyu deserted two twins (Pairs of chromosomes). Atharvaveda, 18-2-33.

The structure of the DNA has been indicated symbolically through Yama, (Chromosome), the Yama's two brindled dogs four eyed watch the path of man, it means they represents two helix of the DNA and the four eyes represents two base pairs, who watch the path of man. It means the life of a individual get regulated by the information, which is there in the coded language of the DNA, simultaneously the deeds of a individual get recorded in the coded language of the DNA.

The same has been expressed in Ṛgveda. also.

With Trikadrukas it purifies it self six wide ones verily great one; tristubh, gayatri, the meters; all those (are) set in Yama. Ath. 18-2-6.

The activity of the DNA has been indicated through Trikadruka (=,=), who represents a triple bond and a double bond of hydrogen, through, which the complimentary base pairs get joined with each other. 'The six wide ones' represents the three base pairs. 'The verily great one' it indicates a triplet language code of the DNA template, The language of the DNA has been indicated through the tristup, Gayatri and the meters; all those (are) set in Yama, it means the information is in the chromosome. The triple and double hydrogen bond play a major role in the synrhesis of mRNA and DNA duplication, which has been expressed by Trikadruka(=,=).

Originally the DNA has only four types of base, the three bases constitute a genetic code means a code for a aminoacid for synthesis of protein, there are 64 codes. In the Vedas the triplet genetic code has been termed as three-headed demon or Trisira the son of Visvarupa (omniform) Tvastr. Till the triplet code is in association with the DNA, it is unproductive. In the nature one of the strand of the DNA act as a template for the syntheis of mRNA in the presence of the enzyne RNA polymerase. Which have the coded genetic information for the protein synthesis. The base sequence in mRNA is complimentary to the base sequence in parent DNA. The synthesis of mRNa for the protein synthesis is its productive power, which initiates the protein synthesis, the same has been expressed through Trita-triple bond, which is the slaughter of the three headed, six eyed demon (three complimentary base pairs) Rg. 10-99-6.

Well skilled to use the weapons of his father Aptya urged on by Indra fought the battle then Trita slew the foe seven rayed three headed and freed the cattles of son of Tvastr. Rg. 10-10-8.

Lord of the brave Indra cleft him in pieces, who sought to gain strength and deemed him mighty. He smote has three heads from his body seizing the cattle of the ominform Tvastr. Rg. 10-10-9.

The release of the cattles signifies the productivity of the mRNA, who take part in protein synthesis through genetic code, the weapons of the Trita is bond energy, which is being represented by Indra and Aptya- water, the triplet code is being generated due to the separation of triplet base through breaking the two double and s single triplet bond with the help of water, the same has been expressed as seven rays, which get break through Indra and Aptya, as such as code of mRNA get translated for the protein synthesis and accordingly Tvastr's visvarupa (omniform) get expressed.

The structure of the duplicating DNA has been expressed symbolically through Sinivali, who is broad hipped, fair armed, fair fingered. The broad hips represent a loop of the DNA and the two separating strips have been termed as fairs arms and the projecting nucleotides are termed as he fingers. She has been invoked with *Raka, Kuhu, Gungu* and *Sarasvati*, where Raka, represent full moon. It means a loop of the DNA is like a full moon. Kuhu and Gungu represent first day of moon, it means the separating DNA exhibits crescent shaped structures, they looks like first day of the moon. During the separation of the DNA helix in the womb a initiator code get opened which is synonym to a particular though energy of an individual ancestor, which may get bind with the initiator code and accordingly a dead person's thought mentality may get rebirth. So it has been prayed that to a girl, O Sinivali, O Sarasvati grant here good progeny.

Stand firm; viraj art thou; as it were, Vishnu here O Sarasvati; O Sinivali, let her have progeny; may she be in the favour of Bhaga. Ath. 14-2-15.

In the Rgveda, it has been indicated that Tvastr give shapes to the infant in the womb and the germ is get by the Sinivali and Sarasvati in the womb, accordingly Aswin (nucleotide pairs) crowned with lotuses, it means the development of the children will be there according to the hereditary characters, where the Sinivali is the code initiatiator and the Sarasvati (thought energy) opens, the door of a language, which is particular to an individual through Aswin (nucleotide pair). It means the first code, which is present on the single stand of the DNA template, it acts as the language code initiator for a particular mentality, which is particular to a individual, hence even twins have different characters and mentality.

1. May Visnu form and mould the womb, may Tvastr duly shape the forms.

Prajāpati infuse the stream and dhatar lay the germ for thee.

2. O Sinivali set the germ, set thou the germ Sarasvati. May the twain gods bestow the germ. The Aswins crowned with lotuses.

3. That which Aswin the Twain rub forth with the attrition stock of gold that germ of thine we invoke that in the tenth month thou may bear. Rg. 10-184-1,2,3.

Evolution through cross breeding in the living organisms is a fact, which has been indicated through a conversation between the Yama and Yami, who are twin brother and sister they represents different pairs of chromosomes and crossing over between the two different strains, where Yami asks her brother to marry with her so as to produce offsprings and Yama has replied that "to marry with a sister is against the law you go to another person and encircle around him like a climber and produce offsprings." (encircle around him like a climber is a metaphor and a indication for the crossing over during meiotic cell division) Rg. (10-10 and Atharvaveda 18-2). The same is the basis of Darwin's theory of organic evolution and Mendel's Laws of Inheritance.

The YAMA is a king of life and death, which represent a pair of chromosome as we know that in the human beings there are twenty three pairs of chromosomes and out of it one pair, who hears the genetic information about the life and death in its genetic sequence of DNA has been termed as Yama and the different chromosomes have been termed as Yama and Yami, which indicates about crossing over between two different strains of chromosomes. Yama's two brindled dogs with four eyes, who watch the course of life of a man, in the modern terminology they represent two strips of the DNA and four eyes symbolically indicates about two complimentary DNA base pairs (Adenine, Guanine, Thymine and Cytosine). A=T, G=C, who records the deeds and thoughts in its genetic code sequence, which remains active through out the life and the language is imperishable. The chromosomes and the DNA maintain their continuity in the living beings through their offsprings. The coded language of the DNA may get stimulated during the DNA duplication in the womb, through the thought energy of an individual of a dead person. The same has been indicated through Yama's two four eyed dogs, who watch the path of the man.

Verily the generator made us(du) in the womb man and spouse-God Tvastr, Savitar of all forms Atharvaveda; none overthrow his ordinances earth knows us (two) as such, also heaven. 18-1-5.

Yami

Desire of Yama hath come up to me Yami, in order to lying together in the same lair (yoni); I would fain yield (ric) my body as wife to husband; may we whirl off, two chariot, wheels. Atharvaveda. 18-1-8.

Yama

Verily I may not mingle my body with they body, they call him wicked (papa) who approach his sister. That is not consonant with my mind and heart, that I a brother, should lie in a sister's bed.

Yami

A weaking alas, art, thou, O Yama; we have not found mind and heart thine, another woman shall embrace thee, as a girth a harnessed (Horse), as a twinning plant (libuja) a tree. Atharvaveda 18-1-15.

Yama

Another man, truly, O Yam, another man shall embrace thee as a twinning plant a tree, either thou seek his mind or he thine; then make for thy self very excellent concord (samvid). Atharvaveda. 18-1-16.

Run thou past the *two four eyed, brindled dogs* of Sarama, by a happy road; then go unto the beneficent Fathers, who revel in common revelry with Yama. Atharvaveda. 18-2-11.

What two defending dogs thou hast, O Yama four eyed, sitting by the road, men watching, with them. O king, do thou surround him; assign to him well being and freedom from disease. Atharvaveda. 18-2-12.

Broad nosed, feeding on lives, copper coloured Yama's two messengers go about after men (jana); let them give us back here today excellent life (asu) to see the sun. Atharvaveda. 18-2-13.

They hid away the immortal (bio soul) one from mortals; having made one of like colour, they have her to Vivasvant; what that was carried also the two Aśvins; and Saranyu deserted two twins (Paris of chromosomes). Atharvaveda. 18-2-33. Tvastr makes a wedding-car for his daughter; by reason of this, all this creation comes together; the mother of Yama, wife of great Vivashtar, being drawn about, disappeared. 18-1-53.

Tvashtar disposed (vi-dha) the garment for beauty. by direction of Brihaspati, of the poets; therewith let Savitar and Bhava envelop this woman, like Surya, with progeny. Atharvaveda. 18-3-6.

Yama beyond, below Vivsvant- beyond that do I see nothing whatever; into Yama has entered my sacrifice; Vivvasvant stretched after the world (bhu) 15. stand firm;arr thou; as it were Vishnu hare O Sarasvati; O Sinivali, let her have progeny may she be in the favor of Bhaga. Atharvaveda. 18-2-32.

Visvarupa is the son of the Tvastr, is a three headed demon slain by both Trit and Indra.

Vivasvat is the father of the Aswin. 10-17-2. and of Yama 10-14-17.

Tvastr is described as ancestor of human race so far as his daughter Saranyu wife of Vivasvat, become the mother of the primeval twins Yama and Yami. Rg. 10-17-2, it has been said that tvastr adorned all beings with forms known as omniform Visvarupa, who is son of Tvastr.

Sinivali is broad hipped, fair armed, fair fingered, prolific a mistrees of a family, who grant progeny, she is invoked with Sarasvati, Raka as well as Gungu, Kuhu, Raka, represents fool moon and Kuku represent first day of moon. Rg. 2-32; 10, 184.

4

प्राचीन भारत में शल्य-चिकित्सा :

शल्य-उपकरणों के परिप्रेक्ष्य में

डॉ॰ सुमन सैनी, प्राध्यापिका

प्रस्तुत शोधपत्र का विषय है- प्राचीन भारत में शल्य-चिकित्सा: शल्य-उपकरणों के परिप्रेक्ष्य में। शोधपत्र में वर्णित विषय सामग्री को तीन भागों में विभक्त किया गया है। प्रथम भाग में अतिसंक्षेपेण आयुर्वेद एवम् आयुर्वेद के अष्टांगों में सर्वप्रथम और सबसे प्रधान शल्यतन्त्र का इतिहास बताया गया है, जिससे आगामी विषयवस्तु की सम्यग्रूपेण स्थापना की जा सके। दूसरे भाग में शल्यग्रन्थों में वर्णित शल्य उपकरणों का वर्णन है तथा अन्तिम भाग में प्राचीन भारतीय चिकित्सा-जगत् में प्रचलित कुछ स्वस्थ परम्पराओं का निर्देश है, जिनका आज भी पालन किया जाता है। परम्परा के अनुसार आयुर्वेद का प्रथम ज्ञान ब्रह्मा को स्वयं हुआ। इस ज्ञान के आधार पर दस हजार अध्यायों वाली एक लक्षश्लोकात्मक 'ब्रह्मसंहिता' का निर्माण किया गया, जो आज अनुपलब्ध है। सुश्रुत के अनुसार इस संहिता का ज्ञान ब्रह्मा ने दक्ष प्रजापति को दिया था। दक्ष प्रजापति से यह ज्ञान अश्विनी कुमारों को प्राप्त हुआ जिनके द्वारा किया गया चिकित्सा एवं शल्यचिकित्सा का वर्णन वैदिक-संहिताओं में अनेकत्र प्राप्त होता है। उन्हें ''देवानां भिषजौ'' कहा गया है। वे शल्यचिकित्सा में पारङ्गत थे। उन्होंने असाध्य व्याधियों का उपचार किया (ऐसे वर्णन अनेकत्र उपलब्ध होते हैं। जैसे मूत्रघात में शलाका के प्रयोग द्वारा मूत्र निकालने का, अपची में शलाका द्वारा वेधन का, विपश्यला की टाँग युद्ध में काटे जाने पर लोहे की टाँग लगाने का, शल्यक्रिया द्वारा प्रसव कराने का वर्णन, ऋजाश्व के अन्धे हो जाने पर उन्हें आँखे प्रदान करने का, पूषा के टूटे दाँत को ठीक करने का, दधीचि का सिर काट कर पहले घोड़े व पुन: उसका अपना सिर लगाने का वर्णन मिलता है। अश्विनीकुमारों ने ही देवताओं की व्याधियों को नष्ट किया व उन्हें यौवन प्रदान किया। वेदों के प्रसंग केवल वर्णन मात्र है (क्रियाविधि का वर्णन प्राप्त नहीं होता परन्तु इससे यह अनुमान हो जाता है कि तत्कालीन शल्यचिकित्सा उन्नत स्तर को प्राप्त कर चुकी थी। अश्विनी कुमारों से यह परम्परागत ज्ञान इन्द्र को प्राप्त हुआ और उससे आयुर्वेदीय ज्ञानविज्ञान तीन विभिन्न सम्प्रदायों के प्रवर्तक ऋषियों ने प्राप्त किया= (1) धन्वन्तरि (2) भारद्वाज (3) काश्यप, जिनमें से शल्य सम्प्रदाय के प्रवर्तक धन्वन्तरि हैं। ये विष्णु का अंश माने जाने वाले, समुद्रमंथन के समय आविर्भूत धन्वन्तरि से भिन्न काशी के राजा दिवोदास धन्वन्तरि थे, जिनका समय लगभग 1000 ई0 पू0 से पहले का है। धन्वन्तरि ने आयुर्वेद को संक्षिप्त करके आठ अङ्गों में विभक्त किया- शल्यतन्त्र, शालाक्यतन्त्र, कायचिकित्सा, भूतविद्या, कौमारभृत्य, अगदतन्त्र, रसायनतन्त्र एवं वाजीकरणतन्त्र। आयुर्वेद के आठ अंगो में से शल्यज्ञान को ही प्रमुख रखकर यह विद्या सुश्रुतादि शिष्यों को दी और कालान्तर में सभी शिष्यों ने इस पर अपनी-2 संहिताएँ लिखीं, जिनमें से सुश्रुतसंहिता सबसे अधिक प्रचलित हुई।

सुश्रुत ने व्यवस्थित रूप से सूत्रस्थान के 46, निदानस्थान के 16, शारीरस्थान के 10, चिकित्सास्थान के 40, कल्पस्थान के 8 तथा उत्तरतन्त्र के 66 अध्यायों में शल्यचिकित्सा का सम्पूर्ण सर्वाङ्गीण वर्णन प्रस्तुत किया है। सुश्रुतसंहिता के सूत्रस्थान के पञ्चम अध्याय से प्रारम्भ करके नवें अध्याय तक शल्य उपकरणों का वर्णन किया गया है। शल्यकर्म तीन प्रकार के बताए गए है- पूर्व कर्म, प्रधान कर्म, पश्चात् कर्म। सर्वप्रथम शस्त्रकर्म के विषय में बताया गया है, क्योंकि शल्यचिकित्सा शस्त्रकर्मप्रधान है। शस्त्रकर्म ८ प्रकार का बताया गया है- तच्च शस्त्रकर्माष्टविधम्। तद्यथा- छेद्यम्, भेद्यम्, लेख्यम्, वेध्यम्, एष्यम्, आहार्यम्, विस्राव्यम्, सीव्यम्। इन कर्मो के लिए निम्नलिखित उपकरणों की आवश्यकता होती है-

छेदन भेदन के लिए- वृद्धिपत्र, नखशस्त्र, मुद्रिका, उत्पलपत्र, अर्धाधार।

छेदन व लेखन- मण्डलाग्र, करपत्र।

विस्रावण- सूची, कुशपत्र, आटीमुख, शरारीमुख, त्रिकूर्चक, अन्तर्मुख।

वेधन- कुठारिका, व्रीहिमुख, आरा, वेतसपत्र, सूची।

आहरण- बडिश, दन्तशङ्कु।

एषण (अनुलोमन)-एषणी।

सीवनकर्म- सूची।

यन्त्र प्राय: लौहनिर्मित होते थे- तानि प्रायशो लौहानि भवन्ति, तत्प्रतिरूपकाणि वा तदलाभे। डल्हण ने टीका में इसकी व्याख्या करते हुए कहा है- 'लोहा: पञ्चसुवर्णादय: तत्कृतानि लौहानीत्युच्यन्ते।' उनके आभाव में, दन्त, शृङ्ग, दारु आदि के बने यन्त्रों का भी प्रयोग किया जा सकता था- लोहवेणु वृक्ष-शृङ्गवल्कलतृणनिस्नायुप्रतानाश्मानश्चैतेभ्यो यन्त्राणि निष्पाद्यन्ते। यन्त्रों के द्वारा 4 प्रकार के कार्य किए जाने का निर्देश है =

1. रोग को देखना- अर्शो यन्त्र के द्वारा।
2. क्रियासौकर्य- भगन्दररोग में भेदनकर्म के समय एषणी का प्रयोग।
3. शल्यनिर्हरणार्थ- तालयन्त्र या सदंश यन्त्र का प्रयोग।
4. मार्गविशोधनार्थ- व्रणधावनयन्त्र द्वारा मार्गविशोधिनी शलाका से संकीर्ण मूत्रमार्ग का विशोधन।

सुश्रुत संहिता के सप्तम अध्याय में यन्त्रविधि का वर्णन है, जिसमें हाथ को प्रधानतम यन्त्र बताते हुए 101 यन्त्र वार्णित हैं- 'यन्त्रशतमेंकोत्तरं (अत्र हस्तमेव प्रधानतमम् यन्त्रमवगच्छ यस्माद्धस्तादृते यन्त्रणामप्रवृत्तिरेव, तदधीनत्वाद्ययन्त्रकर्मणाम्।' (सु॰सं॰ 7/3) यन्त्र की परिभाषा सुश्रुत ने इस प्रकार दी है- 'तत्र मन:शरीरबाधकराणि शल्यानि। तेषामाहरणोपायो यन्त्राणि। तानि षट्प्रकाराणि तद्यथा- स्वास्तिकयन्त्राणि, संदशयन्त्राणि, तालयन्त्राणि, नाडीयन्त्राणि, शलाकायन्त्राणि उपयन्त्राणि चेति।' इन्हें निम्नलिखित तालिका में देखा जा सकता है।

101 यन्त्र (Blunt Instruments)

स्वस्तिक यन्त्र-24 (forceps)

प्रमाण 18 अंगुल

कार्य -अस्थिनिमग्न शल्यों का उद्धरण

9 व्याल मुख
1. सिंहमुख
2. व्याघ्रमुख
3. वृकमुख
4. तरक्षुमुख
5. ऋक्षमुख
6. द्वीपिमुख
7. मार्जारमुख
8. क्षृगालमुख
9. मृगमुख

15 पक्षिमुख
1. काकमुख
2. कंकमुख
3. कुररमुख
4. चाषमुख
5. भासमुख
6. बाजमुख
7. उलूकमुख
8. चिल्लीमुख
9. श्येनमुख
10. गीधमुख
11. क्रौञ्चमुख
12. भृंगराजमुख
13. अञ्जलिकर्णमुख
14. अवभञ्जनमुख
15. नन्दीमुख

संदंशयन्त्र 2 प्रकार

प्रमाण 16 अंगुल

कार्य-त्वचा, मांस, शिरा, स्नायुगत शल्यो का उद्धरण

- सनिग्रह
- अनिग्रह

तालयन्त्र 2 प्रकार (Scoop)

कार्य- कर्ण नासिका के शल्य का उद्धरण

- एकताल
- द्विताल

नाडीयन्त्र 20 (Tubular Inst.)

कार्य- शल्य उद्धरण, शल्य को देखना, दुष्ट रुधिर अवचूषण

1. भगन्दर यंत्र 2
2. अर्शो यंत्र 2
3. व्रणयंत्र 1
4. वस्तियंत्र 4
5. उत्तरवस्ति 5
6. मूत्रवृद्धि स्रावण 2
7. जलोदर स्रावण 2
8. धूमनेत्र
9. निरुद्ध प्रकाश 1
10. सन्निरुद्धगुदा
11. अलाबु 1
12. शृङ्ग 1

शलाकायन्त्र 28 (Spatula /Probe)

कार्य - Probing, Elevation, Retraction

1. गण्डूपदमुखी 2
2. सर्पफणमुखी 2
3. शरपुंख 2
4. बडिशमुखी 2
5. मसूरदलमुखी 2
6. कार्पासकृतोष्णी 6
7. खल्लमुखी 3
8. जाम्बवमुखी 3
9. अंकुशमुखी 3
10. नासाबुर्दहरण 1
11. अञ्जनमुखी 1
12. मूत्रमार्ग विशोधनी 1

उपयन्त्र 25 (शल्य में सहायक उपकरण)

1. रज्जु
2. पट्ट
3. चर्म
4. वल्कल
5. वेणिका
6. लता
7. वस्त्र
8. अश्मोष्ठक
9. मुद्गांक
10. अंगुली
11. जिह्वा
12. दन्त
13. नख
14. मुख
15. बाल
16. क्षार
17. अग्नि
18. भेषज आदि

यंत्रों के आकार सुश्रुत के द्वारा निम्न प्रकार से वर्णित हैं-

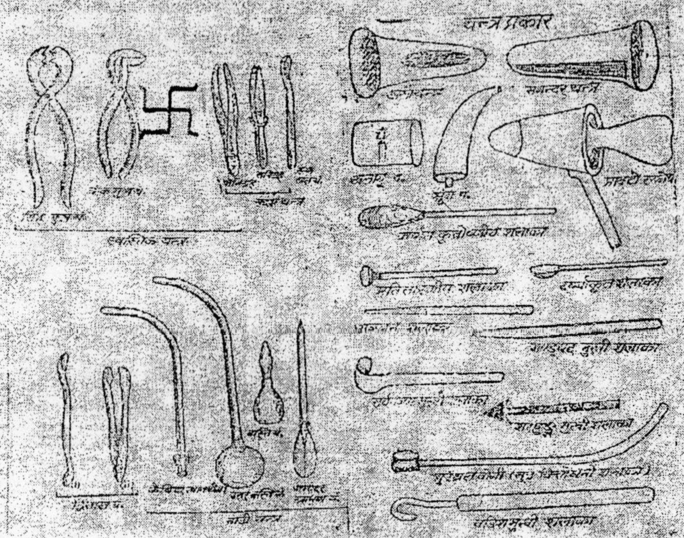

यन्त्रों के वर्णन के उपरान्त शस्त्रों का विस्तृत विवरण प्राप्त होता है। शस्त्र sharp Instruments को कहा गया है। सुश्रुत ने शस्त्रों की संख्या 20 तथा वाग्भट ने 26 बताई है और इनका निम्नलिखित गुणों से युक्त होना आवश्यक माना है- सुग्रहाणि, सुलोहानि, सुधाराणि, सुरूपाणि, सुमुखाग्राणि, अकरालानि। शस्त्र वक्र, कुण्ठ, खण्डित, खरधार वाले, अतिस्थूल, अतितुच्छ, अतिदीर्घ, अतिह्रस्व नहीं होने चाहिए तथा शस्त्रों की धार के लिए क्रमशः पायना, निशातनी और धार संस्थान- ये तीन संस्कार करने चाहिए। पायना (temper) करते समय उन्हें अग्नि में तपाकर तैल, क्षार व उदक आदि द्रवो में बुझाया जाता था। निशातनी संस्कार में काले पत्थर पर घिस कर रोमच्छेदन योग्य बनाया जाता है तथा धार-संस्थान संस्कार में शाल्मली की लकड़ी, चमड़े या मोटे कपड़े पर घिसाया जाता था।

सुश्रुत द्वारा वर्णित 20 शस्त्रों को निम्नाङ्कित वर्गीकृत तालिका में देखा जा सकता है।-

संख्या	शस्त्र		भेद	प्रमाण कार्य	कार्य
1.	मण्डलाग्र	Circular Knife Razor like knife	द्वेसूच्य: तिस्त्र: सूच्य: 1. अग्रभाग गोल 2. अग्रभाग क्षुराकार	षडङ्गु.ल	शुण्डिका Tonsils, बढ़ा दांत, व्रण, मांस का छेदन, मृतगर्भ को काट कर निकालना
2.	कर पत्र	Saw		2/6 अंगुल	अस्थि का छेदन
3.	वृद्धिपत्र	Scalpel	1. अञ्चित/नताग्र 2. प्रयताग्र	7 अंगुल (5½ वृन्त 1½ फलम्)	व्रणशोथ में भेदन
4.	नखशस्त्र	Nail Pairer	1. वक्रधार 2. ऋजुधार	1 अंगुलविस्तार 2 अंगुलफल	नखशल्यभेदन, छेदन, लेखन, प्रच्छत्र व सूक्ष्म शल्य उद्धरण कण्ठगतरोगों में छेदन, भेदन कर्म
5.	मुद्रिका	Finger knife		मुद्रिका तर्जनी में पहन कर उस पर वृद्धिपत्र के समान एक फलक (Blade) फलक 1 X 3 अंगुल	छेदन और भेदन
6.	उत्पलपत्र	Lancet			
7.	अर्धधार	Half edged Knife		8 अंगुल लम्बा 2ग 1 अंगुल फलक	
8.	सूचीशस्त्र	Suturing needle	1.वृत्ताङ्गुलद्वयम् 2.त्र्यङ्गु.लातिस्त्रा 3.धनुवक्रा	2 अंगुल लम्बी 3 अंगुल लम्बी अग्रभाग गोल	1. कम मांस व सन्धियों में सिलने के लिए 2. कठोर व मासंल स्थानो के लिए 3. मर्मस्थान, फलकोष व उदर में प्रयोग
9.	कुशपत्र	Cataract knife		फलक कुशपत्र के समान लम्बा वा पतला वृन्तभाग 3 अंगुल	विस्रवर्ण कर्म के लिए
10	आटीमुख	Curved Scissors		वृंत- 7 अंगुल फलक- 8 अंगुल	रक्त विस्रावण

11.	शरीरमुख	Scissors		10 अंगुल	
12.	अन्तर्मुख	Bistury		6 अंगुल लम्बा अन्दर को मुख किए	रक्त विस्रवण
13.	त्रिकूर्चक			अन्दर की और मुख किए 3 शस्त्र वृंत	सुश्रुत- विस्त्रावर्ण वाग्भट - कुट्टन कर्म
14.	कुठारिका	Axe		वृंत्त- 5½, फलक ½ अंगुल आकार गोदन्त के सामान	व्यधन कर्म सिरावेध
15.	व्रीहिमुख	Trocar		वृंत 2 अंगुल फलक 4 अंगुल	सिरोवेध, जलोदर, मूत्रवृद्धि में वेधन कर्म
16.	आरा	Owl		8 अंगूल लम्बा, 1/3 अंगुल अग्रभाग गोल व पीछे का वृंत भाग चौड़ा	कर्णपाली का वेधन (Punch Biopsy) (पक्वशोथ का विनिश्चय)
17.	वेतसयंत्र	Narrow Blade Knife		अग्रभाग 1अंगुल चौड़ा 4 अंगुल लम्बा	वेधनकर्म वाग्भटकौमुदी-दन्तधार शस्त्र
18.	बडि़श	Sharp look	स्वनत- Full curved नात्यानत Half Curved	मुख तीक्ष्ण अंकुश के समान मुड़ा वृंत 5 ½ _ अंगुल तथा अग्रभाग ½ अंगुल	शुण्ड्का, अर्मादि को पकड़ने के लिए
19.	दन्तशंकु	tooth Scaler		वृंतं 6 अंगुल फलक ½ अंगुल सम तीक्ष्ण, चकोर शंकु के समान मुड़ा हुआ	दन्तशर्करा को हटाना
20.	एषणी	Sharp Probe		8 अंगुल सूत्र के लिए पीछे एक पाश (Eye)	

वाग्भट के अनुसार 6 अन्य शस्त्र हैं-

1. सर्पमुख (कर्ण, नासा के अर्श व अर्बुद के छेदनार्थ)
2. लिंगनाश शलाका (ताम्रमयी द्विमुखी शलाका- लिंगनाश में वेधन कर्म)
3. कूर्च-त्रिकूर्च के समान 7-8 शलाकाकार शस्त्र (उपयोग अर्श में)
4. खज- कण्टक वाला खजकूर्च के समान (छेदनार्थ)
5. कतंरी शरारीमुख के समान तीक्ष्ण अग्रभाग मध्य से जुड़ी (छेदनकर्म)
6. कर्णव्यधान - 3 अंगुल लम्बीसूची - कर्णवेधन के लिए ग्रन्थों में वर्णित

शस्त्रों के आकार निम्नप्रकार से दर्शाए जा सकते है।

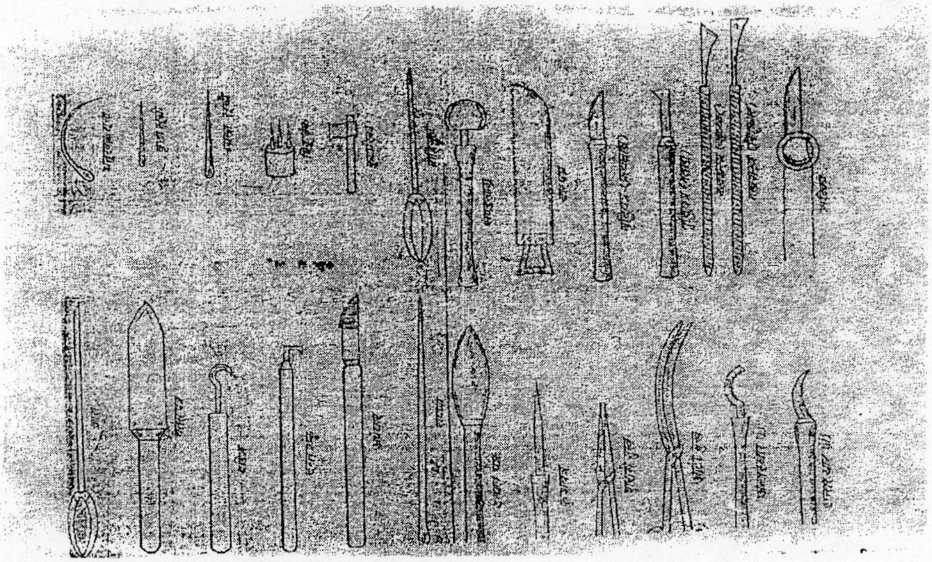

सुश्रुत ने शस्त्र और यन्त्र आदि को रोगी के शरीर पर प्रयोग करने से पूर्व भिन्न-2 वस्तुओं पर अभ्यास करने की बात सूत्रस्थान के योग्यासूत्रीय नामक नवें अध्याय में कही है। आधुनिक शल्य चिकित्सा में इसे Operative Surgery कहा जाता है। आजकल यह अभ्यास Dissection के समय मृत शरीर पर कराया जाता हैं परन्तु सुश्रुत ने अनेक सरलतया प्राप्त होने वाले पदार्थों पर यह अभ्यास कराने की बात कही है। इस अभ्यास के करने में कौन सा शस्त्र शल्यचिकित्सा करते समय कहाँ से पकड़ना है, यह भी अभ्यास हो जाता है। परन्तु सुश्रुत ने सुश्रुत संहिता के शरीर स्थान 49 में भी अच्छी प्रकार से शरीररचना की जानकारी के लिए मृतमानवशरीर को ही उपयुक्त ठहराया है- 'तस्मात्

समस्तगात्रमविषोपहतमदीर्घव्याधिपीडितमवर्षशतिकं निःसृष्टान्त्रपुरीषपुरुषम्' में मृतदेह को preserve करने की विधि भी बताई है। निम्नलिखित तालिका में योग्या की विधि और वस्तुएं निर्दिष्ट हैं-

शल्यकर्म का नाम	कर्माभ्यास के योग्य वस्तु
1. छेदन, ऊर्ध्वकर्तन, अधःकर्तन का अभ्यास (Exicision)	लौकी, तरबूज,ककड़ी, खीरा आदि पर करना चाहिए।
2. लेखन कर्माभ्यास के लिए (Scrapping)	रोम युक्त, फैला हुआ चमड़ा
3. भेदन-कर्माभ्यास के लिए (Incision)	जलमिश्रित कीचड़ से भरी हुई दृतिमशक पर।
4. वेधन अभ्यास के लिए (Puncturing/ tapping)	कमलनाल और मृत पशुओं की सिराओं पर।
5. एषण अभ्यास के लिए (Probing)	घुन लगी लकड़ी, खोखला बाँस, कमल नाल, सूखी हुई तुंबी के मुख में।
6. आहरण-अभ्यास के लिए (extraction)	कटहल, कन्दूरीफल, बिल्वफलमज्जा और मृतपशुओं के दाँत निकालने में योग्या करें ।
7. विस्रावण-कर्म सीखने के लिए (Blood Letting)	मोम लगे सेमल के चपटे काष्ठ-फलक पर। मधुमक्खी के छत्ते से मधु विस्रावण।
8. सीवन-कर्म सीखना हो तो (Suturing)	पतले या मोटे कपड़े के प्रान्तों पर या चमड़े के टुकड़ो पर योग्या करे।
9. बन्धन-योग्या(पट्टी बाँधने का अभ्यास) (Bandaging)	आदमी के पुतले के अंग-प्रत्यंग पर योग्या करें।
10. कर्णसन्धिबन्धन का अभ्यास (Earplasty)	मृदुचर्म, मांसपेशी या कमलनाल पर मृदुमांस खण्ड़ो पर योग्या करे।
11. अग्निकर्म और क्षार कर्म अभ्यास कास्टिक सोड़ा /पोटाश द्वारा	मृदुमांस खण्ड़ो पर योग्या करे।
12. नेत्र प्रणिधान (वस्तिनेत्र का प्रयोग), वस्तिपीडन (व्रणवस्ति पीडन)का अभ्यास (squeezing)	जलपूर्ण पात्र के पार्श्व में छेद करके या अलाबू मुख में अभ्यास योग्या करें।

अन्त में उपसंहार रूप में प्राचीन भारत में शल्यचिकित्सा के सन्दर्भ में उस समय प्रचलित कुछ स्वस्थ परम्पराएँ, जिनका आधुनिक चिकित्सा जगत् में भी पालन किया जा रहा है, का उल्लेख करना मै आवश्यक समझती हूँ-

1. सुश्रुत संहिता प्रथम उपलब्ध ग्रन्थ है, जिसमें तथा वाग्भट आदि अन्य शल्यक्रिया के आचार्यो ने भी शल्य के सभी अंगो का विस्तार से प्रतिपादन किया गया है, जैसे- अस्थि, नेत्र, श्रोत्र, घ्राण, व्रणचिकित्सा, प्रसूतितन्त्र प्लास्टिक सर्जरी आदि जबकि आज प्रत्येक अंग की चिकित्सा शल्य के विशेषज्ञ पृथक-2 है।

2. सुश्रुत के काल में शरीरसंरचना (anatomy) को जानने के लिए शवच्छेद की बात कही है। आज भी मृत शरीर पर अभ्यास करके ही शरीरसंरचना को जानकर चिकित्सक्‍ इस कार्य में प्रवृत्त होते हैं।

3. शल्य से पूर्व अधिपति की अनुमति ग्रहण करनी होती थी। विशेषत: मूढ़गर्भ के उद्धरण के सन्दर्भ में अधिपति की अनुमति भी ली जाती थी है। आज के चिकित्सा जगत् में भी किसी शल्य क्रिया से पूर्व Consent of the patient/ attendant लेना आवश्यक है।

4. सुश्रुत ने चिकित्सा व्यवसाय में सफलता के लिए कई सुझाव दिए हैं और चिकित्सक की शिक्षा सम्बन्धी जो योग्यता बताई गई है। वे सभी आज के समय में प्रासङ्गिक है। सुश्रुतसंहिता के सूत्रस्थान के दशम अध्याय- 'विशिखानुप्रवेशनीयम्' अध्याय में चिकित्सक के लिए शास्त्र का पूर्णतया अध्ययन, शास्त्र के विषय को प्रत्यक्ष देखना और कर्माभ्यास करना, शास्त्र के अध्ययन की क्षमता प्राप्त करना आवश्यक है। अन्तिम योग्यता राज्ञाज्ञातेन बताई गई है- योग्यता प्रदर्शित कर राजा से चिकित्सा व्यवसाय के लिए अनिवार्यत: आज्ञा प्राप्त करना। आज भी चिकित्सक के लिए Medical council of India से पंजीकृत होना आवश्यक है। चिकित्सक की मनोदशा भी व्यवसाय के अनुकूल होनी चाहिए। उसे सुमनसा, कल्याणमिव्याहारेण, अकुहकेन बन्धुभूतेन भूतानां, सुसहायवता वैद्येन भवितव्यम्' कह कर उसे शुद्ध मन से युक्त व्यवहार में कल्याणकारी तथा आचरण में निष्कपट प्राणियों में भ्रातृभाव/दुखियों का सहायक होना चाहिए। ऐसे गुण, आचार विचार वाला शल्य चिकित्सिक ही धर्म, अर्थ, काम, यश को प्राप्त करता है। एक बात अलग कही गई है- स्त्रियों का परहार जो आधुनिक चिकित्सा जगत् से नितान्त भिन्न है क्योंकि आज Nursing care स्त्रियाँ ही करती हैं।

5. प्राचीन ग्रन्थों में कुछ ऐसे भी स्थल प्राप्त होते हैं, जहाँ चिकित्सा वर्णन विधि को एक कारिका में इस प्रकार से बताया गया है, जो आधुनिक शल्यशास्त्रियों के लिए भी उपादेय है। भग्न (Fracture) के रोहण के लिए चिकित्सित स्थान के प्रारम्भ में कही गई ऐसी सम्पूर्ण कारिका है, जिसमें रोगी की मानसिक दशा, वातिक प्रकृति, ज्वर आदि के होने से भग्न का रोहण कठिनाई से होता है-ऐसा बताया गया है- "अल्पाशिनोऽनात्मवतो जन्तोर्वातात्मकस्य च। उपद्रवैर्वजुष्टस्य कृच्छ्रेण सिद्ध्यति।।" यहाँ शल्यक्रिया से पूर्व और उसके पश्चात् भी रोगी की मानसिक दशा का धयान देने की बात कही गई है।

आधुनिक समय में भी बड़े-बड़े हस्पतालों में counselling cells बनाए गए हैं। जो शल्य चिकित्सा से पूर्व और पश्चात् रोगी की मनोदशा का ध्यान रखते हैं और उसे सकारात्मक बनाने का प्रयास करते हैं।

6. शारीरस्थान में शरीर-रचना सम्बन्धी अवयवों को गिनाने से पूर्व सुश्रुत ने चिन्ताशरीर का उपक्रम, अव्यक्त का निरूपण, अव्यक्त से सभी तत्त्वों प्रकृति व आठ विकारों का उत्पत्तिक्रम, ज्ञानकर्मेन्द्रियों का अधिभूत, आध्यात्मिक, अधिदैवत भेद से त्रैविध्य, सभी का त्रैगुण्य से युक्त होना और इस विषय पर आधारित वैद्यक सिद्धान्त जिसमें तीन गुणों से व्याधियों का सम्बन्ध बताया गया है, का वर्णन करते हैं। आज भारतवर्ष में तथा पाश्चात्य चिकित्सा जगत् में भी alternative medicine का विकास हो रहा है, जिसमें हृदय की व्याधियों का इस विधि से उपचार किया जा रहा है जिसमें अध्यात्म की भूमिका को रेकी, स्पर्शचिकित्सा आदि चिकित्सा पद्धतियों में स्वीकार किया जा रहा है।

7. उस समय में शल्यचिकित्सा के लिए सुविधाओं में कमी थी- सुश्रुत ने तीक्ष्ण सुरा से संज्ञाहरण की बात कही है। सौश्रुत काल में शल्य की वेदना हरने के लिए कोई अन्य प्रभावकारी संज्ञाहरण औषधि उपलब्ध नहीं थी, जैसे ईथर, हैलोथीन गैस आदि। साथ ही प्रभावकारी जीवाणुनाशक तथा प्रतिजीवी औषधियों का भी अभाव था, इसलिए बडे-बडे शल्यकर्म उसी अवस्था में किए जाते थे जब शल्य न करने से रोगी की मृत्यु निश्चित थी। सुश्रुत स्वयं कहते हैं= अक्रियायां ध्रुवो मृत्युः क्रियायां संशयो भवेत्।

इस प्रकार शल्यग्रन्थों में दिए गए शल्य उपकरणों के वैविध्यपूर्ण तथा वर्गीकृत विवरण से यह अवश्य पता चलता है कि उस समय में शरीर के प्रत्येक भाग के अंग से सम्बद्ध शल्यचिकित्सा अत्यंत उन्नत अवस्था में थी और आज भी उन विवरणों के विवेचनात्मक अध्ययन से चिकित्सा जगत् लाभान्वित हो सकता है।

5

अथर्ववेदे शरीरविज्ञानम्

डॉ॰ राजेश्वर-मिश्रः

सर्वविधानां ज्ञान-विज्ञानानामाकरः वेद इति प्रसिद्धः। वेदेष्वथर्ववेदस्यायुर्वेदेन सह प्रगाढः सम्बन्धः वर्तते। प्रमाणमस्य स्वयमेवाथर्ववेदे प्राप्यते यतो हि वेदोऽयं 'भिषग्वेद' इत्युक्तम्।[1] 'आयुर्वेदसम्बद्धाः वेदस्यास्य मन्त्रा' इति गोपथ-ब्राह्मणोऽपि स्फुटं प्रमाणयति, तत्र चा'थर्वा' शब्दस्यार्थः भेषजमिति कृतः।[2] शाङ्खायन-श्रौतसूत्रेऽपि वेदोऽयं भेषजमिति निगदितम्।[3] यजुर्वेदस्यैकस्य मन्त्रस्य (11.32) व्याख्या-प्रसङ्गे 'शतपथ-ब्राह्मणे' "प्राणो वा अथर्वा" इति प्रतिपादितम्।[4] अतः प्राणविद्या जीवनविद्या वाथर्वणविद्यास्तीत्यनेन प्रमाणितम्। अथर्ववेदस्यापरं नाम ब्रह्मवेद इत्यप्यस्ति।[5] वेदस्यास्य ब्राह्मणेन गोपथेन ज्ञायते यत् 'ब्रह्म'-शब्दः भेषजः भिषग्वेदस्य चापि बोधकः। उक्तमपि "येऽथर्वाणस्तद् भेषजं यद् भेषजं तदमृतं यदमृतं तद् ब्रह्म" इति।[6] अतः अनेन प्रतीयते यदुभावपि समानार्थकाविति। 'पुनरङ्गिरसामायुर्वेदेन शरीरविज्ञानेन च सह सम्बन्धो वर्तत' इत्यपि गोपथ-ब्राह्मणेन ज्ञायते। अत्र च प्रतिपादितं वर्तते यदङ्गानां रसा अर्थात् तत्त्वानि यस्मिन् वर्णितान्यसावाङ्गिरसो वेदः।[7] गोपथब्राह्मण एवान्यत्र रसायन-विज्ञानमेवाङ्गिरस इत्यभिहितम्।[8] छान्दोग्यबृहदारण्यकोपनिषद्भ्यामपि शरीरविज्ञानं रसायनविज्ञानञ्चाङ्गिरो विद्येति स्वीक्रियते।[9] पुनश्च प्राणविद्याप्यङ्गिरो विद्येति संज्ञितम्।[10] प्राणा अङ्गानां रस इति। यतोहि यथाविधिं प्राणसंचरणेनैव जीवनं धार्यते; अतः

1. अथर्व॰, 11.6.14 : ऋचः सामानि भेषजा यजूंषि ।
2. गो॰ ब्रा॰, 1.3.4: येऽथर्वाणस्तद् भेषजम् ।
3. शां॰ श्रौ॰ सू॰, 16.2.1 : अथर्ववेदो वेदः सोऽयमिति भेषजं निगदेत् ।
4. शत॰ ब्रा॰, 6.4.2.2 ।
5. अथर्व॰, 15.6.8 : तमृचश्च सामानि च यजूंषि च ब्रह्म च ...।
6. गो॰ ब्रा॰, 1.3.4 ।
7. तत्रैव, 1.1.7 :एतमद्धरसं सन्तमद्धिरा इत्याचक्षते ।
8. तत्रैव, 1.3.4 : येऽद्धिरसः स रसः ।
9. छां॰ उप॰, 1.2.10 : एवाद्धिरसं मन्यन्तेऽद्धानां यद्रसः ।
   बृह॰ उप॰, 1.3.19 : सोऽयास्य अद्धिरसोऽद्धानां हि रसः
   प्राणो वा अद्धानां रसः प्राणो हि वा अद्धानां रसः ।
10. शत॰ ब्रा॰ 6.1.2.28: प्राणो वा अद्धिरा ; तुल॰ बृह॰ उप॰, 1.3.19 ।

'शतपथ-ब्राह्मणे' (6.1.2.28), 'बृहदारण्यकोपनिषदि' (1.3.19) च "प्राणो वा अङ्गिरा" इत्युक्त्वा प्राणविद्या जीवनविद्या वाङ्गिरोविद्येति प्रतिपादितम्। चरक-सुश्रुतयोरुभयोः संहितयोरप्यायुर्वेदमथर्ववेदस्योपाङ्गमुपवेदो वा स्वीकृतः। एष आष्टाङ्गिकश्चेति मतम्।[1] अतः स्पष्टं वक्तुं शक्यते, यदथर्ववेद आयुर्वेदस्य मूलमिति।

अथर्ववेदस्य कतिपयैर्मन्त्रैर्ज्ञायते यत्तदानीन्तना जनाः शरीरविषयकं विज्ञानमपि सुष्ठुरूपेण जानन्ति स्म, यतो हि तस्मिन्काले सर्वरोगभैषज्यकर्मार्थं कुशलैर्वैद्यैर्मानवदेहस्य सर्वेषामवयवानां सर्वासां पर्वग्रन्थीनां नामान्यवगच्छन्ति स्म। अथर्ववेदस्यैकस्मिन्सूक्ते[2] मानवदेहस्य सर्वावयवजान् रोगान् बाह्वादिपर्वसु जातं व्याधिञ्च दूरीकर्तुं शरीरस्याङ्गानि, विभिन्नानि अस्थि-संस्थानानि च परिगणितानि सन्ति। तानि चेत्थं भवन्ति- अक्षिणी, नासिके, कर्णौ, चिबुकं (ओष्ठस्याधः प्रदेशः), शिरः मस्तिष्कं (शिरशोऽन्तरवस्थितो मांसविशेषो), जिह्वा, ग्रीवाः (अत्र बहुवचननिर्देशात् ग्रीवाया अवयवभूतानि चतुर्दश-सूक्ष्माण्यस्थीनि उच्यन्ते यस्य संकेतः शतपथ-ब्राह्मणे[3] प्राप्यते), उष्णिहः (ऊर्ध्वं गताः कण्ठस्य धमन्यः, नाड्यः), कीकसा (जत्रुवक्षोगतास्थिः), अनूक्यम् (समवयन्त्यस्थीन्यस्मिन्निति मुख्यास्थि यत्र त्रयस्त्रिंशद् द्वात्रिंशद् वास्थीनि[4] समवयन्ति), अंसौ (स्कन्धौ), बाहू, हृदयम्, क्लोमा (हृदयस्य दक्षिणपार्श्वे समीपस्थो वर्तमानः मांसपिण्डविशेषः), हलीक्ष्मम् (हृदयस्य वामपार्श्वे स्थितः मांस-पिण्डविशेषः, फेफड़ा इति भाषायाम्), मतस्ने (उभयपार्श्वसम्बन्धं वृक्यं, गुर्देति भाषायाम्), प्लीहा (उदरपार्श्वस्थितः श्येनाकारः मांसपिण्डः, तिल्लीति भाषायाम्), यकृत (हृदयसमीपस्थैतत्संज्ञकः कालखण्डः, जिगरेति भाषायाम्), आन्त्राणि, गुदा (आन्त्रसमीपस्थमलमार्गः), वनिष्ठुः (स्थविरान्त्रम्, उदरं), कुक्षी (उदरस्य दक्षिणोत्तरभागौ), प्लाशिः (मूत्राशयः, मलपात्रम्); नाभिः, उरू, अष्ठीवन्तौ (जानू), पार्ष्णी (पादयोरपरभागः), प्रपदौ (पादाग्रौ), भसस् (कटिप्रदेशः), श्रोणी (कट्योरधरभागौ), भसर् (गुह्यस्थानम्), अस्थीनि, मज्जा, स्नावः (सिराः), धमनयः, पाणी, अङ्गुलयः, नखाः, अङ्गं (मांसपेशीः), लोमानि, पर्वाणीति च। एतद्व्यतिरिक्तम् अथर्ववेदे केचिदन्येऽप्यवयवाः उल्लिखिताः प्राप्यन्ते। ते सन्ति-मांसं, गुल्फौ

---

1. सुश्रु॰ सं॰, 1.6 : आयुर्वेदम् अष्टाङ्गमुपाङ्गमथर्ववेदस्य ; चरक सं॰ 1.30.20-21 ; सुश्रु॰ सं॰, 1.1.5 : इह खलु आयुर्वेदं नामोपाङ्गमथर्ववेदस्य ।
2. अथर्व॰, 2.33. 1-7 ; 20.96.17-23 ।
3. शत॰ ब्रा॰, 12.2.4.10 : ग्रीवाः पञ्चदशः । चतुर्दश वाएतासां करुकराणि वीर्यं पञ्चदशम् ।
4. शत॰ ब्रा॰, 12.2.4.14 : अनूकं त्रयस्त्रिंशः । द्वात्रिंशद् तस्य करुकराण्यनूकं त्रयस्त्रिंशम् ।

(टखनेति भाषायाम्), रोमकूपः, उच्छ्लौ (अधःपादभागौ), कबन्धः, उरो, स्तनौ, कफोडौ, पृष्टिः, मुखं, हनू, ललाटं, कृकाटिका (ग्रीवायाः पृष्ठभागः), रेतः, कपाल[1], त्वकोदरं दन्ताश्च।[2]

अथर्ववेदकालिकाः भिषजः देहे विद्यमानानां विभिन्ननाडीनां बिषयेऽपि जानन्ति स्म। 'एताः नाड्यो देवकृता' इति तेषां धारणा।[3] वेदेऽस्मिननेकाः नाड्य उल्लिखिताः। तासु-तीव्राः (श्वेतवर्णाः कफवाहिन्यः नाड्यः); अरुणा (किञ्चित् रक्तवर्णा वातवाहिन्यः नाड्यः), लोहिन्यः (रक्तवर्णाः रक्तवाहिन्यः नाड्यः), ताम्रधूम्राः (नीलवर्णाः पित्तवाहिन्यः नाड्यः) प्रमुखाः सन्ति।[4] अथर्ववेद एकस्मिन्मन्त्रे गवीन्योः नाड्योः उल्लेखः प्राप्यते।[5] सायणाचार्यस्य मते एत आन्त्रेभ्यो विनिर्गतस्य मूत्रस्य मूत्राशयप्राप्तिसाधने पार्श्वद्वयस्थे नाड्यावित्युच्येते।[6] अत्रैव बस्ति (मूत्राशयः), मेहनः (मूत्रनालः), बस्तिबिलम् (मूत्राशयद्वारम्) इत्यादीन्यङ्गान्यपि वर्णितानि सन्ति।[7] अन्यस्मिन्मन्त्रे शरीरे प्रवहमानानां प्राणापान-व्यानोदान-समानादीनां वायूनामक्षितिः (अक्षीयमाणा ज्ञानशक्तिः), क्षितिश्च (क्षीयमाणा क्रियाशक्तिः) शक्तिद्वयस्य, वाङ्मन-इन्द्रियद्वयस्याप्युल्लेखः प्राप्यते।[8] एतद्भिन्नमथर्ववेदेऽन्यत्रापि शरीराङ्गान्युल्लिखितानि सन्ति[9] यत्र रेतोऽपि परिगणितम्।

शरीरे विद्यमानानां शिराणां धमनीनाञ्च विषयेऽपि ऋषिभिर्सम्यग्विचारितमथर्ववेदे। तासां संख्याऽपि निर्दिष्टाऽत्र। वेदेऽस्मिन् धमनीनां (हृदयगतानां प्रधाननाडीनां) संख्या शतं शिराणां शाखानाडीनाञ्च सहस्रमिति स्वीकृतम्।[10] कठोपनिषद्यपि हृदयस्य प्रधाननाडीनां संख्याएकशतं निर्दिष्टम्।[11] एताः शिराः रक्तवाहिन्यः नाड्य इत्युच्यन्ते[12] देहस्य चोर्ध्वाधोमध्यादिषु भागेषु सर्वत्रैताः परिव्याप्ता अपि चातिशयेनाल्पाः स्थूलतराश्च सन्तीत्यपि जानन्ति स्म ऋषयः।[13] अथर्ववेदे

1. अथर्व०, 10.2.1-17 ; 9.7.1.
2. तत्रैव, 11.2.5-6 ।
3. तत्रैव, 6.138.4 : ये ते नाड्यो देवकृते . . . ।
4. तत्रैव, 10.2.11 : तीव्रा अरुणा लोहिनीस्ताम्रधूम्रा ऊर्ध्वा अवाचीः पुरुषे तिरश्चीः।
5. अथर्व०, 1.3.6 : यदान्त्रेषु गवीन्योर्यद्वस्तावधि संश्रितम् ।
6. तत्रैव, द्रष्टव्यः सायणभाष्यः ; तुलनीय अथर्व०, 1.11.5 : . . . वि योनिं गवीनिके ।
7. तत्रैव, 1.3.6-8 ।
8. तत्रैव, 11.10 (8). 4 : प्राणापानौ चक्षुः श्रोत्रमक्षितिश्च क्षितिश्च या । व्यानोदानौ वाङ्मनस्ते वा......॥; तत्रैव, 11.9 (7). 25 ।
9. तत्रैव, 10.9.13-25 ; 11.8.11-29 ।
10. तत्रैव, 1.17.3 : शतस्य धमनीनां सहस्रस्य हिराणाम् ; तत्रैव, 7.35.2 : शतं हिराः सहस्रं धमनी : ॥
11. कठोप०, 3.16 : शतं चैका च हृदयस्य नाड्यस्तासां मूर्धानमभिनिःसृतैका ।
12. अथर्व०, 1.17.1 : हिरा लोहितवाससः ।
13. तत्रैव, 1.17.2 ।

काश्चिन्नाड्य: नाम्नोल्लिखिता:, यथा-अवरा-मध्यमा-कनिष्ठिका-सिकतावती-धनूर्बृहतीति।[1] सायणाचार्यस्य मते धनू: मूत्राशयो नाडीविशेषोऽस्ति।[2] एकस्मिन्मन्त्रेऽष्टौ मन्य: (धमन्य:) संकेतिता: सन्ति या: सायणाचार्य: कण्ठगता: नाडीविशेषा: स्वीकरोति[3]। अनेन प्रकारेण शरीरेऽस्मिन् शतं शिरा: सहस्रं वा धमन्यश्च सन्ति।

शरीरे मूलतत्त्वरूपे विद्यमानानां वायु-पित्तं कफश्चेति त्रिधातू-नामूल्लेखोऽप्यथर्ववेदे प्राप्यते। अनेकेषु मन्त्रेषु पित्तस्योल्लेख: समुपलभ्यते।[5] अन्यत्रैकस्मिन्मन्त्रे वातपित्तश्लेष्मविकारजनितानां कासादिरो-गाणमुन्मूलनप्रसङ्ग स्फुटमेते त्रयो धातव: अभ्रजा:, वातजा: शुष्म इत्यभिधानैरभिहितम्।[6] त्रयाणां धातूनां विकारेण शरीरे नानाविधा: रोगा उत्पद्यन्ते।[7] सुश्रुतसंहितानुसारं वात-पित्तश्लेष्माणएव शरीरस्योत्पत्तिहेतव: सन्ति।[8] अथर्ववेद एते धातव: वायु-अर्क-रयीत्यभिधानैरपि संकेतिता:।[9] चरकसुश्रुत-संहितयोरप्यैतेषां नामान्येतान्येव।[10] अस्माकं शरीरमपि सप्तधातूनां सङ्घात इति। अष्ठाङ्गहृदयग्रन्थे स्वीकृतेष्वष्टधातुषु (रस-रक्त-मांस-मेदास्थि-मज्जा-शुक्राणीति) केषाञ्चिदुल्लेखोऽप्यर्थवेदे प्राप्यते। ते अस्थि-स्नायु-मांसं-मज्जेति सन्ति।[11]

अथर्ववेदे चक्षु:-श्रोत्र-घ्राण-रसना-मुख-त्वक्-शिश्नादीनीतीन्द्रियाणि सप्तप्राणा इत्युक्तानि।[12] ब्राह्मण-ग्रन्थेष्वपीन्द्रियाणि प्राणेत्यभिहितानि सन्ति।[13] वेदेऽस्मिन् चिद्रूपिणी बुद्धिरेव शरीरस्य संचालिकेत्यत्र सम्यक् निरूपितम्।[1] एवं शरीरविज्ञानस्य सम्यक् विवेचनमथर्ववेदे समुपलभ्यते।

---

1. तत्रैव, 1.17.2-4 ।
2. अथर्व०, 1.17.4: द्रष्टव्य: सायणभाष्य:-धनू:धनुर्वद्वक्रो मूत्राशयो नाडीविशेष: ।
3. तत्रैव, 2.12.7 : द्रष्टव्य: सायणभाष्य:-अष्टौ मन्य: धमन्य: कण्ठगता नाडीविशेषा: ।
4. तत्रैव, 20.83.1 : त्रिधातुशरणं त्रिवरूथम् . . . . . ।
5. तत्रैव, 1.24.1 : पित्तम आसिथ; तत्रैव, 18.3.5 : अग्ने पित्तमपामसि ।
6. तत्रैव, 1.12.3 : यो अभ्रजा, वातजा यश्च शुष्मो . . . . ।
7. चरक सू०, 1.57 : वायु: पित्तं कफश्चोक्त: . . . . ।
8. सुश्रुत सू०, 21.3 : वातपित्तश्लेष्माणएव देहसंभवहेतव: ।
9. अथर्व०, 18.2. 29 ।
10. चरक सू०, 12.11-12 ; सुश्रुत सू, 21.8 ।
11. अथर्व०, 11.8.11 : अस्थि स्नाव मांसं मज्जामाभरत् ।
12. अथर्व०, 2.12.7 : सप्त प्राणा: ; अत्र द्रष्टव्य: सायणभाष्य:- सप्तसंख्याकान् प्राणान् शीर्षण्यान् चक्षुरादीन्द्रियाणि ।
13. तै० ब्रा, 1.2.3.3 : सप्त वै शीर्षण्या: प्राणा: ; तां० ब्रा०, 2.14.2 : प्राणा इन्द्रियाणि ; तत्रैव, सप्त शिरसि प्राणा: ; शत० ब्रा०, 9.5.2.8 : सप्त शीर्षन् प्राणा: ।

अत आधुनिकचिकित्सायां प्रयुज्यमानानां यन्त्राणामभावेऽपि वैदिकऋषीणां देहस्यास्याङ्गोपाङ्गविषयकं विज्ञानं नितरां वैशिष्ट्यं धारयतीत्यथर्व-वेदोल्लिखितस्य शरीरविज्ञानस्य परितनेन विवेचनेनेदं सुतरां वक्तुं शक्यते।

1. अथर्व॰. 11.8.17 : सास्मिन् वर्णमाभरत् ।

6

भारतीयचिकित्साविज्ञाने वनस्पतीनामौषधीयं महत्त्वम्

(डॉ॰ मदन लाल वर्मा)

वैज्ञानिकेतिहासस्याध्ययनेन ज्ञायते यद् आदौ एकैकस्य विज्ञानस्योद्भवः आवश्यकतायास्तस्य चोपयोगितायाः कारणेन समभवत्। वनस्पतिविज्ञानार्थमिदं वृत्तं विशेषेण सत्यम्।

मानवजीवनमतिदुर्लभम्। जीवने स्वस्थताया अत्यावश्यकता संवर्तते। शरीरं स्वस्थं न भवति चेत्तर्हि जीवनस्य वास्तविक आनन्दो लब्धुं न शक्यते। अतो मानवजीवनेन सह वनस्पतीनां भैषज्यानां चाखण्डनीयः सम्बन्धः। 'शरीरमाद्यं खलु धर्मसाधनम्' भणितिरियं मानवजीवनेऽतिमहत्त्वपरा।

स्वस्थो मानव उत्तमं स्वास्थ्यं कामयते रुग्णश्च रोगमुक्तिं वाञ्छति। अत उत्तमस्वास्थ्यस्य कृते रोगमुक्त्यै च वानस्पतिकौषधीनां विशिष्टं महत्त्वमस्ति।

आयुर्वेदानुसारेणाचार्यैर्वनेषु भ्रमद्भिः प्राणिनां प्रकृतेराधारे तेषां रोगस्य निर्मूलनहेतूनां वनस्पतीनामुपयोगो विहितः। इमे वनस्पतयः प्राकृतिकोपहाराः सन्ति प्रकृतिरेव च स्वस्थताया आधारः। अतो भारतीयचिकित्सा-विज्ञानानुसारेण मनुष्यस्य प्रकृतिः वात-पित्त-कफेष्वाश्रिता शरीरं च पञ्चतत्त्वैर्निर्मितम्।

यथा -	वातः	वायुः
	पित्तम्	अग्निः
	कफः	जलं, पृथ्वी
	शरीरस्य रिक्तस्थानम्	आकाशः।

अतः शरीरे रुग्णे सति अस्य चिकित्सा पञ्चतत्त्वैरेव उचितरूपे सम्भवा।

उक्तं च - 'सर्वं द्रव्यं पाञ्चभौतिकम्'।

इत्थं वनस्पतयोऽपि पञ्चभौतिका एव। अतो वनस्पतीनां भारतीयचिकित्साविज्ञाने महत्त्वपूर्णं स्थानम्।

भारतीयचिकित्साविज्ञाने वनस्पतीनां महत्त्वम् आयुर्वेदविशेषज्ञेषु सम्मतमेव। आदौ वनस्पतीनां खाद्य-लाभप्रद-हानिप्रदश्रेणीषु वर्गीकरणं विहितम्। एतदनन्तरमनुभवेन परीक्षणेन च विदितं यत् कतिपयबालतरुषु आहोस्वित्तेषां क्वाथे पीड़ायाः न्यूनत्वस्य व्रणरोहणस्य च गुणाः विद्यन्ते। अनुभवेन इदमपि ज्ञातं यत् केचिद् वृक्षकाः आनन्दप्रदानुभूतीः प्रददति, केषांचिच्च निद्रायमाणः प्रभावो जायते।

अतिप्राचीनकाले यथा सृष्टेः सकलान् पदार्थान् प्रति मानवस्य हृदये जिज्ञासा समुत्पन्ना तस्याः पूर्तिश्च तत्कालीनपरिस्थित्यनुसारेण विहिता, तथैव वनस्पतिजगदपि मानवाय कुतूहलस्य विषयो बभूव। पुरातनाः मानवाः वननिवासहेतुना प्रकृतेर्निकटसम्पर्के आसन्, वनस्पतयश्च तस्य सखायः सहचराः सम्बभूवुः।

आयुर्वेदस्य वनस्पतिसम्बद्धप्राचीनग्रन्थानां प्रणेतारः प्रारम्भे उपयोगितावादिन आसन्। उदाहरणार्थं-चरकसंहितायाः सुश्रुतसंहितायाश्च रचयितारौ-इत्यादिः। एतादृशेषु ग्रन्थेषु तेषामुद्भिज्जानां वनस्पतीनां च वर्णनं प्राप्यते, ये औषधीनां कृषेरुद्यानकलायाश्च कृते प्रयुज्यन्ते स्म। उच्यते यत् पंचविंशतिसहस्रपूर्वं तक्षशिलाविश्वविद्यालयस्य आचार्यो भिक्षुः अजेयः स्वकीयमेकं शिष्यं जीवनं कथयाञ्चकार यत् -

"विश्वविद्यालयं परितश्चतुर्योजनविस्तारे उत्पन्नानां क्षुद्रपादपानां संग्रहं कुरु तेषां च गुणान् वर्णय।"

अयमेवासीत् स जीवनो यः कालान्तरे राज्ञो बिम्बिसारस्य चिकित्सको बभूव। धन्वन्तरिनागार्जुनादिकाः प्राचीनाः भारतीयचिकित्सकाः औषधसम्बद्ध-गुल्मानां गुणैः सुपरिचिता एव न आसन्, प्रत्युत तेषां विशेषज्ञा अवर्तन्त।

अनेकेषां वनौषधानामभिधानानामुल्लेखो वेदेषु लभ्यते। ज्वर-यक्ष्मा-अश्मरीप्रभृतिषु एकाधिकेषु रोगेषु कार्यकर्त्रीणामौषधीनां नामान्यपि वेदेषु प्राप्यन्ते। एवं च पृश्निपर्णी-सहदेवी-अपामार्ग-कुष्ठ-गुग्गुलु-पिप्पली-मृगशृंग-अश्वत्थ-सोम-यष्टीमधुप्रभृतीनां वानस्पतिकौषधानां वर्णनमथर्ववेदे लभ्यते।

वेदवाङ्मयेन, रामायणेन, महाभारतेन, उपनिषद्भिः, पाणिनेर्व्याकरणेन, जातकग्रन्थैः, मेगस्थनीजस्य लेखादिकैः ज्ञायते यत् प्राचीनभारते औषधविज्ञानं पूर्णतया विकसितम् आसीत्।

आयुर्वेदस्य द्रव्यगुणविज्ञाने औद्भिदं द्रव्यं चतुर्विधमासीत्-वनस्पतिः, वीरुद्, वानस्पत्यः, औषधं च। येषु दृश्यानि पुष्पाणि भवन्ति, ते वनस्पतयः कथ्यन्ते। यथा-वट-उदुम्बर-प्लाक्षादयः। लताः गुल्माश्च वीरुधः। यथा-प्रसारिणी कुशादयः। वानस्पत्यास्ते सन्ति येषु पुष्पाणि फलानि-उभयानि स्पष्टरूपेण प्ररोहन्ति। यथा-आम्र-जम्बू-आदयः। ओषधयश्च ताः सन्ति, याः फलागतानाम् उत वा पक्वानामनन्तरं विनश्यन्ति। यथा-गोधूम-यव-दूर्वादयः।

प्राचीनभारते वनस्पतीनां विषये वर्णयद्भिरस्माभिः स्वीकरणीयं यदस्माकं पूर्वजैस्तानि बहूनि भोज्यवस्तूनि अलभ्यान्यासन्, यानि अद्यत्वे सर्वथा सामान्यानि सन्ति। यथा-पीपीकरः (पपीता), सीताफलम्, दृढबीजम् (अमरूद) च। तदा-उपस्करेषु शृङ्गवेरं, कृष्णमरिचम्,एला, अम्लिका, पलाण्डुः, लशुनं च प्राप्यवस्तून्यासन्।

चरकमहर्षिणा भणितमपि -

"औद्भिदं तु चतुर्विधम्। वनस्पतिस्तथा वीरुद्वानस्पत्यस्तथौषधि:।
फलैर्वनस्पति: पुष्पैर्वानस्पत्य: फलैरपि।
ओषध्य: फलपाकान्ता: प्रतानैर्वीरुध: स्मृता:।।"[1]

सुश्रुतेन वानस्पत्यार्थे 'वृक्षो' भणित:। वस्तुत: वनस्पतयो लम्बवृक्षा: सन्ति।

"वनानां पाता पालयिता वा[2]।"

वानस्पत्या: मध्यमप्रमाणा: वृक्षा:।

"वृक्षो व्रश्चनात्। वृत्वा क्षां तिष्ठति[3]।"

वीरुध: गुल्मा: लताश्च। ओषधयश्च लघुक्षुपा:।

"ओषध्य ओषद्धयन्तीति वा। ओषत्येनाधयन्तीति वा। दोषं धयन्तीति वा[4]।"

अत्र 'वनस्पति' शब्दो व्यापकमर्थं गृह्णाति।

प्राचीनसंस्कृतवाङ्मये आम्र: पिकस्य मदनस्य च आवास: कथित:। एनमभिलक्ष्य नगरस्यैकस्य नाम आम्रपुरं (आमपुर), एकस्य राज्ञो नाम आम्रगुप्त: (आमगुप्त)एकस्य पर्वतस्य च नाम आम्रकूट: (आमकूट) संजात:। पूर्वकाले जनै: शुभावसरेषु आम्रस्य शाखा: जलभरितघटेषु न्यस्य शकुनरूपेण प्रयोक्तव्या आसन्। कदलोऽपि रसालवत् महत्त्वपूर्ण आसीत्। अस्योल्लेखो बौद्धसाहित्ये प्राप्यते।

'राउल्फिया' नामकस्य बालतरोरपि प्राचीनेषु लेखेषु वर्णनं लभ्यते। योऽधुना विश्वस्मिन् जगति विश्रुतो जात:। चरकस्य विशदग्रन्थेऽस्य गुल्मस्य प्रसंग:, अस्य क्षुपस्य संस्कृतभाषायां 'सर्पगन्धा' नामकस्यागदस्य रूपे आगतोऽस्ति य उन्मादरोगापहरणस्य हेतोरवगत:। अस्य वृक्षकस्य महत्त्वपूर्णभेषजस्य स्वीकरणे चिकित्साविज्ञानस्य त्रिंशद्शताब्द्यो लगिता:। अयं गुल्म उच्चरक्तचापार्थे मानसिक्यै-अव्यवस्थायै च-उत्तमभैषज्यरूपे संसारे विख्यात:।

आधुनिकवनस्पतिशास्त्रे यत् 'क्लोरोफिल' इति नाम्ना प्रख्यातं, तस्य मूलम् अथर्ववेदे 'अवि' इति रूपेण लभ्यते। तत्र लिखितम् -

"अविर्वै नाम देवता ऋतेनास्ते परीवृता।
तस्य रूपेणेमे वृक्षा: हरिता: हरितस्रज:[5]।।"

महाकवेर्माघस्य 'अष्टाङ्गहृदयनामके' ग्रन्थे भारतीय-चिकित्सा-विज्ञानक्षेत्रे वनस्पतीनां महत्त्वं परिलक्ष्यते।

अथर्ववेदस्तु जीवनदायिनीनां रोगविनाशिनीनां च ओषधीनां तथा च वनस्पतीनां वर्णनै: पूरितांऽस्ति।

वनस्पतीनां महत्त्वविषये ऋग्वेदस्य अरण्यानीसूक्तस्य विशेषेण उल्लेख: करणीय:[6]।

अरण्यानीएव वनस्पतिपूर्णमरण्यम्। तत्र विविधशब्दैः वनान्तं महत्त्वमुपवर्णितम्। तस्मिन्नेव प्रसंगे अन्तिमे मन्त्रे निगदितं यत् -

'अरण्यानी प्रशस्यते, यतः सा विविधवनस्पतीनां सौरभेण परिपूरिता, तत्र च विविधानुलेपः परिमलः प्रसरति।

वनस्पतयो वातावरणेन प्रभाविताः भवन्ति। होमकुण्डे यत् किञ्चिदपि हविः क्षिप्यते, वनस्पतयः सूक्ष्मभूतं तत् सर्वं जलेन पवनेन च सार्धं तस्यास्वादनं कुर्वन्ति, आस्वाद्य च वृद्धिं यान्ति विशिष्टगुणान् च धारयन्ति।

"वनस्पतिः शमिता देवो अग्निः स्वदन्तु।
हव्यं मधुना घृतेन[7]।।"

अथर्ववेदे वनस्पतीनां नामभिः गुणैश्च बाहुल्येन पृथगुल्लेख उपलभ्यते, आयुर्वेदस्य च दृशा तत्र तेषां वैशिष्ट्यं लब्धुं शक्यते।

भैषज्यगुणैः साकं वनस्पतयः कृमिभ्योऽपि आवासं यच्छन्ति।

वजपिंगनामकयोर्द्वयोर्वनस्पत्योरुल्लेखो गर्भहानिकराणां कृमीणां विनाशाय प्राप्यते।

"ये सूर्यात् परिसर्पन्ति स्नुषेव श्वशुराधि।
वजश्च तेषां पिंगश्च हृदयेऽधिनिविध्यताम्[8]।।"

वनस्पतयो विविधरूपैर्मंगलकारकाः भवन्ति। पुंसवनाय आहोस्वित् पुत्रलाभाय शमीलतायां समुत्पन्नोऽश्वत्थः सहायको भूयात् - इदमपि उल्लिखितमस्ति -

"शमीमश्वत्थ आरूढस्तत्र पुंसवनं कृतम्[9]।"

पिप्पल्याः सामान्यभेषजरूपे उत वा सर्वरोगभेषजरूपे वर्णनं लभ्यते[10]।

तत्रेदमप्युक्तं यत् पिप्पली यस्य मनुष्यस्य शरीरे प्रविशति, तस्य शरीरं कदापि हिंस्यं न भवति[11]।

वनस्पतयो वातावरणं शोधयन्ति। अतएव वाजसनेयिसंहितायां व्याहृतं यद् व्यापकपरमेश्वरेण 'वरुणेन' वनेषु विस्तृतम् अन्तरिक्षं प्रसारितं पर्वतप्रदेशे च सोमादिभेषजानां स्थानं निर्धारितम्।

"वनेषु व्यन्तरिक्षं ततान--------वरुण-----सोममद्रौ[12]।"

भारतीयचिकित्साविज्ञाने विविधानां वनस्पतीनां प्रयोगोऽनेकविधरोगेषु रामबाणरूपेण सफलः परिलक्ष्यते।

अत्र कतिपयरोगाणां वानस्पतिकौषधानां चोल्लेखो विधीयते।

अरिष्टः - अरिष्टस्य समूलांगानि भैषज्यगुणैः परिपूर्णानि सन्ति। सर्वतोभद्रस्य पत्राणि सम्यक् चूर्णयित्वा स्फोटे संयोजनेन पिटको नश्यति। कुष्ठरोगे निम्बस्य पत्राणानुत्क्वाथेन जलेन स्नातव्यम्। निम्बस्य तैलं गोदुग्धे पानेन महिलायाः रक्तप्रदररोगो विनश्यति। प्रतिदिनं प्रातः काले

निम्बस्य पत्राणां चर्वणमुदररोगो विनश्यति। प्रतिदिनं प्रातः काले निम्बस्य पत्राणां चर्वणमुदररोगेषु चर्मरोगेषु चातिलाभकरं भवति।

कारवेल्लः - आयुर्वेदग्रन्थे 'अष्टांगसंग्रहे' कारवेल्लस्य विषये व्याहृतं यत् कारवेल्लको रसे ईषत् कटुः, जठराग्निं प्रदीपकः, परमकफनाशकः। आयुर्वेदानुसारेण कारवेल्लो रूक्षस्तिक्तो रुचिकरो बुभुक्षावर्धापकः, पाचको रक्तशोधकः शोथनाशको लोचनोपयोगी, आर्तवजन्यः स्तन्यशोधको गुल्मः प्लीहा-शूलनाशकश्च।

आमलकः- आमलके खाद्यौजं सर्वाधिकं विद्यते। अस्य नियमितप्रयोगेण भारो वर्धते।

बहुगन्धम् (दालचीनी) - बहुगन्धस्य प्रयोगेण उदरविकारेभ्यो रक्षणं जायते। गन्धवल्कस्य प्रयोगो ज्वरातिसारेऽपि भवति।

पलाण्डुः - पलाण्डुप्रयोगेण-आमाशयस्य विद्रध्याः (कैंसर से) रक्षणं भवितुं शक्यते। सार्धद्वयतोलकमात्रायां पलाण्डुः, सप्त कृष्णानि मरिचानि च शिलावटके उत वा औषधमर्दनभाजने खल्ले प्रक्षिप्य - इयत् पेषणीयानि यद् विशोधनस्यावश्यकता न स्यात्। पुनस्तस्मिन् तावन्मात्रायां जलं मिश्रितं करणीयं यावन्मात्रायां रुग्णः पातुं शक्नुयात्। पानेनैव विषूचिकाया उद्विग्नतायां, पिपासायां, वमनेऽतिसारेषु च लाभो भवति।

अश्वगन्धा - अश्वगन्धायाः प्रयोगेण विद्रध्यां लाभोऽवाप्यते। अश्वगन्धा-गन्धवल्क-नागरमुस्ता-वाराही कन्दैः सिद्धजलक्वाथश्चतुर्षु पंचसु दिनेषु ज्वरातिसारं शमयति।

आढकी (अरहर) - मुखे जिह्वायां छालाः भवेयुः, भोजने कष्टं स्यात्, तर्हि आढक्या दालीं जले क्लेदयित्वा अनेन जलेन चलुकं करणीयम् आहोस्विद् वृत्तबीजायाः पत्रकाणां पौनः पुन्येन चर्वणं विधाय निष्ठीवनं कर्तव्यम्। अनेन प्रकारेण छालाः नंक्ष्यन्ति। आढक्याः शल्कानि धूमपानचषके उत वा धूमपानयन्त्रे पूरयित्वा पानेन वारं वारमायान्ती हिक्किका दूरीभवति।

स्थूलैरण्डः (पपीता) - भारतीयचिकित्साविज्ञाने महापंचांगुलस्य प्रभूतं महत्त्वमुक्तम्।एतस्य नियमितसेवनेन अनेके गभीरा असाध्याश्च रोगाः नश्यन्ति। उच्चरक्तचापरोगिभिर्नियमबद्धरूपेण स्थूलैरण्डः सेवनीयः। अस्मिन् 'कारपेन' नामकमेकं क्षारीयं तत्त्वं भवति यद् रक्तचापं नियन्त्रयति।

महाज्योतिष्मती (मालकांगनी) - महाज्योतिष्मतीएकः प्रसिद्धो वनस्पतिः। आयुर्वेदमतानुसारेण विश्लेषणेऽस्याः गुणस्तीक्ष्णः रसः कटुः, विपाकः, कटुकः, प्रभावश्च मेध्यम्। त्रिदोषसिद्धान्तानुसारेण विश्लेषणे इयं कफवातप्रशमनी। विभिन्नेषु रोगेषु अस्याः प्रयोगो विधीयते। वातव्याधिचिकित्सायां बाह्यप्रयोगार्थेऽस्याः प्रयोगो लाभप्रदः।एतस्याः सेवनेन होराद्वयत्रयानन्तरं वेगेन स्वेद आयाति ज्वरश्चापक्षीयते। ज्वरे अस्यास्तैलस्य सेवनं क्रियते। अस्याः सेवनेन क्षुधा वर्धते।

कपोतवंका (ब्राह्मी) – आयुर्वेदस्य प्राचीनतमग्रन्थचरकसंहितायाः सूत्रस्थाने आरण्यौषधानामुल्लेखो वर्तते।एतेष्वेवैकस्मिन् प्रसंगे 'ब्राह्मी' इत्यस्यारण्यौषधस्य वर्णनमप्यवाप्यते। अस्या उपयोगो मस्तिष्क-दौर्बल्ये, अपस्मारे, योषापस्मारे, रक्तविकारे, ज्वरे, कासे विधीयते। परमस्या उचितमात्रायामेवोपयोगः करणीयः। बहुतरमात्रायामुपयोगेन उद्विग्नता, शिरोवेदना किचित्कालार्थं मूर्च्छाऽपि भवितुं शक्नोति।

निर्गुण्डी – इयमायुर्वेदशास्त्रे वर्णितैका ईदृशी वनौषधिर्वर्तते, या सर्वविधशोथापहरणे रामबाणवत् कार्यं करोति। 'धन्वन्तरिनिघण्टु' ग्रन्थे सिन्धुवारस्य गुणधर्माणामुपयोगस्य च विषये लिखितं यत् –

"निर्गुण्डी कटुतिक्तोष्णकृमिकुष्ठज्वरापहा वातश्लेष्मप्रशमनी।"

अनेन प्रकारेण सिन्धुवारस्योपयोगः सर्वविधशोफे, उदरपीड़ायां, प्लीहावृद्धौ, कफविकारे, फुप्फुसयोः शोथे, अर्द्ध-शिरोवेदनायां, विषाक्तजीवजन्तूनां दंशने च सर्वसिद्धएव।

अशोकः – अशोको वनस्पतिर्वृक्षो वा भारते, चीनदेशे, मलेशियादेशे, अन्येषु च दक्षिणपूर्वेशियादेशेषु लभ्यते। अस्य 'केंकोली' इति नामापि संस्कृतभाषायां वर्तते। अस्य वनस्पतेः प्रयोगः- अर्शोरोगे लाभकरः।एतस्य पत्राणि रक्तशोधकानि भवन्ति। अशोकबीजानां प्रयोगोऽस्थिभग्ने विधीयते।

बालकानां रोगेष्वपि वानस्पतिकौषधानां प्रभूतं महत्त्वमस्ति। यथा-शाङ्र्गधरमतानुसारेण कासश्वासे मधुयष्ट्यादिक्वाथः प्रदेयः। शय्यामूत्ररोगे पंच कृष्णमरिचानि, बिम्बीमूलं षण्माषकमात्रायामेकस्मिन् षट्टंके जले पिष्ट्वा प्रातः सायं प्रतिदिनं सप्ताहद्वयं यावत् प्रदानेन रोगोऽयं नश्यति।

कार्पासबीजानां (बिनौलों की) – अष्ठ्याः (गिरी का) चूर्णं तोलकद्वयमात्रायां दुग्धे क्षीरं साधयित्वा खादनेन स्तनयोर्दुग्धं वर्धते।

सुखपूर्वकं केवलं **'जिमीकन्द'** सेवनेन गुदांकुरो (बवासीर) विनश्यति।

सार्धचतुर्माषकमात्रायाम् 'अरिष्टं' (रीठा) अर्धपादे जले क्लिन्नं विधाय फेनिलं करणीयम्। पुनस्तज्जलं वस्त्रेण विशोध्य रुग्णं पाययेत्। कतिपयदिनानि यावत् पानेन पुराणाः अतिसाराः समाप्यन्ते।

धत्तूरस्य पत्राणामासवः कुट्टयित्वा निष्कालनीयः। इममासवं शिरसि संयोजनेन यूकाः म्रियन्ते।

बर्बुरस्य (बबूल के) पत्राणि निशाकाले चूर्णयित्वा वटयो निर्मातव्याः। इमाः वटीः दुःखिते नेत्रे न्यस्य उपरि पट्टिकां बद्ध्वा शयनीयम्। प्रातः यावन्नेत्रयोः लालिमा पीड़ा च विनंक्ष्यतः।

अश्वत्थस्य द्वे पत्रे पिष्ट्वा इक्षुसारे (गुड़ में) मिश्रिते विधाय खादनेन-उदरस्य पीड़ा दूरीभवति।

न्यग्रोधस्य पत्राणि अतस्याः (अलसी के) तैले ईषत्तापेन प्लुष्य प्रातः सायं खल्वाटतायां मर्दनेन कतिपयदिवसेषु खल्वाटे शीर्षे केशाः प्ररोहन्ति।

इन्द्रायणस्य मूलं सूक्ष्मतया पिष्ट्वा वस्त्रेण विशोधनीयम्। माषकत्रयमात्रायां चूर्णे तोलकैकमात्रायां पुराणं गुड़ं मिश्रितं विधाय प्राबल्येन कुट्टनीयम्। भूयो **जपाजलस्य** (गुलाब जल की) पृषतैः क्लेदयित्वा मृदुलं कृत्वा षट्वटिकाः निर्मातव्याः। प्रतिदिनं द्वे वटिके उष्णजलेन सह सेवनेन उदरस्य कृमयो विनश्यन्ति।

पलाशस्य बीजानि खदिरश्च - समभागे जले पिष्ट्वाऽवलेपनेन कतिपयेष्वेव दिनेषु दद्रुरोगो विलीयते।

वृन्दायाः (तुलसी के) पत्राणां रसे **जम्बीरस्य** रसं मिश्रितं कृत्वा आनने मर्दनेन प्रतिच्छायाफलं (झाँई) मुखस्य बिन्दवः (मुहांसे) आदिकाः दूरीभवन्ति।

एकपादमात्रायां **जपाजलेएक**पादमात्रायां **दाड़िमबीजानि** क्लेदयित्वा रात्रिपर्यन्तं स्थापनीयानि, प्रातस्तन्मर्दनं कृत्वा विशोध्य पेयम्। अनेन सर्वविध-उदरविकारे, अर्द्धशिरोवेदनायां नेत्ररोगेषु च लाभो भवति।

पंच तोलकमात्रायां **शतपुष्पा** (सौंफ) पंचतोलकमात्रायां **बिल्वं,** दशतोलकमात्रायां शर्करजा-एतानि त्रीणि वस्तूनि कुट्टयित्वा पिष्ट्वा षण्माषकमात्रायां चूर्णं नवजलेन सह प्रातः सायं ग्रहणेन-अर्शो (पेचिश) उदरवेदना वा दूरीभवति।

अपक्वलघुमूलकसेवनं यकृत्रोगे प्लीहारोगे च लाभकरं भवति।

काशीफलस्य खण्डानि जले उष्णानि करणीयानि। तस्योष्णजलस्य चलुकौषधेन कण्ठमालारोगे लाभो जायते।

मधुमेहरोगे **जम्बुफलस्य** गुटिकानां, **शतपुष्पाणां मेथिकानां** च बीजानां चूर्णस्य मुखे जलेन सह निक्षेपणं प्रातः सायमतिलाभकरं भवति।

पेयपदार्थेषु- **आम्र-दाडिम-नारिकेल-जम्बीर-जम्बु-द्राक्षादिकानां** फलानामासवाः स्वास्थ्यप्रदाः सन्ति।

एवं प्रकृतिर्मानवायएतादृशानमूल्यानमोघान् वनस्पतीन् दत्तवती, येषां सेवनेन स पूर्णे जीवने स्वस्थः सुखी च वस्तुं शक्नोति।एतेषु **गोधूमः** प्रमुखः। अस्मिन् बालतरौएतादृशा अनेके गुणाः विद्यन्ते, ये मनुष्यं नित्यं सुस्थं न्यस्यन्ति। भारतीयचिकित्साविज्ञाने वनस्पतिविद्भिः विहितेन गोधूमस्य रासायनिकविश्लेषणेन इदं ज्ञातं यदास्मिन् वृक्षकेऽन्यैर्वनस्पतिभिः सदृशमाधिक्येन शक्तिवर्धकाः रोगनाशकाः प्रतिबन्धकगुणाश्च भवन्ति। उदाहरणार्थमस्मिन् खाद्यौजं (विटामिन), प्रोभूजिनं (प्रोटीन) 'कार्बोहाइड्रेट' आदीनि प्रचुरमात्रायां भवन्ति। गोधूमगुल्मानां रसोऽतिबलवर्धको जायते। गोधूमस्य क्षुद्रपादपो भारतीयैः प्रायेण 'यवांकुरः'-इत्यवगम्यते। अस्य बालतरोर्निरन्तरं सेवनं निद्रानाशात्,

पाण्डुरोगाच्चर्मरोगात्, सन्धिवातात्, प्रदररोगात्, केशानामसमयपतनात्, विद्रधिरोगात्, प्रतिश्यायात्, अश्मरीरोगात् (पथरी)-आदिभ्यो रक्षति।

कर्णपूरस्य (कानुपरस्य) संग्रहालये प्राचीनमेकं पुस्तकं लब्धुं, यस्मिन् चतुर्दशाधिकैकशतं रोगाणां वानस्पतिकोषधैः उपचारस्य वर्णनं प्राप्यते। पुस्तकमिदमायुर्वेदस्य भारतीयचिकित्साविज्ञानस्य च दृशाऽतिमहत्त्वपूर्णम्। अस्मिन् शीतला-क्षयरोग-विद्रधि-इत्यादिकरोगाणां वर्णनं वर्तते।

हिमानीनवरत्नतैलस्य आधुनिकभारतीयचिकित्साविज्ञाने सविशेषं महत्त्वमस्ति। तैलमेतत् **आमलक-ब्राह्मी-कर्पूर-पुदीन-पुष्प-'केसूत'** -आदिकानां मिश्रणेन निर्मितमस्ति, यत् शिरोवेदनायै, अनिद्रायै, केशानां कृते च रघुवंशकारस्य तां सूक्तिं चरितार्थां करोति-'केवलोऽपि सुभगो नवाम्बुदः, किं पुनस्त्रिदशचापलाञ्छितः[13]"।

सर्वासां चिकित्सापद्धतीनां चिकित्सकाःएकस्यैव शरीरविज्ञानस्य शरीररचनायाः विज्ञानस्याध्ययनं कुर्वन्ति। सर्वौषधस्य च प्रयोगस्य रीत्यास्तस्य च गुणानामवगुणानां विषये परिचयो भैषज्येन साकं दीयते। अतःएतस्मिन् विषये आयुर्वैदिकानामगदानां महत्त्वं वानस्पतिकमभिलक्ष्य सर्वाधिकम्। इत्यत्र नास्ति काऽपि विचिकित्सा-इति शम्।

सन्दर्भाः

1. चरकसंहितायां सू॰ 1
2. यास्कनिरुक्ते 8/3
3. यास्कनिरुक्ते 12/29
4. यास्कनिरुक्ते 9/29
5. अथर्ववेदः 10.8.31
6. द्रष्टव्यः ऋग्वेदः 10.146
7. ऋग्वेदः 10.110.10
8. अथर्ववेदः 8.6.24
9. अथर्ववेदः 6.11.1
10. द्रष्टव्यः अथर्ववेदः 6.109.1
11. द्रष्टव्यः अथर्ववेदः 6.109.2
12. वाजसनेयिसंहितायाम् 4.3
13. द्रष्टव्यं रघुवंशमहाकाव्यम्

Section V

Different Healing Therapies in Sanskrit Shastrais

1

AYURVEDIC APPROACH TO MENTAL ILLNESS

J.L.N Sastry[1]

Key Words- **Ayurveda, Yoga, Roga, Medhya Rasayanas**

Ayurveda bestowed means of positive health to human kind. The very definition of health according to Ayurveda is "complete physical, mental and spiritual well being-ness" (*SAMA DOSHAH SAMAGNISCHA SAMA DHATU MALA KRIYAH PRASANNATMENDRIYA MANAH SVASTHA ITYABHIDHIYATE*). This indicates that Ayurveda is different from modern medicine. As a matter of fact, modern medicine is two-dimensional viz., *soma* (body) and *psyche* (mind) where as, Ayurveda is three-dimensional i.e., body, mind and *soul*.

In the present times, all the three aspects are found faulty with majority of population. The most important factor for these illnesses is change in life-style resulting in stress. Therefore, the role of Ayurveda and Yoga are more emphasized today to deal with these life-style diseases and stress disorders. The author did try to lucidly present the Ayurvedic perspective of Mental Diseases and their management in this paper.

Embryological Considerations in Manasa Rogas:

Ayurveda deals with *Atma* and *Manas* right from the embryological stage. Human embryo comprises of several factors, including *Atmaja bhavas* (subtle factors related to Atma) & *Sattva bhavas* (subtle factors related to Manas).

Atmaja Bhava: *Indriya Jnana* (perception of motor & sensory organs), *Vijnana* (IQ), *Ayu* (life), *Sukha* (comfort), *Duhkha* (discomfort) etc. (*INDRIYANI JNANAM VIJNANAMAYUH SUKHA DUHKHADIKAM CHATMAJANI* - S.S.Sa.3/33).

Sattvaja bhava: *Prakriti* (constitution), *Nidra* (sleep) etc. (S.S.Sa.4.).

Anatomical Considerations in Manasa Rogas:

---

1 Adviser-Ayurveda. Dabur India Limited, Dabur Dhanwantry Hospital, Shri Dhanwantry Ayurvedic College, Sec-46B, Chandigarh U.T, India

Physically it is difficult to express the site of *Manas*. For general understanding, most of the Ayurvedic descriptions quote *Hridaya* as the site for *Manas*. Earlier Acharyas of Ayurveda might be indicating at the kinetic and potential energy of heart muscle as *Atma* and *Manas* respectively. Though situated at Hridaya both Atma and Manas act all over the body.

However, Harita quoted *Sirah* (head) as the site for *Manas* - *SIRAH TALVANTARA GATAH MANAH*. He also mentioned that *Majja* (bone marrow) is *Vata sthana* (site for Vata dosha). The author is of he opinion that the *Manas* described between *Sirah* (scalp) and *Talu* (palate) is nothing but the brain. To support the view, it is also important to realize that *Kapala Asthi* also possess *Majja* (marrow) which may be the brain.

Physiological Considerations in Manasa Rogas:

Manas is responsible for *Jnana* (mediation of proper knowledge) or *Ajnana* (inability to gain knowledge). In other words, it is the connecting link between *Atma, Indriya* and *Indriyartha*. (LAKSHANAM MANASO JNANASYABHAVO BHAVA EVA CHA. SATI HYATMENDRIYARTHANAM SANNIKARSHE NA VARTATE. VAIVRITTYANMANASO JNANAM SANNIDHYATTACCHA VARTATE.)

Manas is characterized by *Anutva* (subtle in nature) and *Ekatva* (involves with one issue at onetime). (ANUTVAMATHA CHAIKATVAM DVOU GUNOU MANASAH SMRITOU.)

Manas has control over the Indriyas and differentiates between *Hita* (good / useful) and *Ahita* (bad / useless) as well as *Ooha* (guess) and *Vichara* (discretion). Finally the decision taken by mind is expressed in the form of *Buddhi*. (INDRIYABHIGRAHAH KARMA MANASAH SVASYA NIGRAHAH. OOHO VICHARASYA, TATAH PARAM BUDDHIH PRAVARTATE.)

Manas is one among the *Karana dravyas* along with *Atma*. *Atma* is considered to be *Chetayita* (which initiates action) while Manas is *Karayita* (which performs an act). Actually, *Manas* is said to be *Achetana* (which does not have any activity but performs an act) while *Atma* is said to be *Chetana* (which represents activity but cannot perform any action by it self). (ACHETANAM KRIYAVACCHA MANASCHETAYITA PARAH. YUKTASYA MANASA TASYA NIRDISYANTE VIBHO KRIYA. CHETANAVAN YATASCHATMA

TATAH KARTA NIRUCHYATE. ACHETANATVACCHA MANAH KRIYAVADAPI NOCHYATE.)

Pathological Considerations in Manasa Rogas:

Sushruta classified diseases in to three categories viz., *Adhyatmika rogas* (which are linked with Atma), *Adhi daivika rogas* (which are linked with heavenly factors) & *Adhi bhoutika rogas* (which are linked with physical world).

Adhyatmika Rogas are sub-divided into three groups viz., Adibala pravritta (hereditary / genetic diseases), Janmabala pravritta (congenital diseases) & Doshabala pravriita (general diseased conditions).

Adhi daivika Rogas are sub-divided into three groups viz., Kalabala pravritta (diseases lined to season, age etc.), Daivabala pravritta (diseases related to infections or natural calamities) & Svabhavabala pravritta

Adhi bhoutika Rogas are considered under one group viz., Sanghatabala pravritta (head injuries resulting in memory loss etc.)

Vidyudasani krita (lightening & thunderbolts resulting in mental illness) & Pisachadi krita (infections leading to mental illneness).

Pathological involvement of Manas involves three Srotas viz., Hrit Srotas (Apasmara), Sanjnavaha Srotas (Nidra nasa) and Manovaha Srotas (Unmada).

According to Ayurveda diseases are classified as two types viz., *Shariraka roga* (physical illness) & *Manasa Rogas* (mental illness). The mental illness may be sub-divided in to three categories:

1. ***Kevala Manasa Roga*** (mental illness with exclusive psychic symptoms):

There are certain mental diseases in which the organic pathogenesis in the body is not specifically traceable and the symptoms of such diseases are mostly psychological e.g.: *Unmada* (schizophrenia / insanity), *Dhriti nasa, Kama, Krodha, Lobha, Bhaya, Soka, Chinta* etc.

2. ***Manasika Roga with Sharira Lakshanas*** (mental illness with predominantly physical symptoms):

The etiology of these diseases will be mainly psychological nature but the symptomatology of the disease is somatic in expression e.g.: *Apasmara* (epilepsy), *Murcha* (syncope), *Sanyasa* (coma), *Apatantraka* (hysteria), *Nidra nasa* (insomnia) etc.

3. ***Sharira Roga with Manasika Lakshanas*** (physical illness with predominantly psychic symptoms):

There are several somatic illnesses, which are clinically present with the mental (psychological) symptoms. They also need attention of *Manasa roga chikitsa*.

a. *Manasika Jvara* (psychic fever)

b. *Bhayaja & Sokaja Atisara* (psychic diarrhoea)

c. *Vataja Grahani* (psychic enteropathy)

d. *Vata Vyadhi* (psychiatric conditions related to CNS)

Therapeutic Considerations in Manasa Rogas:

Charaka mention that *Manasa Rogas* to be treated by means of *Jnana, Vijnana, Dhairya, Smriti* and *Samadhi* - "*MANASO JNANA VIJNANA DHAIRYA SMRITI SAMADIBHIH*" (C.S.Su.1/58).

According to Ayurveda treatment of diseases is three-folded viz.,

Daivavyapasraya: This treatment includes *Mantra-Oushadha, Mani, Mangala, Bali, Upahara, Homa, Niyama, Prayaschitta, Upavasa, Svastyayana, Panipata, Gamana* etc. - "*TATRA DAIVAVYAPASRAYAM MANTROUSHADHA-MANI-MANGALA-BALI-UPAHARA-HOMA-NIYAMA-PRAYASCHITTA-UPAVASA-SVASTYAYANA-PRANIPATA -GAMANADI*" (C.S.Su.11/55).

Yuktivyapasraya: This treatment includes *Ahara* and *Oushadha*- "*YUKTIVYAPASRAYAM PUNARAHAROUSHADHA DRAVYANAM YOJANA*"(C.S.Su.11/55).

Sattvavajaya: This treatment mainly includes counseling- *SATTVAVAJAYAH PUNARAHITEBHYO ARTHEBHYO MANO NIGRAHAH* (C.S.Su.11/55).

Role of Medhya Rasayana:

The *Rasayana therapy* which improves / enhances *Medha* (intelligence) is known as *Medhya Rasayana*. Several herbs possessing this property are repeatedly used in the management of psychiatric disorders. Some of them are: *Brahmi* (Bacopa monnieri), *Mandukaparni* (Centella asiatica), *Kushmanda* (Benencasa cerifera), *Indravaruni* (Citrullus colocynthis) etc.

Other herbs like *Tagara* (Valeriana wallachi), *Jatamansi* (Nardostachys jatamansi), *Sarpagandha* (Rauwalfia serpentina), *Ashwagandha* (Withania somnifera), *Vacha* (Acorus calamus) have been screened for their psychotropic activity and are found to be effective in the management of various mental illnesses.

REFERENCES:

Brandt KD (2001): in Harrison's Principle of Internal Medicine, ed. by Wilson JD, Braunwald E, Isselbacher KJ, Peterdorf RG, Martin JB, Fauci AS, Root RK; published by McGraw-Hill. Inc., New York, 15th ed, chapter 321(Osteoarthritis), pp-1987-1993.

Chunekar K.C, Pandey G.S. (1993): Bhavprakash Nighantu, Chaukhambha Bharati Academy, Varanasi.

Christopher H., Chilvers E.R., Hunter J.A.A. & Boon N.A.; Davidson's Principles and Practice of Medicine. London: Churchill Livingstone, 1999. pp.802-805.

Ramu G.; An Approach to Mental Examination Based on Ayurvedic Concepts. New Delhi: CCRAS, 1981.

Karel Nespor & Singh R.H.; The Experiences with Ayurvedic Psychotherapy 'Satvavajaya' in Europe. Ancient Science of Life, Vol. V, No. 3, January 1996, page 154-155.

Ramu M.G.; The Role of Shodhana in Manasika Roga and Jatrurdvagata Viakra.

Gupta S.P.; Psychiatry in Indian Medicine. 1976.

Yadavji Trikamji; Manasaroga-Vijnana. Baidyanath Publications, 1983.

2

MĀNAS ROGA- A CONNECTING LINK BETWEEN ĀYURVEDA & YOGA

-Dr Aayam Gupta[1]

INTRODUCTION

ĀYURVEDA= 'ĀYU' + 'VEDA'

AYU= Sharira + Indriya + Sattva (Manas) +Atma[2]

VEDA=Realistic Knowledge (unlike nescience that means wrong knowledge).

Here presence of 'manas' is mandatory. As perception/ acquiring 'jnana' (knowledge) depends entirely on 'manas'[3]

Ayurveda has been influenced by Samkhya philosophy & hence it recognizes the theory of evolution of Samkhya having the root as prakruti (Jada) & purusha (Chetana), which interact with each other to form `avyakta'. Mansika Dosh as envisaged in Ayurveda i.e. Sattva Rajas & Tamas are inseparable part of `avyakta'.[4] Later in the course of evolution 'trividha ahamkara' (sattvaj, rajas & tamas) have been described. These are the subtlest parts of the subtle 'Manas' that later manifest in the grosser form. Perhaps that is why 'sthul sṛṣti' cannot be uninfluenced by these Mansika Gunas (at the subtlest level) & thence by manas (at the subtle/psychic level) leading to psycho-physiological changes that are manifested at the grosser/physical level as also reflected in our behaviour.

In Ayurveda the term 'Manas' has been used in several context i.e. manas

---

1 An Ayurvedic Physician & Yoga Therapist by profession having clinical experience in psychiatry. Ex-Resident Medical Officer cum Lecturer at Kaivalyadhama Yoga Institute, Lonavla, Maharashtra. Presently working as a 'Yoga Consultant' in the 'Yoga Therapy Unit & Diet Clinics' at Dabur Dhanwantry Hospital, Shri Dhanwantry Ayurvedic College & Hospital, Sec-46B, Chandigarh. His permanent residential address is- H. No- 1665/7, Urban Estate, Kurukshetra, Haryana- 136118, Ph.- 01744222387. E-mail- draayamgupta@yahoo.com, aayamaayam@rediffmail.com

2 Shariraindriyasattvatmasamyogodharijivitam/
nityagauchanubandhauca paryayaerayuruchyate// (Charak Samhita Sutra Sthana 1/2)

3 Lakshanam manaso jyanasyabhava bhava eva ca// (Charak Samhita Sharir Sthana 1/18)

4 Sarvabhutanama karanamakaranama sattvarajastamolakshanamashtarupamakhilasya jagatah sambhavaheturvyaktamnama / (Sushrut Samhita Sharira Sthana -1/3)

as a 'dravya'[1] & as a part of 'chaturvimshati purusha / chikitsya purusha' in the context of 'chikitsa' & manas as Ubhae indriya (as it works like both jnanendriya & karmendriya i.e. it aids in both sensory & motor functions). Thus the concept of Manas as given in our scriptures is very subtle yet having a broad connotation & modern psychology or therapeutics perceives it as 'Mind'& its components-Id, Ego & Super Ego & hence has understood it in a quite narrow sense.

Brief Concept of Mind in Modern Psychology

According to Freud's Psycho-analytic Theory, Manas/Mind is assumed to have two extremities 1) Physical (Brain & Spinal Cord) 2) Consciousness. The Psychic apparatus which is present between these two phenomena has 3 parts that are interrelated with each other-

1) Super-Ego (Idealistic Principle/Moral Principle)
2) Ego (Reality principle/ Adjustment principle) - based on logic & organized.
3) Id (Pleasure Principle/ biological principle) - all inborn characteristics (instincts, reflexes, motivation), devoid of any ideal, disorganized.

This Psychic apparatus works at 3 levels-

1) Conscious level- includes all thoughts etc. that an individual is conscious of.
2) Pre-Conscious - Those memories that can be brought to the conscious level through conscious efforts.
3) Unconscious - The memories are dormant & inactive still greatly influence our overall behaviour. They are brought to the surface of mind by dreams, hypnosis, projective tests & accidentally.

In contrast the concept of 'Sattva-anurupa kaya' & 'Manas Prakruti' concept in Ayurveda shows the extreme importance given to the 'Manas' by our ancient seers. Ayurveda has given due consideration to both physical & mental constitution of the body & has stressed on Mental hygiene for mental fitness.[2]

---

1 Khadinyatma manah kalo dishasca dravyasamgrah// (Charak Samhita Sutra Sthana 1/48) Yatrashritah karmagunah karanam samvayi yata tada dravyam//

2 Shariram hiyapi sattvamanuvidhiyate, sattvam ca shariram// (Charak Samhita Sharir Sthana 4/36)

CONCEPT OF ĀYURVEDIC PSYCHIATRY

In Ayurvedic texts we find description & treatment of psychiatric disorders under one of its branches called-'Bhuta Vidya' (Ayurvedic psychiatry). The basic source of this concept comes from Atharvavediya Atharvangirasa.

The term 'Bhuta' can be interpreted as akin to mental disorders or unknown microorganisms. According to Dr P.V. Sharma all disorders (Idiosyncratic) with unknown cause & etiologies comes under 'Bhuta Vidya' whether physical/mental. Mostly Psychiatric disorders have non-organic nature & are idiosyncratic in origin. Hence this interpretation of the term 'Bhuta' is most probably used in the sense of mental disorders. The description of mental disorders in Ayurvedic texts is not exhaustive & rather scattered but still if we read between the lines we can find a very practical approach of dealing with mental ailments which is lacking in Modern Psychiatry i.e. the consideration of Atma Tattva/Manas.

MENTAL DISORDERS AS MENTIONED IN AYURVEDIC TEXTS[1]

A) Mental disorders with impact on Body (psychosomatic) e.g Shokaj atisara,

B) Physical disorders with impact on Psyche e.g. Apasmāra - Cikitsā Sthāna - 10, Nidāna Sthāna -8 & Indrīya Sthāna 5/22-23.

C) Mental disorders with purely Psychic Origin e.g.

a) Ummada - Cikitsā Sthāna - 9, Nidāna Sthāna - 7 & Indriya Sthāna 5/18-21.

b) Atatvabhinivesa- Cikitsā Sthāna - 10 & Nidāna Sthāna -8.

c) Smriti bhramśa - Śārira Sthāna -1/101.

d) Mad & Murcha - Sūtra Sthāna -24/25-27.

e) Nidranasha - Sūtra Sthāna -21/52-57.

Some references of mental disorders also found in other Ayurvedic texts- Sutika Unmada, Pralapa & Nidranasa (Kasyapa samhita, Khila Sthana-11)

FOUNDING PRINCIPLES OF ĀYURVEDA & YOGA THERAPY

AYURVEDA- ***Tridosa Theory*** -Vata, Pitta, Kapha (Physical doshas) & - Raja & Tamas (Mansika Doshas)

1 References from Charak Samhita (Tripathi, Brahmanand, 1999)

Panchamahabhutas (Ether, fire, air, water, earth)

Atma Tattvas, Hridaya, Manas

YOGA- ***Heya (Dukha/Roga)*** - Adhyatamika (Intraorgasmic), Adhibhautika (Extraorgasmic) & Adhidaivika (celestial/due to previous birth),

Hetu (Cause of Dukha),

Hana (A State without any Dukha),

Hanopaya (Various methods to achieve the Hana)

INDICATION OF NORMALITY IN ĀYURVEDA & YOGA

AYURVEDA - Samdoṣah samāgniśca samadhātu malakriyaḥ/

Prasannātmendriyamanaḥ svastha ityabhidhīyate//

(Suśruta Samhitā Sūtra Sthāna 15/42)

YOGA- Yogaśchittavṛttinirodhaḥ (Pātanjali Yoga Sutras 1/2)

......The perfect psycho physiological & balanced state of mind is indicated here.

INDICATION OF ABNORMALITY IN AYURVEDA & YOGA

AYURVEDA - Vitiation of 'Mansika Doṣa -Rajas & Tamas' leading to the development of Mental disorders. In chronic & long standing cases psyche ultimately affects the 'Śārīrika Doṣa -Vāta, Pitta, Kapha' as well & leads to physical ailments as well.

YOGA- Citta Vrittis.[1]

CAUSE OF ABNORMALITY IN ĀYURVEDA & YOGA

AYURVEDA -Main cause -Prajnaparadha[2]

Rogastu doṣa vaiṣamyam doṣasāmyamārogatā

(Aṣtānga Hridaya 1/20-21)

YOGA- Avidyā asmitā rāga dveṣa abhiniveṣāh kleśah//

(Pātañjala Yoga Sutra 2/3-9).

Basic cause is 'Avidya'[1]

1. Vrittayah Panchatayyah klistaklistah/
Pramana viparyaya vikalpa nidrasmritayah.. (P.Y.S 1/5-11)
2 Prajnaparadha (Cause of mental illness)- Sharira Sthana -1/102-108.

TYPES OF ABNORMALITY IN AYURVEDA & YOGA

AYURVEDA [2]- Mental Disorders are of two types- 'Nija' (Due to Dosha vitiation) & 'Agantuja' (external trauma)

Nija is further divided into three types-

Adibalapravritta (Hereditary),

Janmabalapravritta (Congenital),

Kalabalapravritta (Due to environmental influences-internal/external)

YOGA- Term 'DUKHA' in Yogic text encompasses all types of mental & physical disorders.

PATHOGENESIS OF MENTAL DISORDERS

AYURVEDA - Ayurvedic view of mental fitness is based on 'Triguṇa Concept', i.e. the state of perfect balance of three mental humors (Sattva, Rajas, Tamas). Sattva never gets vitiated. Vitiation of other two leads to disequilibria & results in emotional disintegrity & ultimately leads to development of mental disorders.

Examples-

Viṣādo roga vardhanānāma, ṣokah ṣośanānāma, daurmanasyam avriśyanāma/

(Charaka Samhitā, Sūtra Sthāna 25/40)

Matrayāpyabhyavahritam pathyam ca annam nā jiryati/
Chintā śoka bhaya krodha dukha śayya prajāgarah//

(Charaka Samhitā, Vimāna Sthāna 2/5)

Irśyā bhaya krodha parikṣatena lubdhena rugdainyanipiditena/
Pradveśa yuktena ca sevyamānamannam na samyag pariṇāmam eti//

(Suśruta, Sūtra Sthana 46/501)

---

1 Avidyakshetaram uttreshamaprasuptatanu vicchinodaranama/
Anityashuchi dukhaṇatmasu nitya shuchi sukhatma khyatiravidya// (P.Y.S 2/3-4)

2 Rogastu dosha vaishamyam dosha samyamarogata/
Nijagantuvibhagena tatra roga dvividha smritah//20//
Teshama kaya mano bhedada adhithanamapi dvidha/
Rajastamashca manaso dvau ca doshavudahritau//21// (Ashtanga Hridayam 1/20-21)

Murchā Pralapo vamathuh prasekah sadanam bhramah/

Updravah bhavantyete māranam cāpyajirnatah//

(Suśruta Sūtra Sthāna 46/504)

YOGA- Vyādhi styān samasaya pramādālasyāvirati bhrānti darśanālabdhabhumikatva anavasthitattvāni cittavikśepāste antarāyah/ (P.Y.S 1/30)

Dukha daurmanasya aṅgamejayatva śvāsa praśvāsa vikṣepa sahbhuvah/ (P.Y.S 1/31)

A Reference from GITA[1] (A treatise based on Bhavana Yoga with its components- Jnana Yoga, Bhakti Yoga & Karma Yoga)

Overindulgence in sensual pleasures → Attachment → Desires → Anger
(Abnormal Psyche) (First sign of Psychic Derangement)

↓

Complete disintegration ← Insanity ← Psychotic features ← Neurotic features

MANAGEMENT OF MENTAL DISORDERS IN ĀYURVEDA & YOGA

AYURVEDIC MANAGEMENT[2]

Spiritual Therapy (daiva vyapāśraya)[3]	Therapy Based On Reasoning (yukti vyapāśraya)[4]	Psychic Therapy (sattvāvajaya)[5]

Apart from this strictly following Ayurvedic lifestyle (Sadvritta pālana) ensures max healthy life span- Jivem śārada śatam

1. dhyayato vishyana purusah sangah teshu upajayate/
sangat sanjayate kamah kamat krodho abhijayate//
krodhat bhavati sammoha sammohat smriti vibhramah/
smriti bhramat budhinasho budhinashat pranaśyati// (62-63)
2. Trividhama aushadhama iti- daivavyapashrayam yuktivyapashrayam Sattvavajayashcheti / (Charaka Samhita Sutra Sthana 11/54)
3. Daivavyapashrayam-mantraaushadhi manimangala-bali uphara homa niyama prayashchitttopavasa svasthyayan pranipata gamanadi/ (Charaka Samhita Sutra Sthana 11/54)
4. Yuktivyapashrayam -punarahara aushadha dravyanama yojana/ (Charaka Samhita Sutra Sthana 11/54)
5. Sattvajayah punarhitebhyoh arthebhyomanonigraha/(Charaka Samhita Sutra Sthana 11/54)
Manaso jnanavijnana dhairya smriti samadhibhi (Charaka Samhita Sutra Sthana 1/58)
Manasanama tu ashabdadrishto vargah sukhavaha/ (Sushruta Samhita Sutra Sthana 1/35)

Astānga Hrdayam (Sūtra Sthāna 1/26) mentions- Dhi dhairyātmādi vijnānam manodoṣa aushadham param//

YOGIC MANAGEMENT

- Heyam Dukham Anāgatam[1] (Prevention)
- Aṣtānga Yoga[2] (Control & Rehabilitation)
- Kriyā Yoga[3]

Cause of mental fitness is 'sattva guna'. To develop 'Sattvik' temperament & maintaining normal psyche (Prasannātmendriya manah) several methods are mentioned in P.Y.S apart from more easy methods as mentioned above. These are As follows-

A) **Concept of Pratipakṣa bhāvana**[4](P.Y.S 2/33-34)

B) **Concept of Citta Prasādana**[5] (P.Y.S-1/39). It would be interesting to mention here that in Patanjala Yoga Sutras the heart is considered the abode of citta- Hridaye cittasamvita (P.Y.S-3/34) that means through samyama on the heart (arises) the knowledge of the citta. In Ayurvedic texts also

---

1. Kleshamula karmashayo drishtadrishta janma vednïyah/
   Sati mule tada vipako jatycyurbhogah/
   Te hrilada paritapaphalah punyapunyahetutvata/
   Parinama tapa samskara dukhagyanananta phala agunavrittivirodhata ca dukhameva sarva vivekinah/
   Heyam dukham anagatam /(P.Y.S. 2/12-14)
2. Yoganganushnhanadashudhikshaye jnanadiptïravivekakhyateh/
   Yama niyama asana pranayama pratyahara dharana dhyana samadhyoashtavangani/
   (P.Y.S 2/17-18)
3. Tapah svadhyayeshvara pranidhanani kriyayogah/
   Samadhi bhavnarthah klesha tanu karnarthashca/ (P.Y.S. 2/1-2)
4. Vitarka badhane pratipaksha bhavanam/
   Vitarka himsadayah krita karita anumodita lobha krodha moha purvaka mridumadhyadhimatra dukha ajnana anantaphala iti pratipakshabhavanam// (P.Y.S. 2/33-34)
5. Maitri karuna muditopekshanama sukhadukha punyapunya vishayanama bhavnatashcitta prasadanam/33/
   Prachardana vidharnabhyama va pranasya/34/
   Vishayvati va pravrittirutpanna manasah sthiti nibandhani/35/
   Vishoka va jyotishmati/36/
   Vitaragavishayam va chittam/37/
   Svapna nidra jnanalambanam va/38/
   Yathabhimata dhyanatva/39/

frequently the word 'Hridaya' is used as a synonym of 'Manas/citta'. The concept of 'manas' in relation to 'citta' is still controversial. According to modern understanding the region of functioning of citta-mind in the body would be head or more precisely the brain & the nervous system linked with it, provided it is assumed & accepted that a region is there in the body, where the citta functions.

C) Concept of Iśvara praṇidhāna[1] (P.Y.S-1/13-19)..

INTEGRATED APPROACH OF AYURVEDA & YOGA

In Āyurvedic texts we find references regarding concept of yoga, powers of yogis etc in Śārir Sthāna-1/138-141 &

Yoga is a powerful medium to achieve the concept of perfect health as given in Suśruta Samhitā i.e.

Samadoṣah samagniścha............ svastha iti abhidhïyate// (Suśruta Sûtra Sthāna 15/42)

In Yogic texts also we find references of Āyurveda-

Vaidya Śāstrokta vidhinā Kriyām Kurveet Yatnatah/

Kuryātadyogacikitsām ca sarvarogeśu rogvit// (Haṭha Pradîpikā 5/22)

Yogic texts clearly mention that Yoga therapy is only applicable to those disorders that are caused by wrong practices of yoga[2], they also mention that in case of all other types of disorders, Ayurvedic approach to treatment can be adopted along with judicious application of yoga practices to handle the crisis (H.P 5/22).

Another link between yoga & Ayurveda is indicated by the following verse with reference to Patanjali Muṇi-

Yogen cittasya paden vāchām malam śarirasya ca vaidyakena/

Yo apākarottam pravaram munināṃ pātanjali prānjalirānatoasmi//

---

1. Ishvara pranidhanatva/13/
   Kleshakarmavipakashair aparamrishtah purusha vishesha ishvarah/14/
   Tatra niratiśayam sarvajna beejam/15/
   Purveshamapi guruh kalenanavachedata/16/
   Tasya vachaka pranavah/17/
   Tata japah tadartha bhavanama/18/
   Tatah pratyakchetana adhigamo api antaraya abhavashca/19/
2. Pramadi yujyate yastu vatadistasyajayate...(H.P 5/1)

Thus we may assume that āyurvedic & yogic principles are essentially similar in nature & application of both these traditional sciences can be instrumental in managing most of the day to day problems encountered these days.

STRENGTHS OF ĀYURVEDA & YOGA AS A SCIENCE OF HEALING MENTAL DISORDERS- A DISCUSSION

STRENGTHS OF ĀYURVEDA AS A SCIENCE OF HEALING MENTAL DISORDERS

1) Ayurveda is a science of perfect living in tune with the nature. It lays emphasis on treating the person & not just curing the disease.

2) Prevention is the best remedy for any disorder. For this Āyurvedic approach to disciplined lifestyle i.e. strict mental discipline & `āchāra rasāyan' i.e. strict adherence to moral values & proper socio-behavioural conduct can be very effective. Even if someone develops symptoms of any mental disorder it can be effectively taken care of by salubrious use of Auśadha (mental tonics), Ahāra (sātvik diet) & vihāra (spiritual practices). Ayurveda says body & mind are both abodes of disease likewise of well- being; morbidity of body is remedied by medication & of mind by spiritual knowledge, philosophy, fortitude, remembrance & concentration.

3) There are certain findings where Āyurvedic approach has shown positive results in the management of psychiatric disorders. For example, animal studies & clinical studies by Singh et al, Appa Rao et al & Kupu Rajan et al between 1973 to 1980 on āyurvedic mental tonics (Medhya Rasāyan) for assessing their psychotropic effects as also their intelligence enhancing property were done (Abhang, 1987). Drugs chosen for research were-*Hydrocotyl asiatica* (Maṇḍūkparṇi), *Convolvulus pluricaulis* (śankhapuśpi), *Withania somnifra* (Aṣ wagandhā), *Bacopa monnieri* (Brāhmi), *Tinospora cordifolia* (Guduchi) & *Glycyrrhiza glabra* (Yaṣtimadhu). These drugs were found to have tranquillizing & antidepressant action, barbiturate hypnosis potentiation effect.

4) Same drugs were found efficacious in quantitative reduction of anxiety, maladjustment & disability leading to improved mental functions & also reduction in level of urinary vanillylmandelic acid (VMA)[1] & corticoids were noted in clinical studies. Their medhya effect was found to be a secondary effect to anti-anxiety effect. *Hydrocotyl asiatica* was found to have positive effect on intelligence & general adaptive behaviour in mentally retarded children.

---

1 VMA (3-methoxy-4-hydroxymandelic acid)- Approximately 90% of catecholamines epinephrine &
norepinephrine are metabolized to VMA & are secreted in the urine.

5) Out of all medhya rasāyanas only Maṇḍūkparṇi (Hydrocotyl *asiatica*) & Mālakāngani (*Celastrus panniculatus*) have been tested to see their effects on human intelligence even in the absence of anxiety (Abhang et al).

STRENGTHS OF YOGA AS A SCIENCE OF HEALING MENTAL DISORDERS

1) Yoga talks of man as a part of total cosmos & it tries to establish the relation between microcosm & macrocosm. It is a normative science that speaks of virtuous thoughts & actions.

2) It is applied in nature & strives to understand, predict & control human behaviour.

3) It is essentially experiential & experimental in nature. In fact it is a system of mento-spiritual training for realizing the true nature & the creative potential of transcendental consciousness (eternal self) & its freedom from the yoke of conscious & unconscious impulsions, reactivity's & conditionings of phenomenal (empirical self) consciousness (Bhogal, 1995).

4) Patanjali's concept of Antarāyas (Acquired mental afflictions. So called as they develop in various developmental stages of yoga sādhanā), Citta Vikṣepa (Fickleness of mind), Kleśās (Inborn mental afflictions) & Vikśepa Sahabhuvah (Incidental afflictions of mind & body) may be considered akin to mental disorders though they have been mentioned as an obstacle in the path of self-realization for a full time yogi.

5) Various yoga techniques are an outcome of systematic observation/ experimentation/ documentation of results of experiments by a long line of sages.

6) The yogic concept of 'Mokṣa' or self-realization in the context of therapy is akin to attaining highest psycho physiological state of consciousness.

7) Yoga works toward better personal & interpersonal adjustments.

8) It strives to improve upon the life & living of human individuals.

9) It is concerned with Cognitive (thought), Conative (action) & Affective (feeling) aspects of human behaviour & even transcending these as one proceed with higher yoga practices.

10) The norms of behaviour i.e. the Yamas (codes of social conduct) & Niyamas (Codes of personal conduct) do not undergo any change. They are independent of the demands of the given time.

11) Scientific researches in yoga have indicated a better physical & emotional stability in its practitioners. The hostility, anxiety, neuroticism & various personality traits have been measured using psychological tools/experiments & a total personality development has been evident in many such attempts.

12) Above all yoga gives purposefulness in life. The greatest antidote for any disorder is happiness, which is provided by yoga to all poor & rich alike.

CONCLUSION

In light of the facts as listed above & many more which could not be highlighted in the present paper, we may conclude that both Āyurveda & Yoga although independent disciplines in themselves can actually complement each other in therapy & much better results can be anticipated in the long run. Modern Scientists, Physicians & Researchers has to still learn a lot from the texts of traditional medicine & yoga, but an open & broad outlook is required with a keen urge to benefit the mankind.

BIBLIOGRAPHY

1) Abhang R. Y (June 1991). A study to evaluate the effect of a micro (Sūkṣma) medicine from a Medhya Rasāyan on intelligence of mentally retarded children using psychological & biochemical parameters. J.R.A.S., CCRAS, New Delhi, 13(1-2): 35-47

2) ---------------- (April 1987). Medhya Rasāyan - past, present & future. Deerghāyu International, Vol. 4: 8-12

3) Achal Ayodhya Prasad (1976). Pracheen bharatiya Manovikar Vijnana. Ayurvedic & Tibbia Academy, Lucknow, Uttar Pradesh

4) Bhatnagar Rajendra Prasad (1987). Abhinav Manas Roga Vijnāna - Samanvayatmak adhyan. Surya Prakashan Samsthan, Udaipur, Rajasthan.

5) Digambarji, Swami (1998 Ed.). Hatha Pradipika of Svatmaram, Kaivalyadhama, S.M.Y.M Samiti, Lonavla (Pune), Maharashtra.

6) Gupta Aayam (2003). Yoga Therapy in Psychiatric Disorders. A paper presented in the International Conference Nov. 13-15, 2003 on 'Dimensions of Mental, Spiritual & Physical Health', Yoga Centre, University of Jammu, Jammu & Kashmir.

7) Gupta Aayam (2002). Modern Psychiatry, Yoga & Ayurveda- An Overview & Research Issues. A paper presented in the IVth International conference (Dec. 28-31, 2002) on "Yoga Research & Value Education", Kaivalyadhama, Lonavla, Maharashtra.

8) Gupta S.P. (1976) Psychiatry in Indian Medicine.

9) Jain Rajkumar (1986). Yoga Evam Ayurved. Anekant Sahitya Shodha Samsthana, Itarasi, Madhya Pradesh.

10) Karel Nespor & Singh R.H. (Jan 1996). The Experiences with Ayurvedic Psychotherapy 'Satvāvajaya' in Europe. Ancient Science of Life, Vol. V (3): 154-155.

11) Karembelkar P.V (2002). Pātanjala Yoga Sūtras. Kaivalyadhama S.M.Y.M Samiti, Lonavla (Pune), Maharashtra

12) Kuvalyānanda, Swami & Vinekar S.L (1994). Concept of disease in yoga & principles of yogic treatment. In 'Yogic Therapy Its Basic Principles & Methods.' Central Health Education Bureau (Directorate general of Health Services), Ministry of Health & Family Welfare, Govt. of India, New Delhi. PP: 1-5

13) Pathak Balakrishna Amarji (1983). Manas Roga Vijnana -I. Shri Baidyanath Ayurved Bhavan Limited, Nagpur, Maharashtra.

14) Ramu G (1981). An Approach to Mental Examination Based on Ayurvedic Concepts, CCRAS, New Delhi.

15) Shastry Ambikadutt (2000). Suśruta Samhitā - I. Chaukhamba Samskrit Samsthan, Varanasi.

16) Srikantamurthy K.R. (1996) Clinical Methods in Ayurveda. Chaukhamba Orientalia, Varanasi.

17) Tripathi, Brahmanand (1999). Charak Samhita. Vol-I & II. Chaukhamba Surbharati Prakashan, Varanasi.

18) Upadhyaya Govind Prasad (2000) Ayurvediya Manas Roga Chikitsa. Chaukhamba Surbharati Prakashan, Varanasi.

19) Vahia, N.S; Vinekar, S.L & Doongaji, D.R (1966). Some ancient Indian concepts in treatment of psychiatric disorders. Brit. J. Psychiat, 112: 1089-1096.

20) Yadavji Trikamji (1983). Manasaroga- Vijnana. Baidyanath Publications.

3

Indian Origin of Reiki and Prāṇic Healing

Dr. M.C. Bhartiya

'Reiki' is a Japanese word composed of two words *'Rei'* and *'Ki'*. The word *'Rei'* means, *'Universal'* and *'Ki'* means 'Life force energy' which in Sanskrit is called *prāṇa*. Reiki is thus the 'universal life force energy'. According to another interpretation *'Rei'*, means 'spiritual consciousness' or 'divine consciousness'. Thus 'Reiki' means 'the life force energy emanating from the divine consciousness i.e. God.'

Reiki is a healing therapy where life force energy is imparted to a patient through hands by touching or by keeping hands at a distance of about two-three inches. This energy is capable of healing many types of physical, mental and emotional ailments. The present originator of this therapy is Dr. Mikau Usui, a Japanese Doctor, who flourished in Japan in later half of the nineteenth century and first quarter of the twentieth century (1865 to 1926 A.D.). From Japan this therapy has travelled to America and from America to the other parts of the world including India.

'Pranic healing' is a similar therapy where also the life force energy is imparted through hands. But instead of giving this energy to the physical body, it is given to the 'aura' which is the 'energy body' of the patient. This aura or the 'energy body' works as a sort of causal body for the physical body and any ailment existing in the physical body also exists in the aura in the form of negative energy. This negative energy is first manually removed from the aura and then the gap in the aura is filled with fresh life force energy which heals the ailment. The healing occurring first in the aura is precipitated to the physical body. The present propounder of this therapy is Mr. Choa Kok Sui, who belongs to Philippine. Choa Kok Sui however admits that he has not discovered this science; it is an ancient science.

Till recently, history of Reiki was not very clear and was based on the verbal statements of Mrs. Hawayo Takata, a desciple of Mr. Chujiro Hayashi, who in turn was a desciple of Dr. Usui. According to Mrs. Takata's version, Usui was a Christian and though he knew that touch healing was a Buddhist technique, he could not find a teacher for it and he himself re-discovered it during a twenty-one day meditation and fast at a Mountain named Kurama Yama. But recently Usui's own notes have been discovered and translated into English[1]. According to Usui's own version, he got the technique from a Buddhist Tantra named *Vajra-tantra*. The technique was of course complicated filled with Buddhist religious rituals. Usui simplified the technique, got rid of its religious taint and presented it in a form capable of being practiced by any person without any religious prejudice. Usui called it *Manchos Reiki*. *Manchos* is a Tibeten term meaning *Bhaisajya-dharma* i.e. 'a healing technique', The present technique of Reiki is based on this *Manchos Reiki,* and the present form was given by Usui's disciple Chujro Hayashi and Chujiro Hayashi's desciple Mrs. Hawayo Takata.

The History of *Vajra-tantra,* the original basis of Reiki, goes back to Shakyamuni Buddha, the historical Buddha. Shakyamuni Buddha, during his teaching career of some 50 years, is said to have spoken the sutra of *Bhaisajya - guru* (The Buddhist archetype of healing). The teaching was passed from Shakyamuni Buddha to a physician known as Ratnagarbha. The teaching passed to Ratnagarbha was known as *Vajra-tantra.* Through an unbroken lineage of more than a thousand years, this *tantra* was passed on from teacher to disciple. During this period this *tantra* made a journey from India to Tibet, from Tibet to China and finally to Japan in the seventh country A.D. by a Buddhist Master Kobo Dashi. The teaching continued to be handed down throughout time from teacher to desciple until it reached a Buddhist monastery in Hokkaido, Japan. There it layed untouched for some three hundred years. The uneducated Buddhist monks at Hokkaido, not understanding its value, sold it to a book dealer who took it to Kyoto where Mikao Usui came into its contact and purchased it. He studied the tantra and after much diligence and practice grasped its essential teachings which formed the basis of Dr. Usui's Reiki system.[2] The Buddhist *Vajra-tantra* is thus at the root of the present Reiki Therapy.

Reiki, as we know, is a hands - on healing, where the universal life force energy or *prāṇa* is imparted to the patient through hands. A reference in *Atharva-veda,* where hands are eulogized for healing all types of diseases, shows that the technique in some form was prevalent at that time. The mantra runs thus :

अयं मे हस्तो भगवान् अयं मे भगवत्तरः ।
अयं मे विश्वभेषजोऽयं शिवाभिमर्शनः ।।
हस्ताभ्यां दशशाखाभ्यां जिह्वा वाचः पुरोगवी ।
अनामयित्नुभ्यां हस्ताभ्यां ताभ्यां त्वाऽभिमृशामसि ।।[3]

"This my (one) hand is glorious, this my (another) hand is even more glorious. This my (hand) is the healer of all (ailments); it is the bestower of health through touch. The tongue precedes the voice.[4] With these two hands, having ten branches (i.e. fingers), removers of all ailments, I touch thee."

It was sometimes thought that these mantras refer to hypnotism[5], but one who is acquainted with hands - on healing like Reiki or *Pranic*, can clearly see this healing in them. This Atharvavedic hands-on healing refers to the healing through *prāṇa* i.e. life force energy, is also confirmed by a fore running mantra of this very hymn:

आ वात वाहि भेषजं वि वात वाहि यद्रपः ।

त्वं हि विश्वभेषज देवानां दूत ईयसे[6] ।।

"O wind (i.e. *prāṇa*) ! bring forth the healing and remove every ailment. You are indeed the healer of all (ailments) and come as an envoy of gods."

It is to be noted that the Sanskrit word *vāta* stands for both air and life force energy.

Upanisads are filled with various statements eulogizing the importance of *prāṇa*. *Chhandogya, Kausitaki* and *Prasnopanisad* refer to a story which establishes the superiority of *prāṇa* over all other sense organs and motor organs like speech (*vāc*), perception (*cakṣu*), hearing (*śrotra*) and mind (*manas)* etc. All the senses started quarrelling among themselves on the subject of relative superiority. It was later decided to let every sense leave the body one by one and see whether the body is sustained without it. Each one of the organs, except *prāṇa,* left the body one by one. There was of course some weakness in the body, but the body continued and did not fall. But as soon as the *prāṇa* started leaving the body, all the senses felt as if they all became lifeless. All the senses thus accepted the superiority of *prāṇa. Kausitaki upanisad* says that a being lives only till *prāṇa* stays in the body :

यावत् यस्मिन् शरीरे प्राणो वसति तावदायुः ।[8]

We here give some quotations from upanisads stressing the extraordinary importance of *prāṇa.*

स ईक्षांचक्रे। कस्मिन्नहमुत्क्रान्त उत्क्रान्तो भविष्यामि कस्मिन्वा प्रतिष्ठिते प्रतिष्ठास्यामि।

स प्राणानसृजत । (*Prasnopanisad*, VI, 3-4)

At the time of creation God deliberated. 'As a result of whose departure ṣhall I (also) depart and as a result of whose continuance shall I continue ? (And) He created *prāṇa*.

प्राणं देवा अनुप्राणन्ति मनुष्याः पशवाश्च ये। प्राणो हि भूतानामायुः। तस्मात्सर्वायुषमुच्यते।

(*Taittiriya up*, iii, ii, 1)

"All gods, human beings and animals act following *prāṇa.* Since the life of all beings depend on *prāṇa,* it is therefore called the life of all.'

प्राणो ब्रह्मेति व्यजानात्। प्राणाद्ध्येव खल्विमानि भूतानि जायन्ते।
प्राणेन जातानि जीवन्ति। प्राणं प्रयन्त्यभसंविशन्तीति।

(*Taittiriya up*., III, iii, 1)

He (Bhrgu) came to know Brahman as *prāṇa,* because it is *prāṇa* whence all beings are born and after being born live through *prāṇa*. (After death) they move towards and enter into *prāṇa*.

यो ह वै ज्येष्ठं च श्रेष्ठं च वेद ज्येष्ठश्च ह वै श्रेष्ठश्च भवति। प्राणो वाव ज्येष्ठश्च श्रेष्ठश्च।

(*Chadogya up*., v, i, ii, *Brhadaranyaka* up. vi, ii, 1)

'One who indeed knows the greatest and the best becomes greatest and best. *Prāṇa* is indeed the greatest and the best.'

'All the above quotations show that Vedic seers knew well the importance and superiority of *prāṇa* in the body. The quotation from Atharva-veda shows that they knew how to impart *prāna* though hands for the purpose of healing and as we have seen above, *Vajra-tantra* has been the source of Reiki. This all goes to establish that Reiki and Praṇic healing are of Indian origin.

Notes :

1. Lama Yeshe : *Medicine Dharma Reiki,* Full Circle, Delhi, 2001.
2. Lawrence Ellyard : *The Tao of Reiki* (Full circle, Delhi 2001), pp 22-23
3. *Atharva - Veda*, iv, 13, 6-7
4. This probably refers to the chanting of some mantra with hands-on-healing.
5. See Hindi commentary on these mantras by Pt. Sripad Damodar Satavalekar.
6. *Atharva - Veda*, IV, 13, 3
7. *Chhandogya,* 5,1, *Kausitaki,* 2,14
   Prasnopanisad, 2,1-4
8. *Kausitaki up*., 1,2.

4

Tantra as Science of Mental Healing

Dr. H.S. Sinha

Tantra psychology has been recognised now by the Western Psychologist as an effective therapy of curing mental ailments viz., anguish, anxieties, tension and stress. Despite tremendous advancement in the field of empirical sciences, man has not been able to find any solution of his mental agonies and psychic troubles like the fear of the unknown. He still carries an immeasurable load of ambiguities and imponderables. A sense of insecurity is hauting human psyche. The increasing incidence of hypertension and stress are the manifestation of some inner psychological trauma.

The intricate mechanism of mental disorders has puzzled the human mind from the very beginning. This has been a problem since the dawn of human history, as to how one can find a release from his inner torture and where one can seek solace against the vagaries of the unknown forces which are highly inaccessible but create trouble from within the form of dreadful dreams, fear of the unknown. Previously, however, the phenomenon of hidden distress was attributed solely to mysterious forces. The reason was that whatever data the introspection could furnish was only of the mento-emotional nature the '*animum fluctuation*' and the resultant vibrations in the mind. The mysterious forces revealed themselves in the unconscious processes i.e. the process of concealment, disguise and Archetypal symbols like *'dark Mother and Father symbols'*, called Śiva and Śakti according to ancient healing discipline of Tantra. Due to the misunderstanding of the pithy symbolism of 'Phallus Serpentine (kuṇḍalinī), cakra and maṇḍalas, the Ṭantra were branded as Black Magic. But after the interpretation given by Sir John Woodroffe and C.G. Jung to these symbols, it has been realized that these symbols are pregnant with deep meanings and if they are interpreted in a medico-psychological frame of reference, this hence-to-forth neglected science of healing can be effectively developed into a useful branch of psycho-pathology and of modern therapeutic disciplines, within the area of medical sciences. We can rather say that the Tantra contains the therapeutic techniques used from the earliest known history of mankind for curing the neurotic disorders and mental ailments of neurasthenic as well as of Psychasthenic types.

Interestingly, the modern man still exploits the religious systems like yoga and technique of meditation developed thousands of years ago to seek solace and mental satisfaction. West has evolved no treatment of such anxieties except prescribing tranquilisers which contain opiates and opium derivatives. The religious idea of tranquility was based on the starting premises that the restlessness which every individual feels under the pressure of his psycho-social environment is mainly due to the waves of anxiety (cintā-taraṅgacapala)[1] interfering with the amorphous peace of mind. The goal of tantric training was also therapeutic i.e. to extricate man from the trammels of internal conflict, mental tension and agony and therapy to get him freed from all sorts of neurogenic stresses. Tantra psychology provides a way to liberate a sufferer patient from the vagaries of his own morbid dispositions and tendencies (Saṁskāras). A reflection on Tantrāloka of Abhinavagupta, Śrīmṛgendra Tantra and Kamikāgama shows that they have conceived the human personality as 'Coincidentia oppositorum' (i.e. the transformation of the opposites) a higher synthesis of the unconscious and conscious. Surprisingly enough C.G. Jung, an eminent psychologist of 20th century has also used the same symbolism to represent them. The unconscious is represented by the infra-personal female (Śakti in tantra, is Anima Complex in Jungian psychology) and conscious is represented by Suprapersonal masculine Śiva which is the 'animus' of Jung. In this synthesis, masculine principle acts as psychopompos and by guiding and accompanying the movements and transformations of the dark-complexioned Śakti (anima-in-shadow of Jungian Psychology) it introduces a rhythmic order into the latent instinctual chaos. The resultant harmonizing process is the blissful dance of Śiva with his consort Kālī. This is again very akin to the concept of individuation, used by Jung. The feminine emotional aspect (the dark mother) co-ordinates actively with the conscious and the process of individuation takes place by the communion of Śiva and Śakti in their secret mutual enjoyment (*annyonya viharana of Śiva Śakti*).[2]

The fundamental postulates of Tantra are also more or less identical to those of the post-Freudian psycho-therapy. According to both these systems, the mental disturbances are primarily of emotional character and, therefore, irrational and eccentric. They require a principle of rationality for their amelioration. In psychology Freud called it 'nirvana principle' because the unification of personality brings a positive relief from the intense neurotic anxiety and from deranged excitability.

---

1. Kamakalavitasa, (verse 55) commentary by Natanandana-natha (tran) by Arthur Avlon.
2. Ibid., verse 2.

In the light of this, Tantra psychology is a scientific search for the intra-psychic determinants of abnormal behaviour and a deep penetration into the dark recesses of human personality, which are still inaccessible to behavioural sciences. Tantra like all other branches of yoga lays stress on the introspective analysis of self which was misunderstood as a religious practice by the Westerners. But the intensive researches into the pathology of civilized communities have made it imperative on them to admit that after a long period of outward expansion men should look within themselves once more.[1] That is why Jung declared that he had no objection if psycho-therapy belonged to the field or religion.

The act of ceremonial circumambulation i.e. dancing round in a circle to seek the communion of Śiva and Śakti for the birth of self (Individuation), which was once considered to be a purely a Tāntric ritual, has now been adopted as a highly potent and widely used therapeutic symbol for all those, whose minds are disturbed by countless Pathogenic conditions. Jung has found many incontestable evidences to show that individuation, on the synthesis of contra-sexual polarities, is realizable through the practice of circumambulation. Thus the Indian ancient technique of Tantra surpasses modern psychiatry in dealing with certain psychiatric disorders. The practice of circumambulation, i.e. walking or dancing around the Śiva has succeeded in treating psychic disorders like anxiety, neurosis, tension (mental as well as muscular), phobias and even hysteria. The early symptoms of these mental ailments appear in the form of 'anhedonia' (absence of joy and pleasure), Rumination i.e. exaggerated guilt feelings about some past mistakes, chronic fatigue, depressed sexual activity, hallucinations (specially dreadful).

Dancing around the symbol of Śiva in Tantra, which is the symbol of perfect calm and tranquility even when there are strong waves of churning ocean, was recognized as ideal therapy for the treatment of emotional and psychological disorders. The patient was inspired or aroused to dance on the sound of a percussion instrument called 'ḍamaru' or drumlet (Mṛḍangam). Guru speaks some Mantras which sustain the rhythmic movements of the patient. This practice is continued till patient gains rhythm in his movements, which restores the harmony that had been disturbed. Now the same technique is being used by the Jungians like Coster, by replacing the Śiva-liñga (the unmoved psychopompose) by Guru and by Doctor. The Guru or Doctor sits in the middle and patient takes dancing steps around him. Dance may just be the ideal therapy for the treatment of emotional and psychological disorders faced by most people in the fast paced life of today.

---

1. Jung C.G. Modern Man in Search of Soul, p. vii.

After months of experimenting, researchers at the University of Bonn have found that just as primitive man expressed his emotions, needs, worries and aspirations, through dance, the civilized man also needs precisely this outlet if he is to develop a sound personality devoid of neurosis.

"Most people in the modern world turn their emotions inward, thus bottling up their feelings or reactions, "says Mrs. Gorebel of Bonn University who has developed a new therapy which makes her patients leap, kick, hop, clap, swing and move their hands rhythmically.

The basic idea is to make them dance away their aggressions and suppressed emotions. The wide and brisk movement of whole range of isolated muscles produces such intensive concentration that the dancer is compelled to come out of his shell.

To bring out a few characteristic similarities between the postulates of Tantra and those of modern psycho-therapy, it may first of all be pointed out that Tantra conceives the human personality as the psychic totality of two antithetical beings symbolized as the massing together of Śiva (Śiva-Śakti mithunapiṇḍa).[1] It is a circumscribed functional complex of conscious (prakāśa) and unconscious (vimarśa). In the final stage of individuation the two opposites, which appear independent of each other in a patient, push forward from the opposite directions to form a whole. This is symbolized in the corresponding picture depicting a 'conjuntis' (Devī Śivonmukhi)[2] between two contra sexual principles, and the Tantra prescribes a practical code for the reunification of these two. Similarly, the modern therapy has come to realize that in the psyche of modern man, the conscious side has been over-emphasized and this is the cause, why unconscious threatens to burst so often. As a remedial step, a need is always felt to integrate the unconscious. For example, Jacobi advocates[3] that the wholeness of personality is achieved when the main pairs of opposites are linked together in a living relation'. Similar is the pronouncement in the opening verse of Saundarya Lahari that is only when Śiva is united with Śakti that he is able to exert his power as Lord. In Kāilasa Saṁhitā, Vimarsa is prefixed by 'Antaralina' connoting inwardness. It means that Śakti is the latent inner shadow. Śakti is symbolically female which symbolizes the latent emotionally and catholic power. More precisely Śiva stands for agile thought pattern and Śakti for exuberant eroticism.

---

1. Satcakranirupana verse, 49.
2. Ibid.
3. Jacobic Jalande 'Psychology of Jung', p. 102.

Through this technique the anima, elevated from unconscious into conscious, loses its tremendous load of libido. It becomes depotentiated and loses the power of possession. The conflicting elements of conscious and unconscious come to harmony in the self. This integration of personality through the emergence of unified Śiva and Śakti is found in the art and religion of several people, especially of the Far East. It is represented by the Mandala symbol, in the form of a square or a circle, with a central point. Maṇḍala, according to Jung represents the reconciliation of opposites, the fusion of the conscious and the unconscious. In describing the characteristics of the primordial Śakti, almost all Āgamas refer to Śakti as Kamakala Vilāsa (the primal sexuality) having its three aspects in which the gross (sthūla) or the physiological aspect is also included.[1] Since Kala is that which moves anybody (prasaranam preranam sa kurvati tam sa kala),[2] Devi is the primary motor force, the creative erotic energy, from which all psychological faculties are activated. This Saundarya Lahari, is the archaic on which personality is built. In Jungian language, it is the foundation an the regulator of all psychic existence. The use of word Kamakala for Devi, also signifies that primary pleasure function (ānanda) is akin to the sexual or libidinal factors prefigured in the psychological structure of man as described by Freud. The latent libidinal factors remain unconscious and that is why tantra has symbolized them as 'dark' (Kāli).

It can, thus, be concluded that the psychiatric postulates of Tantra psychology and its symbolism, if correctly interpreted in its therapeutic context, can provide basic postulates for all leading theoretical schools of personality research which investigate the dynamics of mental life.

* * *

1. Arthur avlon op.cit., p.1.
2. Dass Gupta, S.N. History of Indian Philosophy, Vol. 23.

5

संस्कृत-शास्त्रों में आयुर्विज्ञान

डॉ. वसन्त कुमार

संस्कृत-शास्त्र परम्परा में आयुर्विज्ञान के विषय में विस्तार से चिन्तन हुआ है। वेदों में आयुर्विज्ञान का संक्षिप्त विवरण मिलता है। आयुर्विज्ञान के विभिन्न पक्षों का जो बीजन्यास वेदों में हुआ था, अर्वाचीन शास्त्रों में उसी का वृक्ष रूप दृष्टिगोचर होता है। प्रायः सभी प्रकार के शास्त्रों में आयुर्विज्ञान का न्यूनाधिक उल्लेख मिलता है। लेकिन कुछ शास्त्रों में तो केवल आयुर्विज्ञान विषय का ही चिन्तन हुआ है। इस प्रकार के विस्तृत चिन्तन ने मानवता का बड़ा उपकार किया। जीवन में यदि मानव को इस विषय का ज्ञान नहीं होता तो वह अपने जीवन के लक्ष्य को प्राप्त नहीं कर सकता था। लक्ष्य तो दूर, निरापद एवं सुखी जीवन की भी कल्पना नहीं कर सकता था। प्राचीन मनीषियों ने ज्ञान-विज्ञान कला-कौशल के क्षेत्र को जो अमूल्य योगदान किया, वह आयुर्विज्ञान पर ही आश्रित था। यदि हमें अपनी आयु, शरीर, इन्द्रियाँ आदि का ज्ञान नहीं होगा तो हम कुछ भी नहीं कर सकते। वेद में इस बात का स्पष्ट उल्लेख है कि हमारा जीवन कैसा हो? जीवन का लक्ष्य क्या हो?–

पश्येम शरदः शतं जीवेम शरदः शतं शृणुयाम शरदः शतं प्रब्रबाम
शरदः शतमदीनाः स्याम शरदः शतं भूयश्च शरदः शतात्॥[१]

मात्र प्राण धारण, भोजन, शयन, जागरण ही जीवन नहीं है। अदीन होकर अपनी इन्द्रियों को स्वस्थ रखें और उनसे उत्तमोत्तम कल्याणकारी कार्य सम्पादित करते रहना ही जीवन है। धर्मार्थकाममोक्ष पुरूषार्थ चतुष्टय की सिद्धि मानव जीवन का चरम लक्ष्य है। लेकिन यह भी आयुर्विज्ञान के बिना सम्भव नहीं। क्योंकि आरोग्य ही जीवन का मूल-मन्त्र है। चरक मुनि कहते हैं–

धर्मार्थकाममोक्षाणामारोग्यं मूलमुत्तमम्। च. सं. १.१५

आयुर्विज्ञान का प्रयोजन है-

स्वस्थस्य स्वास्थ्यरक्षणम्,
आतुरस्य च बिकारप्रशमनम्। च. सं. ३०.२६

---

1. यजु. 36.24 2. च. सं. सूत्र. 1.41

संस्कृत-शास्त्रों में आयुर्विज्ञान के लिए आयुर्वेद शब्द का प्रयोग बाहुल्य से हुआ है। आयुर्विज्ञान अथवा आयुर्वेद की क्या परिभाषा है? इसका उत्तर इसी विद्या के आचार्यों ने स्वयं दिया है। जैसे कि चरकमुनि लिखते हैं-

हिताहितं सुखं दुःखमायुस्तस्य हिताहितम्।

मानं च तच्च यत्रोक्तमायुर्वेदः स उच्यते।।[१]

एक अन्य आचार्य सुश्रुत भी आयुर्वेद की परिभाषा करते हैं-

आयुरास्मिन् विद्यतेऽनेन वा

आयुर्विन्दन्ति इत्यायुर्वेदः।[२]

आयुर्वेद का पठन-पाठन, श्रवण, मनन, आचरण ही मानव को आयु हितायु और सुखायु दे सकता है। यह हितायु-सुखायु की प्राप्ति ही धर्म-अर्थ आदि का साधन है। ऐसा कथन आयुर्वेद के आचार्य वाग्भट का है।[3]

आयुर्विज्ञान को इस महत्ता को अंगीकार करके ही संस्कृत के आचार्यों ने इस पर पर्याप्त प्रकाश डाला। अब हम प्रकृत विषय पर आगे बढ़ते हुए विचार करेंगे कि हमारे शास्त्रों में आयुर्विज्ञान के किन-किन पक्षों पर बल दिया है। आज की पाश्चात्य चिकित्सा-पद्धति से प्राचीन पद्धति भिन्न थी या समान? आयुर्विज्ञान में चिकित्सा का क्या स्थान है? आयुर्विज्ञान तथा चिकित्सा पर्याय है या इनमें कोई भेद है? इन बिन्दुओं पर विचार करने के बाद विभिन्न शास्त्रों में वर्णित आयुर्विज्ञान के विभिन्न पक्षों- आहार-विहार, ऋतुचर्या, काय एवं शल्य-चिकित्सा पर विचार करेंगे।

भारतीय-आयुर्विज्ञान श्रेष्ठ : अनेक अवसरों पर प्रायः यह सुनने को मिलता है कि पाश्चात्य-चिकित्सा पद्धति उत्तम है। इसमें प्रायः सभी रोगों को तत्काल प्रशमन करने की सामर्थ्य है। जबकि आयुर्वेद में ऐसी सामर्थ्य नहीं है। इसी कारण पाश्चात्य-चिकित्सा (एलोपैथी) अधिक प्रचलित है। इस विषय में मेरा विचार है कि आज आयुर्वेद का पूर्णरूप सबके सामने नहीं है तथा उसके विकास प्रचार-प्रसार एवं शोध के लिए उतना प्रयास भी नहीं हो रहा। शल्य चिकित्सा के क्षेत्र में इसकी श्रेष्ठता अधिक दृष्टिगोचर होती है। जब आयुर्वेद में आचार्य सुश्रुत

---

1. सु. सूत्र. 1.15
2. आयुः कामयमानेन धर्मार्थ सुखसाधनम्।
   आयुर्वेदोपदेशेषु विधेयः परमादरः।। अष्टा. सू. 1.2
3. छान्दोग्य क: 7.26.2

का शल्य शास्त्र बहुत उत्तम कोटि का है। इसमें अंग प्रत्यारोपण, प्लास्टिक सर्जरी आदि की भी विधियां हैं। लेकिन मूल-अन्तर पर रखा है। हमारे यहाँ वेद से लेकर अर्वाचीन आचार्यों तक ने इस तथ्य पर अधिक बल दिया है कि **हम कभी रोगी ही न हों।** हमारा आहार-विहार ऐसा होना चाहिए कि शरीर में विजातीय द्रव्य सञ्चित ही न हो सकें। हमारे जीवन में चिकित्सा का स्थान ही नहीं रहा। यदि किसी कारणवश मनुष्य रोगी हो भी जाता था, तो वह उस शरीर को आत्मा के अयोग्य मानकर शरीर त्याग कर देता था। ऐसे अनेक उदाहरण मिलते हैं। मनुष्य रोगी होना अशुभ माना जाता था। रोगी होने का अशुभ विचार मन में आने ही नहीं देते। उसकी चिकित्सा की बात तो बहुत दूर है। जिस औषध-विज्ञान एवं शल्य-विज्ञान के आज गुण गाए जा रहे हैं, उसकी हमारे पूर्वजों को आवश्यकता ही नहीं पड़ती थी। उनके जीवन में औषध का स्थान ही नहीं था। चिकित्सालयों की तो बात ही क्या है। अरबों रूपये के व्यय से बने आज के चिकित्सालय तो उनके लिए श्मशान तुल्य हैं। आज भी ऐसे उत्तम आहार-विहार का पालन करने वाले मनुष्य हैं, जो इन चिकित्सालयों की तरफ मुख भी नहीं करते। उन्होंने जीवन में कभी औषध का सेवन ही नहीं किया। उनकी जीवन-शैली भारतीय आयुर्विज्ञान पर आधारित है। प्राचीनकाल में प्रायः सभी मनुष्य इस का पालन करते थे। शनैः शनैः इसका ह्रास हुआ और हम रोगी होने लगे। जिसके परिणामस्वरूप औषध एवं चिकित्सालयों की आवश्यकता पड़ी। अब हम निरामय-संसार के आधार-भूत तत्त्वों पर विचार करेंगे, जो कि भारतीय आयुर्विज्ञान की विश्व को अमूल्य देन है।

आहार-विहार : सात्विक भोजन, दिनचर्या, ऋतुचर्या, आदि आयु के आधारभूत तत्त्व हैं। हमारे प्रत्येक क्रिया-कलाप का अन्तर्भाव आहार-विहार में ही हो जाता है। इन्हीं को हमारे जीवन का मूल आधार माना है। आहार के विषय में उपनिषद् वाक्य है-

आहार शुद्धौ सत्त्वशुद्धिः सत्त्वशुद्धौ ध्रुवा स्मृतिः।[१]

आहार की शुद्धता तथा सात्विकता से मन शुद्ध होता है। मन ही हमारे शारीरिक स्वास्थ्य की आधार शिला है। इसके बिना स्वस्थ शरीर की कल्पना भी असम्भव है। क्योंकि मन ही शरीर की प्रमुख संचालिका शक्ति है। मन से बुद्धि प्रभावित होती है, जिसके विकृत होने से मनुष्य पतित होकर नष्ट हो जाता है। महर्षि चरक ने भी कहा है-

प्रज्ञापराधो हि मूलं रोगाणाम्।

इस प्रकार हमारा जीवन उत्तम आहार पर ही आश्रित है। क्योंकि मानवशरीर के लिए जो

1. मात्राशीस्यात्। आहारभात्ता पुनरग्नि बला पेक्षिसणी।। 7.26.2

भोज्य पदार्थ हैं, वे ही उसकी औषध हैं और जो औषध है वही भोज्य है। सुश्रुत कहते हैं-

अन्नमूलं बलं पुसां बलमूलं हि जीवनम्।

हमारा आहार हितकारक तथा अल्पमात्र में हो, साथ ही नियतकाल पर ही सेवन किया जाना चाहिए- स्वास्थ्य सूत्र भी है-

'हिताशी स्यान्मिताशी स्यात् कालमोजी जितेन्द्रियः।'

महर्षि चरक ने सूत्र स्थान के पञ्चम अध्याय में इस सिद्धान्त को पुष्ट करते हुए मिताहार पर बहुत बल दिया है। इसी विषय में जैमिनि-दर्शन में एक जनश्रुति बहुत प्रचलित है। एक बार महर्षि जैमिनि अपने आश्रम बैठे थे। उस समय वृक्ष की शाखा पर बैठा एक पक्षी बोला- **'कोऽरूक्'** कौन मनुष्य निरोग है? उत्तर में महर्षि जैमिनि बोले- **'हितभुक्'**। जो हितकर भोजन करता है। वह पक्षी पुनः बोला **'कोऽरूक्'** जैमिनि ऋषि ने पुनः उत्तर दिया **'मितभुक्'** परिमित आहार करने वाला। वह तीसरी बोला **'कोऽरूक्'** इसका ऋषि ने उत्तर दिया- **'ऋतभुक्'** अपने सत्याचारण का आहार करने वाला कभी रोगी नहीं होता। इस प्रकार आयुर्विज्ञान के आचार्य समाज को अपने उपदेशों से दीर्घायु का रहस्य बताते थे। जिससे मनुष्य शारीरिक एवं मानसिक रूप से स्वस्थ रहकर अपने जीवन के लक्ष्य को प्राप्त कर सके। पथ्य-सेवन रोग निवृत्ति का सबसे उत्तम साधन माना गया है। अतः कहा है-

विनापि भेषजैर्व्याधिः पथ्यादेव निवर्तते।
न तु पथ्यविहीनानां भेषजानां शतैरपि॥

गीता में भी आयुर्विज्ञान की चर्चा करते हुए श्रीकृष्ण अर्जुन को सात्विक आहार के लाभ तथा राजसिक-तामसिक की हानियां बताते हैं :

आयुः सत्वबलारोग्यसुखप्रीतिविवर्धनाः।
रस्याः स्निग्धाः स्थिरा हृद्या आहाराः सात्विकप्रियाः॥

सात्विक भोजन से मनुष्य दीर्घायु, बलवान्, नीरोग, बुद्धिमान् एवं सर्वप्रिय बनता है। राजसिक[1] एवं तामसिक[2] भोजन इसके विपरीत गुण-धर्म वाला होता है।

---

1. कट्वम्ललवणात्युष्णातीक्ष्णरूक्षविदाहिनः।
आहारा राजस्येष्टा दुःखशोकामयप्रदाः॥ गीता 17-9

2. यातयामं गतरसं पूति पर्युषितंचयत्।
उच्छिष्टमपि चामेधयं भोजनं तामसप्रियम्॥ गीता 17-10AA

आहार-सेवन विधि : हमारे यहाँ भोजन खाने के लिए ही तैयार नहीं किया जाता। जब शुद्ध-सात्विक भोजन तैयार हो जाए तो देवता, अतिथि, गो आदि पशु-पक्षियों को पहले उसका सेवन कराया जाता है। उसके पश्चात् शेष भोजन मनुष्य स्वयं सेवन करता रहा है। अपने लिए पकाकर खाया गया भोजन पापमय माना गया है। उसका लाभ शरीर-मन एवं बुद्धि को नहीं होता। अपितु रोगोत्पत्ति कारण ही बनता है। वेद में कहा है-

मोघमन्नं विन्दते अप्रचेता सत्यं ब्रवीमि

वधा इत् स तस्य नार्यमणं पुष्यति नो सखायं

केवलाघो भवति केवलादी।। ऋग् - १०-११७-६।।

मनु महाराज ने मनुस्मृति[1] में तथा श्री कृष्ण जी ने गीता में इसी तथ्य का उल्लेख कर यज्ञ शेष की महत्ता तथा केवल अपने लिए पकाए गये भोजन के दोष बताए हैं। इस प्रकार हमारे यहाँ आयुर्विज्ञान का इतना सूक्ष्म-चिन्तन होता था कि भोजन जिस भावना से पकाया जाता है उसका उतना ही अच्छा-बुरा प्रभाव शरीर पर पड़ता है।

भोजन के विषय में आयुर्वेद के प्रसिद्ध आचार्य चरक मुनि ने भी लिखा है- 'भोजन उष्ण, स्निग्ध तथा उचित मात्रा वाला होना चाहिए। पहले वाले भोजन के पच जाने पर अगला करें। जिसमें खाद्य पदार्थ वीर्य विरूद्ध न हो, उनके गुणधर्म मिलते हों। उत्तम उपकरणों के द्वारा भोजन के योग्य स्वच्छ स्थल पर सामान्य रीति से भोजन करें। न अतिशीघ्र, न अति विलम्ब से। भोजन के समय हास-परिहास कदापि न करें। यहां तक की परस्पर वार्तालाप भी न करें एकाग्र मन से अपनी पाचन-सामर्थ्य के अनुसार भोजन करें। ऐसा गुण-सम्पन्न उक्त रीति से किया गया भोजन आयु बुद्धि करने वाला होता है। रोग-उत्पत्ति का अवसर हीन ही होता।[2] ऐसा व्यक्ति सौ वर्ष तक जीवित रहता है।[3]

आहार-विहार ही शरीर मन तथा बुद्धि का आधार है। ऐसा प्राय: सभी आचार्यों का चिन्तन रहा है। उपयुक्त आहार-विहार के बिना योग-सिद्धि तो कदापि सम्भव नहीं है। अपितु जो मनुष्य

---

1. मनु. 3.118
2. षट्त्रिंशतं सहस्राणि रात्रीणां हितभोजन:।
   जीवत्यनातुरो जन्तुर्जितात्मा सम्मत: सताम्।। चरक. सू. 27.348AA
3. विगत धाननिशीथे प्रातरूत्थाय नित्यं
   पिबति खलु नरो यो ध्राणनन्ध्रेण वारि।।
   सभवति मतिपूर्णश चक्षुष्ण ताक्ष्र्यतुल्यो,
   बलि पलितविहीन: सर्वरोगैर्विमुक्त:।।

उचित आहार-विहार वाला होता है, उसके सभी दुःख दूर हो जाते हैं। गीता में श्री कृष्ण कहते हैं-

नात्यश्नतस्तु योगोऽस्ति चैकान्तमश्नतः।

न चाति स्वप्नशीलस्य जाग्रतो नैव चार्जुन॥

युक्ताहारविहारस्थ मुक्तचेष्टस्य कर्मसु।

युक्तस्वप्नाववोधस्य योगो भवति दुःखहा॥ ६.१६.१७॥

आहार इसके अतिरिक्त नियमित दिनचर्या भी आयुवर्धक मानी गई है- आयुर्वेद के प्रसिद्ध आचार्य बाग्भट लिखते हैं-

ब्राह्मे मुहूर्ते उत्तिष्ठेत् स्वस्थो रक्षार्थमायुषः।

शरीरचिन्तां निर्वर्त्य कृतशौचसमाहितः॥ अष्टा. ह. सू. २.१॥

मनु महाराज ने तो सर्वप्रथम इस तथ्य को समझा और इसके पालन का निर्देश मनुस्मृति[3] में किया। इनके अतिरिक्त अनेकानेक आचार्यों ने अपने ग्रन्थों में आयुर्विज्ञान के रहस्यों का सरल-रीति से सूत्रात्मकरूप में उल्लेख किया है। इनको जीवन में चरितार्थ कर हमारे पूर्वजों ने अपनी पूर्ण आयु को प्राप्त किया तथा जीवन के चरम लक्ष्य पुरुषार्थ चतुष्टय की सिद्धि की। अब हम कुछ ऐसे ही आयुष्य-प्रदायक सूक्ति रत्नों पर दृष्टिपात करते हैं। इनके अनुसार आचरण से रोग नहीं होते।

उषा-पान : सर्वरोग विनाशाय निशान्ते तु पिबेज्जलम्।

दुग्ध-तक्र-पान : दिनान्ते च पिबेद् दुग्धं निशान्ते च जलं पिबेत्।
भोजनान्ते पिबेत् तक्रं वैद्यस्य किं प्रयोजनम्।।

आहार-अवधि-मात्र : दिवा पुनर्न भुञ्जीत नान्यत्र फलमूलयोः।।
आर्द्रपादस्तु भुञ्जीत नार्द्रपादस्तु संविशेत्।
आर्द्रपादस्तु भुञ्जानो दीर्घमायुरवाप्नुयात्।।[1]

दिशा : आयुष्यं प्राङ्मुखो भुंक्ते ..।।
यो भुंक्ते वेष्टितशिरा यश्च भुंक्ते विदिङ्मुखः।
सोपानत्कश्च यो भुंक्ते सवनविद्यात् तदासुरम्।।

---

1. मनु. 4.76

मात्रा:	:	कुक्षेर्भागद्वयं भोज्यैस्तृतीयं वारि पूरयेत्। वायोः संचरणार्थाय चतुर्थमवशोषयेत्।।
नेत्र ज्योति लाभ	:	भुक्त्वा पाण्तिलं धृत्वा चक्षुषोर्यदि दीयते। अचिरेणैव तद्वारि तिमिराणि व्यपोहति।।
वाम-शयन:	:	वामदिशायामनलो नाभेरूर्ध्वेऽस्ति जन्तूनाम्। तस्मातु वामपार्श्वे शयीत मुक्त प्रपाकार्थम्।।[1]
जल-सेवन:	:	अत्यम्बुपानाच्च न विपच्यतेऽन्नमनम्बुयानाच्च स एव दोषः। तस्मान्नरो बह्विविवर्धनार्थं न मुहुर्मुहु र्वारि पिबेद् भूरि।।

इन सूक्ति रत्नों जो कहा गया है (उसका पालन आज भी बहुत से संस्कारवान् मनुष्य करते हैं। उन्हें इसका लाभ हो रहा है तथा अन्य जन रोगी हो रहे हैं। संक्षेप में न्यूनतम विधान करते हुए लिखा है:-

वामशायी द्विभुञ्जानो षण्मूत्री द्विपुरीषकः।
स्वल्पमैथुनकारी च शतं वर्षाणि जीवति।।

यदि किसी समय प्रमादवश वात-पित्त-कफ कुपित हो जाए तो उस का भी उपचार बताया गया है-

वमनं कफनाशाय वातनाशाय मर्दनम्।
शयनं पित्तनाशाय ज्वरनाशाय लंघनम्।।

योग : भारतीय शास्त्र-परम्परा में योग पर बहुत बल दिया गया है। योग बल से मनुष्य जन्म-मरण के बन्धन से भी मुक्ति पा लेता है। स्वास्थ्य लाभ और रोगमुक्ति की तो बात ही क्या है। योग के आठ अंग, मुक्ति-मार्ग के सोपान हैं। इनके सेवन से मनुष्य शारीरिक मानसिक एवं बौद्धिक रूप से पूर्ण स्वस्थ रहता है। रोग का जीवन में स्थान ही नहीं है। आज हम देख रहे हैं कि योग के नाम पर जन सामान्य को आसनों का ज्ञान कराया जा रहा है। कुछ विद्वान् प्राणायाम पर भी बल देने लगे हैं। जबकि यम-नियम इनसे पहले के सोपान हैं। इनके पालन से हम बौद्धिक तुष्टि को प्राप्त होकर रोगों से मुक्त रह सकते हैं। मन की अशान्ति एवं बौद्धिक दूषण के कारण ही आज मनुष्य अधिक रोगी हो रहे हैं। परिवार समाज, राष्ट्र तथा विश्व में अविश्वास, अव्यवस्था, अराजकता अनाचार का साम्राज्य उत्तरोत्तर बढ़ रहा है। आसन-प्राणायाम के साथ साथ

---

1. भाव प्रकाश

हमें अन्य योगांगों का भी पालन करना चाहिए। जिससे हम अपने पूर्वजों की भांति निरामय, निरापद एवं सौहार्दमय जीवन-यापन कर सकें।

जप-सन्ध्या : प्रातः सायं सन्ध्या, जप करना दीर्घ आयु का कारण होता है। प्रज्ञा, यश, बल तथा ब्रह्मवर्चस की प्राप्त होती है।[1] ईश्वर की न्याय-व्यवस्था को मानने का व्रत लेते हैं। जिससे ईर्ष्या द्वेष की भावना मन में उदित नहीं होती। यह भावना ही रोग तथा दुःख-दारिद्र्य का मूल है।[2]

ब्रह्मचर्य : विषय भोगों से मन को निवृत्त करके ही ब्रह्मचर्य का पालन कर सकते हैं। देवताओं ने ब्रह्मचर्य से ही मृत्यु पर विजय प्राप्त की थी।[3] भोगी व्यक्ति विषयों में फंसकर विनाश को प्राप्त होता है।[4]

इसके अतिरिक्त जो मनुष्य इन्द्रियों के द्वारा केवल शरीर यात्रा हेतु राग-द्वेष से रहित होकर विषयों का सेवन करता है, उसको शारीरिक मानसिक सुख तो प्राप्त होता ही है, अपितु वह परमात्मा की परम कृपा (प्रसाद) को भी प्राप्त कर लेता है। रोग-आदि की तो बात ही क्या है?

आसन-प्राणायाम : इनके द्वारा शरीर-शुद्धि तो जगत्प्रसिद्ध है। प्रायः सदाचारी जन इनका अभ्यास करते ही हैं। मनु महाराज ने प्राणायाम द्वारा इन्द्रियों के दोष (विकार) नष्ट करने को कहा है-

दह्यन्ते धामायमानां धातूनां च यथामलाः।। प्राणायामैर्दहेद्दोषान् ६-७२

तथेन्द्रियाणां दह्यन्ते दोषाः प्राणस्य निग्रहात्।। ६-७१.७२।।

योग दर्शन के रचयिता महर्षि भी लिखते हैं कि प्राणायाम करने से तो शरीर की ही नहीं अपितु बुद्धि की भी शुद्धि होती है। इससे ज्ञान-रूपी प्रकाश को ढकने वाला अज्ञान का आवरण हट जाता है। यदि किसी कारण से जीवन में प्रमाद आ जाए और वह रोगोत्पत्ति का कारण बन भी जाए तो विभिन्न अंगों की शुद्धि (रोग-निवृत्ति) के लिए अलग-अलग आसनों का विधान है। इनके अतिरिक्त नेति, धौति, न्यौली, अग्निसार, त्राटक आदि यौगिक क्रियाओं का विधान है। इनके

1. ऋषयदीर्घसन्धयत्वाछीर्धमायुखाप्नुयुः।
   प्रज्ञां यशश्च कीर्तिं च ब्रह्मवर्चसमेव च।। मनु. 4.94
2. योस्मान् द्वेष्टि यं च वयं द्विष्मस्तं वो जम्भेदधमः।। अथर्व 3.27.1
3. ब्रह्मचर्येण तपसा देवा मृत्युमपाहनत।। अथर्व 11.5.19
4. ध्यायतो विषयान् पुंसः संगस्तेषूपजायते। संगात्-कामात् क्रोधः।
   क्रोधात्-सम्मोहः, सम्मोहात्-विभ्रमः। बुद्धिनाशः ततः प्रणश्यति।। गी. 2.62-63

यथाविधि करने से शरीर में रोगोत्पत्ति होती ही नहीं। इन सब यौगिक क्रियाओं का विधि-विधान 'हठयोग प्रदीपिका, घेरण्डसंहिता आदि ग्रन्थों में है। श्वेताश्वतर उपनिषद् में ऋषि डिण्डिम घोष कहते हैं-

न तस्य रोगो न जरा न मृत्युः प्राप्तस्य योगाग्निमयं शरीरम्।[1]

चिकित्सा-आधारित-आयुर्विज्ञान

हमारे यहाँ प्राचीन काल से यही मान्यता रही है कि मनुष्य को निरामय होना चाहिए। रोग-उत्पत्ति से शरीर विकृत होता है। इसी कारण देवत्व-वृत्ति के मनुष्य तो विकृति-प्राप्त शरीर का स्वेच्छा से परित्याग कर देते थे। वाल्मीकि रामायण में शरभंग-ऋषि ने श्रीराम के सामने अपना शरीर-त्याग किया था।[2] एक जनश्रुति के अनुसार सिकन्दर के काल में भी एक साधु ने इसलिए शरीर-त्याग कर दिया था कि उसको ज्वर हो गया था। जीवन में प्रथम बार शरीर को रोग-ग्रस्त पाकर उस साधु ने उसे आत्मा के अयोग्य माना और त्याग दिया। देवत्ववृत्ति से भिन्न जो मनुष्य अपने शरीर को निरामय नहीं रख पाते, उनके लिए चिकित्सा-आधारित आयुर्विज्ञान की आवश्यकता हुई। इस का आरम्भ ब्रह्म से माना जाता है। इनसे यह ज्ञान दक्ष-प्रजापति, अश्विनी कुमार, इन्द्र भारद्वाज, भृगु, अंगिरा, अत्रि, अगस्त्य आदि मुनियों को प्राप्त हुआ।[3] आज हमें लगभग 2500 वर्ष (दो सहस्र पांच सौ वर्ष) पुराने चिकित्सा शास्त्र के ग्रन्थ उपलब्ध हैं। इन में कुछ प्रति संस्कृत हैं, कुछ खण्डित हैं और कुछ संगृहीत हैं। आज उपलब्ध सर्वाधिक मान्य ग्रन्थों में चरक एवं सुश्रुत संहिता प्रति संस्कृत हैं, भेल एवं काश्यप संहिता खण्डित हैं। संगृहीत संहिताओं में अष्टांग हृदय तथा अष्टांग संग्रह प्रसिद्ध हैं। काय चिकित्सा के लिए चरक संहिता तथा शल्य चिकित्सा के लिए सुश्रुत संहिता सर्वश्रेष्ठ मानी गई है। इनसे पूर्ववर्ती आचार्यों में भारद्वाज ऋषि के दो शिष्य बहुत प्रसिद्ध हुए हैं। प्रथम धन्वन्तरि एवं द्वितीय पुनर्वास आत्रेय। धन्वन्तरि शल्यतन्त्र में प्रवीण थे। इनके ग्रन्थों में निघण्टु, वैद्यक चिन्तामणि, रोग निदान आदि प्रमुख थे। पुनर्वसु आत्रेय ने भिषक्-विद्या में प्रसिद्धि प्राप्त की थी। कायचिकित्सा में निष्णात आत्रेय मुनि को हृदयंगम-औषधालय कहा जाता था। इनके छः शिष्य थे जिनके अग्निवेश, भेल, पराशर, हारीत, जतूकर्ण, क्षारपाणि नाम थे। इनके प्रसिद्ध ग्रन्थ भी थे। जिनका उल्लेख अन्य ग्रन्थों में मिलता है। कुछ उपलब्ध भी हैं।

---

1 श्वेताश्वरतर 2.12

2. एष पन्था नर व्याघ्र मुहूर्त पश्य तातमाम्।
यावज्जहामि गात्रणि जीर्णां त्वचमिवोरगः।। बा. रा. अरण्य 4.38-39

3. चरक सं. सूत्र. 1.4-5,26

कांकानय : बाह्लिक देश (अफगानिस्तान) वासी, 'शालाक्य-तन्त्र' के आचार्य थे। इनके अनेक शिष्य भी थे।

कश्यप-उशना : ये दोनों 'अगद्-तन्त्र' के आचार्य थे। इसमें विविध विषों के शमन का उपाय बताया गया है।

भृगु-अगस्त्य : 'रसतन्त्राचार्य' थे। इसके बल पर मनुष्य दीर्घायु को प्राप्त कर सकता है। आचार्य नागार्जुन का नाम रसतन्त्रकारों में आदर से लिया जाता है। 'सिद्ध-नागार्जुनीय, रसरत्नाकर, रसेन्द्र-मंगल, आरोग्य-मञ्जरी आदि प्रसिद्ध ग्रन्थों की रचना कर रसशास्त्र का विकास किया।

वाग्भट : 'अष्टांग-हृदय' की रचना की।

वृद्ध-वाग्भट : 'अष्टांग संग्रह' को ग्रथित किया तथा भावमिश्र ने 'भावप्रकाश' को।

माधव : 'माधवनिदान' के रचयिता हैं।

शार्गंधर : 1. शार्गंधर संहिता, 2. शार्गंधर पद्धति। इसी प्रकार बोपदेव, बंगसेन, मोरेश्वर डल्हण आदि आचार्यों ने भारतीय आयुर्विज्ञान परम्परा को समृद्ध बनाया और मानवता का महान् उपकार किया।

चिकित्सा : शरीर में विषमता को प्राप्त धातुओं को सम-अवस्था में लाया जाए तथा वैषम्य दोष के कारणों को जिस क्रिया के द्वारा रोका जाए उसे चिकित्सा कहा है।[1] यह चिकित्सा चार चरणों पर आश्रित है। 1. गुणवान् चिकित्सक, 2. गुणयुक्त-औषध द्रव्य, 3. गुण सम्पन्न परिचारक, 4. गुणवान् रोगी।

चिकित्सक : शास्त्र-ज्ञान, चिकित्सा कर्म का प्रत्यक्ष-ज्ञान, दक्षताएवं पवित्रता से सम्पन्न मनुष्य चिकित्सक होता है। इसके अतिरिक्त विद्या, वितर्क, विज्ञान, स्मृति, तत्परता एवं चिकित्सा रूप क्रिया ये छः गुण जिसमें हो(वह उत्तम चिकित्सक होता है।

आचार संहिता : चिकित्सा-कार्य का मूल उद्देश्य पीड़ित मानव की सेवा करना है। स्वार्थपूर्ति नहीं जैसा कि आजकल बहुत से स्थानों पर देखने को मिल रहा है। चरकमुनि अपने स्नातक चिकित्सकों को उपदेश देते हुए कहते हैं-

उत्तम : नार्थार्थं नापि कामार्थमथ भूत दयां प्रति।
वर्तते यश्चिकित्सायां स सर्वमति वर्तते।।

अधम : कुर्वते ये तु वृत्यर्थं चिकित्सापण्य विक्रयम्।

---

1. चरक सं. सूत्र. 16.34-36

ते हित्वा काञ्चनं राशि पांशु राशि मुपासते।। चरक चि. 1.4.58.59।।

चिकित्सक अपने उत्तम धर्म का पालन करते हुए रोगियों को पुत्र की भांति माने और उनके रोगों को दूर करे।[1] सर्वत्र अर्थ लाभ के लिए चिकित्सा न करें। क्योंकि इस कार्य में कहीं से धन कहीं से मित्रता कहीं से पुण्य, कहीं से यश, कहीं से कर्म-अभ्यास आदि कुछ न कुछ तो प्राप्त होता ही रहा है।[2]

आयुर्वेद के अष्टांग :

१. **शल्य** : इसमें शल्य-चिकित्साएवं प्रसूति-कर्म का वर्णन है।

२. **शालाक्य** : जत्रु से ऊपर के रोगों (आंख कान नाक मुख आदि के) की शान्ति के लिए शालाक्य शास्त्र विशेषत: आंख के लिए है।

३. **काय-चिकित्सा** : समस्त अंगों में होने वाले रोगों के लिए।

४. **भूत-विद्या** : विभिन्न भूतों के आवेश से दूषित मन की शान्ति हेतू।

५. **कौमारमृत्य** : बाल-चिकित्सा, धात्री-स्तन्य चिकित्सा।

६. **अगद-तन्त्र** : सर्प आदि दंशएवं विविध-विष से उत्पन्न रोग चिकित्सा।

७. **रसायन तन्त्र** : आयु-बल बुद्धि वर्धक रसायन विज्ञान।

८. **वाजीकरण-तन्त्र** : वीर्य वर्धक, शुक्र शोधक औषध वर्णन!

चिकित्सा के प्रमुख दो विभाग हैं 1. काय-चिकित्सा, 2. शल्य चिकित्सा काय-चिकित्सा पर आहार-विहार प्रकरण मे विचार किया जा चुका है। विभिन्न रोगों की चिकित्सा पद्धतिएवं औषध आदि को तत्तत् ग्रन्थों में जाना जा सकता है। अब आयुर्वेद के प्रमुख अंग शल्य चिकित्सा पर विचार करते हैं : **प्रमुख शल्य-यन्त्र** शल्य-चिकित्सा के मुख्य आचार्य सुश्रुत ने प्रमुख यन्त्रों की संख्या 101 एक सौएक कही है तथा वाग्भट[3] ने इन्हें **असंख्य** माना है। संख्या आकृति भेद से इन्हें छ: भागों में विभक्त किया है। **१.** स्वास्तिक यन्त्र 24, **२.** सन्दंश यन्त्र 2, **३.** ताल यन्त्र 2, **४.** नाड़ी यन्त्र 20, **५.** शलाकायन्त्र 28, **६.** उपयन्त्र 25, ये सभी यन्त्र प्राय: लोह निर्मित होते

---

1. भिषगप्यातुरान् सर्वान् स्वसुतान् इव यत्नवान्।
आबाधोम्यो हि संरक्षे दिच्छन् धार्ममनुत्तमम्।। च. चि. 1.4.56

2. वचचिदर्थ: क्वचिन्मैत्री क्वचि)र्म: क्वचिद्यश:।
कर्माभ्यास: क्वचिच्चैषा चिकित्सा नास्ति निष्फला।। 1.4.56

3. अत: कर्मवशात्तेषां इयत्ता अवधारणमशक्यम्।। वा. ग. सू. स्था. अ. 25

हैं। लोहे-अभाव में अन्य सामग्री से भी बनते थे। आकृति के अनुसार इनका माप[1] भी भिन्न रखा गया है। इसी प्रंसग में यन्त्र कर्म तथा दोष भी दर्शाए है।[2]

प्लास्टिक सर्जरी : कर्ण वेध, कर्ण सन्धाान कर्ण-वृद्धि : शल्य-आचार्य लिखते हैं कि बालक का कर्णवेध करना चाहिए। इसके दो लाभ है। १- आन्त्र वृद्धि, २- भूषण धारण। यहां वे धन-विधि का भी वर्णन किया है। यदि कर्ण दोष हो तो उनकी वृद्धि की विधि-औषध आदि का भी वर्णन है। किसी रोगवश अथवा चोट ग्रस्त होने से कर्ण के दो भाग (कटना) हो जाए तो उनकी जोड़ने की विधि दर्शायी है। यदि मांस त्वचा की आवश्यकता पड़े तो गण्ड प्रदेश से उखाड़कर उस का उपयोग किया जाए।[3] इसके साथ ही शल्य क्रिया के बाद बरतने वाली सावधानियों का भी स्पष्ट एवं विस्तृत उल्लेख किया है।

नासिका - औष्ठ का सन्धान[4] : यदि किसी कारण वश नासिका की त्वचा अथवा मांस कटकर अलग हो जाए तो उसे जोड़ने के लिए गण्ड प्रदेश की त्वचा तथा मांस निकालकर उक्त स्थानों पर जोड़ने चाहिए। सूई से सीकर औषध लगा पटी बान्धो। पटी बान्धते समय श्वास मार्ग के लिए नासिका क्षिद्रों को एरण्ड आदि कोमल नालों से ऊंचा उठा कर रखें। इसी प्रकार औष्ठ सन्धान करना चाहिए। यह कार्य कुशल वैद्य ही कर सकता है। सुश्रुत से पूर्व भी शल्य तन्त्र के आचार्य हो चुके थे। स्वयं सुश्रुत ने लिखा है- अश्विनी कुमारों ने शिव के द्वारा छिन्न यज्ञ के **शिर का सन्धान** किया था।[5]

काया-कल्प : अश्विनी कुमार देव-वैद्य थे। शल्य-क्रिया के साथ साथ काया-कल्प विद्या में भी पारंगत थे। रूप-स्वर-हीन वृद्ध च्यवन् ऋषि की काया को पुनः रूप यौवन सम्पन्न किया था। यह विद्या अब भी दक्षिण भारत में प्रचलित है। राजा यपाति ने भी पुत्र के सहयोग से यौवन प्राप्त किया था।[1]

संजीवनी-विद्या : रामायण अनेक प्रसंग मिलते हैं। जिनमें मृत प्राय: वीरों को इस विद्या के विशेषज्ञ वैधों पुनर्जीवित कर विजय प्राप्ति की योग्य विद्या के विशेषज्ञ वैधों ने पुनर्जीवित कर विजय प्राप्ति के योग्य बना दिया।[6] 1. इन्द्रजित् के ब्रह्मास्त से मूर्छित श्री राम लक्ष्मण एवं वानरों

---

1. सु. सू. 16.27-3
2. सु. सू. 7.5-9
3. सु. सू. 16.27-3
4. सु. सू. 16.27-3
5. सु. सू. 16.27-3
6. अश्विनौ देव मिषजौ यज्ञ बाहाविति स्मृति।महा. आ. 84.28-32
   यज्ञस्य हिशिरश्छिन्नं पुन स्ताभ्यं समाहितम्।।

को 2. रावण की शक्ति से मूर्छित लक्ष्मण को सुषेण वैद्य ने। शुक्राचार्य को भी संजीवनी-विद्या का ज्ञान था। अतः देवता असुरों को पराजित न कर सके। क्योंकि असुरों को आचार्य (शुक्राचार्य) युद्ध में मृत असुरों को पुनर्जीवित कर देते थे। अन्त में देवगुरू बृहस्पति के ज्येष्ठ पुत्र कच ने भी बहुत कष्ट सहकर यह विद्या शुक्राचार्य से सीखी थी।[1] महाभारत की जाग कन्या उलूपी भीऐसी विद्या जानती थी। बभ्रु वाहन के साथ युद्ध में मृत प्रायः अर्जुन को **संजीवनी मणि** से पुनर्जीवित किया था।

क्षुत्पिपासा नाशिनी : गुरु विश्वामित्र ने युद्ध काल के लिए श्री राम लक्ष्मण को दो विद्याएं दी 1. बला 2. अति बला। इनके प्रभाव से वे भूख-प्यास, श्रम, ज्वर निद्रा पर विजय पा गये थे। आज भी वैज्ञानिक, अन्तरिक्ष-यात्र, युद्ध-आदिकाल के लिए इस विद्या से ही कैपसूल आदि बनाते हैं। जिन के सेवन से उन्हें सामान्य भोजन (अन्न-जल) आदि की आवश्यकता नहीं रहती।

सूर्य-चिकित्सा : 'आरोग्यं भास्करादिच्छेत्' यह मत्स्य पुराण का वचन है। इस चिकित्सा का वर्णन वेद में भी मिलता है। उदय होता सूर्य मृत्यु पाश (रोगों) को दर भगाता है।[2] इसके कारण ही भारतीय सूर्य नमस्कार, सूर्य-अधर्य आदि क्रियाएं दैनिक करते हैं। हृदय रोग, शिर के रोग, नेत्र, त्वचा आदि के रोगों को सूर्य नष्ट करता है। सूर्य के प्रकाश वाले भाग को अमर लोक माना है।[3]

नाड़ी-परीक्षा :

चरक-मुनि लिखते हैं-

रोगभादौ परीक्षेत ततोऽनन्तरमौषधम्।
ततः कर्म भिषक् पश्चात्, ज्ञान पूर्व समाचरेत्॥ चरक संहिता॥

रोग-परीक्षा में नाड़ी-परीक्षा का महत्त्वपूर्ण स्थान है। यह विज्ञान इसी कारण प्रसिद्धिएवं विकास को प्राप्त हुआ। रावण, कणाद आदि ने इस नाड़ी-विज्ञान पर ग्रन्थ लिखे। योगरत्नाकर में शार्गंधर आचार्य लिखते हैं :

करस्यांगुष्ठमूले या धामनी जीवसाक्षिणी।
तच्चेष्टया सुखं दुःखं ज्ञेयं का यस्य पण्डितैः। शा. पूर्व. ३.१

1. वा. रा. यु. 101.31-32, 44-45
2. बा. रा. बाल. 22.13-18
3. उद्यन् सूर्योनुदतां मृत्युपाशान्॥ अथर्व 17.1.30AA अथर्व 9.8.1, 22

इस प्रकार संस्कृत-शास्त्रों में आयुर्विज्ञान का बहुविध ज्ञान उपलब्ध होता है। चरकमुनि का ग्रन्थ श्रेष्ठतम माना जाता है। उनके काल में ऐसे अनेक आचार्य थे जिनके ग्रन्थ उस समय प्रचलित थे लेकिन आज वे उपलब्ध नहीं हैं। चरक मुनि स्वयं लिखते हैं।

भिषक् बुभूषुः शास्त्रमेवादितः परीक्षेत।
विविधानि शास्त्राणि मिषजां प्रचरन्ति लोके॥ च. संहिता

आयुर्विज्ञान का आरम्भ वेद से हुआ। अथर्ववेद केएक मन्त्र में 360 अस्थियों का वर्णन है। इसीप्रकार धमनी-शिराएं कृ भी ज्वर विष[1] चिकित्सा आदि का वर्णन वेद में उपलब्ध होतो है। अथर्ववेद में रोग-नाशक औषअधियों का भी विस्तार से वर्णन मिलता है। वेद सब का मूल है। वहीं से आयुर्विज्ञान विस्तार को प्राप्त हुआ। इस प्रकार हम कह सकते हैं कि संस्कृत-शास्त्रीय परम्परा में आयुर्विज्ञान का बहुत उत्कृष्ट-रूप है। हमें उसका लाभ उठाना चाहिए। हम आचार्य बाग्भट के निर्देश का पालन करें :

आयुर्वेदोपदेशेषु विघेयः परमादरः॥ अष्टा. सू. १.२॥

यदि हम आयुर्वेद के उपदेशों को जीवन में स्वीकार करेंगे तो हमारी यह प्रार्थना अवश्य सार्थक होगी :

सर्वे भवन्तु सुखिनः सर्वे सन्तु निरामयाः।
सर्वे भद्राणि पश्यन्तु मा कश्चिद् दुःखभाग्मवेत्॥

---

1. अथर्व, 1.17.3.11 अथर्व, 5.29.6 अथर्व, 10.4.22

6

आयुर्वेद का वैज्ञानिक आधार

डॉ. कैलाशचन्द्र विद्यालंकार

भारतीय सभ्यता और संस्कृति में ज्ञान-विज्ञान की अनेक दिशाओं में गहन और व्यवस्थित चिन्तन हुआ है। उस चिन्तन का आधार जहाँ चारों वेद - ऋग्वेद, यजुर्वेद, सामवेद और अथर्ववेद हैं, वहाँ क्रियात्मक जीवन को सुचारु रूप से चलाने के लिए चार उपवेद - आयुर्वेद, धनुर्वेद, गन्धर्ववेद तथा अर्थवेद (अर्थशास्त्र) हैं। **धनुर्वेद** की आवश्यकता मनुष्य समाज में संघर्ष की स्थिति से निपटने तथा आत्मरक्षा के लिए हुई, **गान्धर्ववेद** जीवन के सौन्दर्यपक्ष की अभिव्यक्तिरूप कलाओं से सम्बन्ध रखता है। **अर्थवेद** में जीवन यापन के लिए आवश्यक शारीरिक बल के साथ-साथ समाजमें सुव्यवस्था तथा भौतिक सुखसाधनों के समायोजन का महत्त्व तथा उसके लिए उपायों का निर्देश है। **आयुर्वेद** का सम्बन्ध मनुष्य के जीवन से है, उसके स्वास्थ्य से है। चरक-संहिता के आरम्भ में ही कहा गया है - 'स्वास्थ्य धर्म, अर्थ, काम और मोक्ष रूप चार पुरुषार्थों का प्रमुख साधन है। व्याधियाँ स्वास्थ्य की तथा अच्छे जीवन की विनाशक हैं। इस कारण मानव जाति की प्रगति में बड़ी रुकावट उत्पन्न होगई, तो विचार हुआ 'इस सबकी चिकित्सा का क्या साधन है? यह सोचने के लिए ऋषि समाधि में बैठ गए।[1] ऋषियों को समाधि की अवस्था में औषधियों का स्फुरण हुआ।' वही आयुर्वेद है -- जीवन का विज्ञान है। जीवन सबसे अमूल्य है अत: यह विज्ञान सभी विज्ञानों में उत्तम है।[2] इसीलिए आयुर्वेद को उपवेदों में प्रथम तथा अष्टादशविद्यास्थानों[3] में महत्त्वपूर्ण स्थान प्राप्त है।

आयुर्वेद का सम्बन्ध जीवन से है। 'आयुस्' शब्द का वेद शब्द से समास होने पर आयुर्वेद शब्द बनता है। तदनुसार - आयु: वेदयति, विन्दयति, ज्ञापयति वा इति आयुर्वेद: अर्थात् जो शास्त्र आयु के विषय में ज्ञान कराता है, 'आयु' को प्राप्त कराता है, उसके विषय में सूचनाएँ देता है अथवा उसकी पहचान कराता है वह आयुर्वेद है।[4] 'आयुस्' शब्द का अर्थ अमरकोश में 'आयति' दिया गया है जिसका तात्पर्य 'गतिशीलता तथा विस्तार' होता है। उस गति शीलता का विस्तार जीवन पर्यन्त रहता है, जीवन की समाप्ति के साथ आयु का विस्तार समाप्त हो जाता है, इसीलिए उसे 'जीवितावधि:' ऐसा कहा गया है।[5] इसके अनुसार आयुर्वेद वह विज्ञान है 'जिससे मनुष्य अपने शरीर के अंगसमूह (Organism) को शुद्ध और स्वस्थ रखने के लिए मानव जीवन के प्रारम्भ से ही मार्ग खोजता रहा है और साधन जुटाता रहा है।'

मनुष्य सदा से प्रकृति के सम्पर्क में रहा है और प्रकृति के प्रत्येक पदार्थ- वनस्पति, वृक्ष, लता, गुल्म के मूल से लेकर फल तक सभी अंगों को उसने अपने जीवन के लिए और अपने स्वास्थ्य के लिए उपयोगी पाया है। प्रकृति में प्राप्त धातु, अम्ल, क्षार और लवण आदि सभी में उसने अपने लिए सहायक और उपयोगी तत्त्व ढूंढा है, वायु तो उसका प्राण है, आकाशीय पिण्ड उसको उष्णता, ऊर्जा, प्रकाश आदि देते हैं। भारतीय मनीषियों ने तो प्रत्येक प्राणी अर्थात् पशु, पक्षी, कीट, पतंग, कृमि आदि में भी जीवनोपयोगी तथा दोष निवारक तत्त्वों का अन्वेषण किया है। उनकी यह प्रकृति में मित्र स्वर सहायक तत्त्वों को पाती है। सभी शास्त्रों, काव्यों तथा अन्य ग्रन्थों में प्रकृति मानव की सहचरी के रूप में वर्णित है। इसीलिए यहाँ प्रकृति का दोहन न करके उसे अपने उपयोग के योग्य बनाने तथा उसका संरक्षण करने में प्रयत्न किया गया है।

यह सभी कुछ वैज्ञानिक दृष्टिकोण और वैज्ञानिक प्रक्रिया से हुआ है। यद्यपि आयुर्वेद का साधारण रूप ऋग्वेद[6] और अथर्ववेद[7] में मिलता है। वहाँ वैद्य, अनेक रोगों तथा उनकी निवारक ओषधियों का वर्णन है परन्तु उनके मूल में वैज्ञानिक प्रवृत्ति कार्य करती रही है। आयुर्वेद के प्रसिद्ध ग्रंथ चरकसंहिता में कहा गया है कि 'सिद्धांत विद्वानों द्वारा परीक्षा करके स्थापित किए जाते हैं।[8] सिद्धांत की व्युत्पत्ति दो प्रकार से की जाती है -

1. सिद्धः अन्ते - अंत में सिद्ध, अंत में प्राप्त परिणाम (थ्यदंस बवदबसनेपवद)

2. सिद्धः अन्तः - अर्थात् निरीक्षण, परीक्षण, पुनर्वीक्षण तथा प्रयोगात्मक उपयोग और

विचार विमर्श के द्वारा प्राप्त अंतिम ज्ञान

सिद्धांत की इस प्रक्रिया को आज की वैज्ञानिक प्रक्रिया - Observation, experimentation, again observation, practical use, conclusion and theory - से मिला सकते हैं। विज्ञान की इस प्रक्रिया के लिए विचारात्मक तथा क्रियात्मक विधि जो अपनायी गयी है, वह निम्न प्रकार से है -

1. प्रत्येक वस्तु को अपनी इन्द्रियों से **प्रत्यक्ष** देखा जाता है। इन्द्रियों की सीमित शक्ति को दूरवीक्षण, सूक्ष्मवीक्षण आदि यंत्रों की सहायता से चरम सीमा तक बढ़ाया जाता है। भारतीय मनिषियों ने भौतिक तथा यांत्रिक साधनों की सहायता के अतिरिक्त योग साधना के द्वारा अन्तरिन्द्रिय की सहायता से विश्व को हस्तामलकवत् देखा था।

2. प्रत्यक्ष से प्राप्त ज्ञान के आधार पर व्याप्ति द्वारा वह **अनुमान** करता है।[9]

3. प्रत्यक्ष से प्राप्त ज्ञान की तुलना से जो ज्ञान प्राप्त होता है वह सादृश्य अथवा **उपमान** रूप होता है।

4. प्रत्यक्ष द्रष्टा, सत्य के परीक्षक, चिन्तक, विद्वज्जनों अथवा आप्त पुरुषों के कहे वचन के रूप में संग्रहीत शब्दों से अर्थात् साहित्य से ज्ञान प्राप्त किया जाता है। इसे वाग्भट ने आगमसिद्ध कहा है।[10]

इस प्रक्रिया में प्रत्यक्ष प्रधान है और वह अन्ततोगत्वा मान्य है और सिद्धांत का स्थापक है, अन्य प्रमाण उसके पोषक हैं।

आयुर्वेद के प्रामाणिक ग्रंथ सुश्रुतसंहिता और वाग्भट के ग्रन्थों में इसी के समान स्थापना की गई है कि 'जो कुछ प्रत्यक्ष देखा है जो आप्त जनों से वचनों के रूप में सुना है उन सब में से युक्ति-प्रत्युक्ति के द्वारा संदेह को दूर करके चिकित्सक को क्रिया अर्थात् स्त्रषधि आदि से रोग निवारण की क्रिया करनी चाहिए'।[11] इस क्रिया को **युक्तिव्यपाश्रय** कहा गया है। यह क्रिया **दैवव्यपाश्रय** और **सत्त्वावजय** नहीं है। वह श्रुति, युक्ति और अनुभव पर आश्रित है। इस उपर्युक्त विधि से प्राप्त ज्ञान को ही शास्त्र की संज्ञा दी जाती है। शास्त्र के दो अंग होते हैं- 1. सुपरीक्षित ज्ञान और दूसरा उसका समुचित और यथास्थान प्रयोग। इस प्रकार आयुर्वेद में प्रत्यक्ष के विषय में युक्तियुक्त चिन्तन, उसके विषय में परीक्षण, पुनः चिन्तन के साथ समुचित प्रयोग का भी प्रावधान है। इसी श्रुति, युक्ति और अनुभव पर आश्रित यह आयुर्वेदशास्त्र प्रगति करता है।

स्वास्थ्य का विवेचन करते हुए कहा गया है कि अवस्था, परिस्थिति अथवा परिवेश तथा जीवन शैली के अनुसार शरीरक्रिया की साम्यावस्था स्वास्थ्य कहलाती है। इसमें विषमता रोग बन जाता है। अनुभव के आधार पर वैद्य विकृत या विषम अवस्था, परिवेश और जीवनशैली को साम्यावस्था में लाकर पुनः स्वास्थ्य प्रदान कर सकता है। यह साम्यावस्था ओषधियों के सम्यक् प्रयोग से तथा यथोचित पथ्य से पुनः स्थापित की जा सकती है। इसीलिए चरकसंहिता के प्रारम्भ में ही स्वस्थवृत्त का वर्णन किया है, जिसमें स्थापना की गई है कि खान-पान, रहन-सहन की नियमितता से मनुष्य स्वस्थ रहता हुआ दीर्घायु प्राप्त करता है।[12]

चरकाचार्य इसी प्रसंग में प्रस्तुत करते हैं कि आयुर्वेदएकऐसा विज्ञान है जो यह ज्ञान कराता है कि कौनसा वृत्त या आचरण और कौनसी वस्तुएँ स्वास्थ्य प्रदान करती हैं, आयु को बढ़ाती हैं और कौनसी अस्वास्थ्यकर हैं।[13] द्रव्य से तात्पर्य है विभिन्न पदार्थ, जिनका वर्णन भावप्रकाशनिघण्टु में किया गया है। गुण अर्थात् द्रव्यों के गुण और कर्म अर्थात् क्रिया का प्रकार तथा प्रयोग। इस प्रकार देश, काल और व्यक्ति की दृष्टि से हानिकर द्रव्यों से बचकर, लाभकारी द्रव्यों के सेवन से स्वास्थ्यलाभ होता है।

इस प्रयोग में प्रयोक्ता के विवेक को विशेष महत्त्व दिया गया है क्योंकि शास्त्र तो बोधक होता है। वह द्रव्य के गुण का ज्ञान कराता है, वह कारक नहीं होता। कारकत्व तो

परिवेश, समय, स्थान और व्यक्ति अर्थात् स्वभाव आदि की दृष्टि से प्रयोगकर्ता के विवेक पर निर्भर करता है। शास्त्र से प्राप्त ज्ञान बुद्धि को विवेक देता है, विवेक के आधार पर बुद्धि मन को आदेश देती है, उससे मन क्रियाशील होता है। यह विवेक काल, देश और अदृष्ट (प्रत्यक्षत: न देखी जाती हुई) बातों के आधार पर निर्णय लेता है। वस्तुत: प्रत्येक व्यक्ति के लिए परिवेश, समय तथा अवस्था की दृष्टि से विषमतारूप रोग एक समान नहीं होता। देश, काल और व्यक्ति या स्वभाव के आधार पर बदल सकता है। इस प्रकार शास्त्र में सिद्धान्त देश, काल की दृष्टि से मर्यादित होते हैं, अत: इनका प्रयोग प्रयोक्ता के विवेक पर निर्भर करता है। इस विषय में वाग्भट का कथन है - 'अनेक रूपों वाले असंख्य विकारों से कुपित मल शरीर को कष्ट पहुँचाते हैं, इसलिए एक-एक के हेतु का साधन दिखाते हुए वर्णन करना शक्य नहीं है, केवल सामान्य बातों का वर्णन किया जाता है'।[14]

आयुर्वेद की यह वैज्ञानिकता शरीररचना, शरीरक्रिया और स्वभाव पर निर्भर है। आधुनिक चिकित्साशास्त्र में भी Anatomy, Phisiology का विशेष महत्त्व है।

शरीररचना का आधार पंचभूत सिद्धान्त है और क्रिया या कार्य का आधार त्रिदोष का सिद्धान्त है। **'यथा पिण्डे तथा ब्रह्माण्डे'** के आधार पर जीव के शरीर को Microcosm कहा जा सकता है और विश्व को Macrocosm। आयुर्वेद की दृष्टि से मानवशरीर की रचना पाँच भौतिक तत्त्वों से होती है और वे तत्त्व वैशषिक दर्शन के आधार पर- अग्नि, वायु, जल, आकाश और पृथिवी हैं। इन तत्त्वों के विभिन्न अनुपात और व्यवस्था में मिलने से विभिन्न शरीरों की रचना होती है। दूसरी ओर समस्त विश्व की रचना भी इन्हीं तत्त्वों से होती है।[15]

पाँचों तत्त्वों के गुण क्रमशः - रूप, स्पर्श, रस, शब्द और गन्ध हैं। इन पाँचों तत्त्वों में से हीएक-एक तत्त्व की प्रमुखता से मानव शरीर में क्रमशः चक्षु, त्वचा, जिह्वा, कान तथा नासिका रूप पाँच ज्ञानेन्द्रियाँ हैं। इसीप्रकार पाँचों कर्मेन्द्रियों का भी इनसे सम्बन्ध है।[16] इन तत्त्वों का वांछित सन्तुलन बने रहने से शरीर स्वस्थ रहता हे और इनमें विषमता हो जाने पर रोग बन जाता है, अत: मानवशरीर को स्वस्थ रखने के लिए, उसके समुचित पोषण के लिए, विकार (रोग) को दूर करने के लिए प्रकृति में प्राप्त विभिन्न द्रव्यों अथवा पदार्थों का प्रयोग सर्वथा युक्तियुक्त है।

शरीरक्रिया की व्याख्या सांख्याभिमत त्रिगुण - सत्त्व, रजस् और तमस् की क्रियाशीलता के आधार पर स्थापित त्रिदोष - वात, पित्त और कफ अथवा वायु, अग्नि और जल के अनुसार की जाती है।[17] पंचभूतों में वात का सम्बन्ध वायु से है, पित्त का अग्नि से और कफ का जल से है। वात का कार्य स्नायुतन्त्र (Nervous system) से, पित्त का पाचनसंस्थान (Digestive system) से और कफ का रक्तसंचार (Circularory system) से है। इसी आधार पर रोगों को पैत्तिक अथवा आग्नेय, कफज अथवा सोम्य और वातज अथवा वायव्य तीन श्रेणियों में विभक्त किया गया है। आधुनिक चिकित्साशास्त्र में शरीर में विघटन (Catabolism), संघटन (Anabolism) तथा इनका सन्तुलन (Matabolism) तीन क्रियाएँ होती रहती हैं। इन तीनों प्रकार के कार्यों से शरीर के सभी जैविक (Biological) कार्य नियमित रहते हैं। इन तीनों में से किसी का भी आनुपातिक सन्तुलन न रहने से शरीर के कार्य विकृत हो जाते हैं जिसे रोग कहा जाता है अर्थात् इनके असन्तुलन अथवा अव्यवस्था से जीवन के लिए शत्रुभूत अवस्थाओं का प्रादुर्भाव हो जाता है। चिकित्सा द्वारा इनका सन्तुलन ठीक हो जाने पर रोग ठीक हो जाता है और व्यक्ति स्वस्थ कहलाता है।[18]

रचना और क्रिया के आधारभूत पंचमहाभूतों और तीन गुणों के अनुपात और व्यवस्था के अनुसार प्रत्येक व्यक्ति का अपना **स्वभाव** निश्चित हो जाता है। उस स्वभाव को भी आयुर्वेद में बहुत महत्त्व दिया जाता है। चरकसंहिता[19] में कहा गया है 'स्वभाव सबसे बढ़कर है' 'स्वभाव ही सबसे आगे क्रिया करता है'। इस स्वभाव को ध्यान में रखते हुए ओषधि और अनुपान का विवेकपूर्वक निर्णय लिया जाता है।[20] इसी कारण बाह्यत: एक ही रोग से पीड़ित व्यक्ति को एक ही प्रकार की ओषधि दी जा सकती है, परन्तु उसका अनुपान भिन्न होता है। Homeopathy में भी व्यक्ति के स्वभाव के आधार पर Constitutional remedy का चुनाव किया जाता है।

इसप्रकार मानवशरीर और समस्त जगत् के पदार्थों में रचनात्मक साम्य होने से उन सभी पदार्थों में शरीर की विषमताओं को दूर करने की सम्भावनाएँ होती हैं। दूसरी शताब्दी ई. पू. में तक्षशिला स्थित विद्यापीठ में आयुर्वेद के छात्र जीवक ने अनेक वर्षों के परिश्रम के उपरान्त पाया कि तक्षशिला के आसपास कोई भी ओषधि ऐसी नहीं है जो किसी रोग की चिकित्सा के लिए उपयोगी न हो। चरक संहिता में भी कहा गया है–'जगत् में ऐसा कोई भी द्रव्य नहीं है, जो जीव के शरीर के लिए औषधरूप न हो'।[21]

आयुर्वेद की वैज्ञानिक अन्वेषण परम्परा में ही आयुर्वेद शास्त्र के अनेक विभाग हुए हैं और उनका भी वर्गीकरण करके विशेष अध्ययन और प्रयोग किए गए हैं।[22] शरीररचना तथा शरीरक्रिया के साथ 'स्वस्थवृत्त' का भी अतिमहत्त्व है जो आयुर्वेद की आधारभूत वैज्ञानिकता को प्रदर्शित करता है। चरकसंहिता में बड़े विस्तार से 'स्वस्थवृत्त' का वर्णन है जिसमें भोजन के सन्तुलन[23] के साथ मनुष्य के व्यक्तिगत आचरण ही नहीं, अपितु सामाजिक जीवन में शिष्टाचार का भी उपपादन किया गया है। ऐसा प्रयत्न रहा है कि मानव जीवन इतना नियमित और दोषरहित और औचित्यपूर्ण हो कि वह आधि और व्याधि दोनों से बचा रहे।

रोग होने की अवस्था में रोग के समुचित **निदान** का भी अत्यन्त महत्त्व होता है, आंगलभाषा में इसे Clinical method कहा गया है। वर्तमान चिकित्सापद्धति में यह निदान अत्यन्त यान्त्रिक हो गया है। आयुर्वेद में नाडीपरीक्षा का विज्ञान अतिसूक्ष्म है, वह चिकित्सक की साधना तथा उसके ध्यान पर निर्भर करता है। जो कार्य E.C.G., Thermometer, रक्तचापमापक यन्त्र से किया जाता है, वह नाडीपरीक्षा से किया जाता था। इसके अतिरिक्त शरीर के अंगों का निरीक्षण, परीक्षण, मल, मूत्र, स्वेद, जिह्वा पर स्थित मैल आदि का निरीक्षण तथा परीक्षण सुश्रुत, वाग्भट तथा चरक आदि संहिताग्रन्थों में उपलब्ध होता है।[24]

आयुर्वेद को **अष्टांग** कहा गया है- इस विषय पर वाग्भट के 'अष्टांगसंग्रह' और 'अष्टांगहृदय' नामक दो ग्रन्थ भी हैं। उसमें शल्य (सर्जरी)[25], शालाक्य (शिरोरोग), कायचिकित्सा, भूतविद्या (मनोरोग), कौमारभृत्य (गर्भाधान से लेकर कुमारजन्म और तत्पश्चात् होने वाले उपद्रव इसमें सम्मिलित हैं), अगदतन्त्र (अगद का अर्थ विष भी होता है और रोग का अभाव भी) इसे आज के ओषधि-विज्ञान के अनुसार Doctrine of antidote कहा जा सकता है, इसमें मानवशरीर में अन्तः या बाह्य रूप से उत्पन्न होने वाले सभी विष सम्मिलित किये जाते हैं, आजकल इन्हें Toxins कहा जाता है। रसायनतन्त्र - अनेक द्रव्यों के मिश्रण से बनाए गए उपयोगी योग, वाजीकरण आदि शरीर के शक्तिवर्धक प्रयोग इसकी विशेषताएँ हैं।[26] इनके अतिरिक्त नेत्ररोगविद्, हृद्रोगविद्, अस्थिरोगविद्, उदररोगविद् आदि विशेषज्ञों के वर्णन के साथ तद्विषयक पृथक् ग्रन्थ भी उपलब्ध हैं।[27] महाभारत में भी रोगहर, शल्यहर, विषहर, कृत्यहर, भिषग् अथर्वन् आदि विभिन्न प्रकार की ओषधियों और चिकित्सकों का उल्लेख है, जो आयुर्वेद की विभाजन और वर्गीकरण पर आश्रित विशेष अध्ययन की वैज्ञानिकता को प्रकट करते हैं।

इस प्रकार आयुर्वेद मानवमात्र ही नहीं, अपितु प्राणिमात्र की आयु अर्थात् जीवन की सुरक्षा तथा स्वस्थता के लिए निरीक्षण, परीक्षण, पुनर्निरीक्षण तथा प्रत्यक्ष अनुभव तथा प्रयोग आदि प्रमाणों के विमर्शात्मक वैज्ञानिक आधार को लेकर रचना, क्रिया और स्वभाव को महत्त्व देते हुए, समस्त जगत् के द्रव्यों की जीव के शरीर के हित के लिए उपयोगिता को परीक्षा और अनुभव के आधार पर इस भारतभूमि पर आदि अर्थात् वैदिक काल से लेकर 12वीं तथा 13वीं शताब्दी तक विकास के पथ पर अग्रसर होता रहा।[28] तक्षशिला 2 ई०पू० में आयुर्वेद की शिक्षा का केन्द्र था। जीवक का उदाहरण ही उसकी प्रगति तथा वैज्ञानिक शोधप्रवृत्ति के लिए पर्याप्त है। छठी-सातवीं ई०पू० में नालन्दा के शिक्षाकेन्द्र में यूनान, रोम, चीन, जापान आदि देशों से जिज्ञासु वैद्य यहाँ के ओषधि तथा चिकित्सा विज्ञान को सीखने के लिए आते थे तथा यहाँ के विद्वान् उनके देश में प्रचलित चिकित्सापद्धति से परिचित होते थे।

संदर्भ :

1. चरकसंहिता, सूत्रस्थान, 1.11-14
2. वही, 1.15-16
3. अंगानि वेदाश्चत्वारो मीमांसान्यायविस्तरः,
धर्मशास्त्रं पुराणं च विद्या ह्येताश्चतुर्दश।
आयुर्वेदो धनुर्वेदो गान्धर्वश्चेति त्रयः,
अर्थशास्त्रं चतुर्थं तु विद्या ह्यष्टादशैव तु।।
4. आप्टे : संस्कृत-इंग्लिश डिक्शनरी
5. अमरकोश:
6. ऋग्वेद, 10.97.2,6,22
7. अथर्ववेद, 4.12.7; 4.22; 4.8.13; 5.22.12
8. चरकसंहिता, 1.3.7
9. सुश्रुतसंहिता, 1.40.12-13
10. चरकसंहिता, 1.10.6 तथा वाग्भट: अष्टांगसंग्रह
11. सुश्रुतसंहिता : दृष्टश्रुताभ्यां सन्देहमवापोह्य आचरेत् क्रियाम्।
12. चरकसंहिता, सूत्रस्थान, स्वस्थवृत्त
13. वही, आयुष्याणि अनायुष्याणि च द्रव्यगुणकर्माणि वेदयति इति आयुर्वेदः।
14. वाग्भट : अष्टांगहृदय
15. चरकसंहिता, शारीरस्थान, 1.16
16. खादयश्चेतनाषष्ठाः धातवः पुरुषः स्मृतः। चरक., शारीर., 1.16
17. वही, 1.5;18.49; वाग्भट : अष्टांगहृदय, 1.7; त्रिदोष = वात, पित्त, कफ

These are the three humours of the body, the vitiation or derangement of the three humours of the body is called Tridosha.

In medieval physiology any of the four fluids of the body, i.e. blood, phlegm, choler (yellow bile) the dominance of which was thought to determine a person's character and general health. Accordingly one's disposition may be sanguine, choleric or melancholy. (Cardinal humour).

In Geek medical system also these three humours were considered as the cardinl elements of the body, which decided the nature of a person and the line of treatment. Reader's Digest Universal Dictionary.

17. वायुः पित्तं कफः प्रोक्तो शारीरो दोषसंग्रहः।
मानसः पुनरुद्दिष्टो सत्त्वं रजश्च तम एव च।। चरकसहिता, 1.57
18. 'स्वभावो परमः' 'स्वभावस्तु प्रवर्तते'। चरकसंहिता, निदानस्थान, 1.4
19. सुश्रुतसंहिता, 1.40.12-13 ; 'हेतुर्लिंगौषधिज्ञानम्' चरकसंहिता, निदानस्थान
20. 'नानौषधं जगति किंचिद् द्रव्यमुपलभ्यते'। चरक., 1.26.12
21. चरकसंहिता के आठ स्थान; सुश्रुतसंहिता के छः अंग; वाग्भट ने तो अपने

ग्रन्थों के नाम ही 'अष्टांगसंग्रह' और 'अष्टांगहृदय' रक्खे।

22. 'सर्वमन्ने प्रतिष्ठितम्'। चरकसंहिता, 1.27.249-50

23. चरकसंहिता में चिकित्सास्थान, 28, सुश्रुतसंहिता में निदान अंग, 1.16, माधवकर का माधवनिदान।

24. सुश्रुतसंहिता, 1.6; इसमें शल्यक्रिया का विशेष वर्णन है, शल्यक्रिया से पूर्व रोगी को निश्चेतन करने के लिए सम्मोहिनी ओषधि सुंघाने का उल्लेख है।

25. चरकसंहिता, सूत्रस्थान, 30.28 सुश्रुतसंहिता, सूत्रस्थान, 1.7,

26. काश्यपसंहिता, पृ0 44

27. Girindranath Mukhopadhyay : History of Indian Medicine, Vol. 1-2

28. कविराज सूरमचन्द्र : आयुर्वेद का इतिहास, 1. पृ0 194-195

7

वेदेषु जलचिकित्सा

डॉ० विक्रमकुमारः

वेदाः खलु सर्वज्ञाननिधयः। एषु जीवनोपयुक्तानि सर्वज्ञानानि समाविष्टानीत्यत्र नास्ति संशयावकाशः। वेदमन्त्रा ननु न केवलं शारीरिकम् अपितु मानसिकम् आत्मिकं सामाजिकञ्चापि समाधानं समामनन्ति।

अत्रास्माकं विचारचर्चासरणिर्वर्त्तते-'वेदेषु जलचिकित्सा।' तत् संक्षिप्यात्र किंचिदुच्यते। वैदिकनिघण्टावुक्तम्-'अमृतमित्युदकनाम' (निघण्टु 1.12) जलं खलु अमृतमित्युच्यते। जलपानेन, जलव्यवहारेण च जना रोगमपाकरोति। अथर्ववेद उक्तम्-'अप्स्वन्तरमृतमप्सु भेषजम्' (अथर्व 1.4.4) जलेऽमृतं, जीवनशक्तिः रोगापाकरणबलञ्च वरीवर्त्ति। वस्तुतः प्राकृतिकपदार्थेषु जलमेकं महौषधं यद्रोगमपाकृत्य शरीरे शक्तिं निदधाति। तस्मादेवोच्यते-**अप्सु मे सोमो अब्रवीदन्तर्विश्वानि भेषजा (अथर्व १.६.२) आपः पृणीत भेषजम् (अथर्व १.६.३)**। इत्थं जलं भेषजत्वात् प्राकृतिकचिकित्सायाः परमोपायत्वेन वेदेषु परिगण्यते।

जलं ननु वृष्टितः, पर्वतस्रोतोभ्यः, निर्झरेभ्यः, तडागेभ्यः, सरिद्भ्यः, कूपेभ्यश्च प्राप्यते। बहिरागत्य चेदं 'जीवनम्' इत्यभिधीयते। यदाऽग्निनांगुलिर्वा त्वग् वा शरीरांशो दह्यते तदाऽस्माभिर्ब्रूयते-**अहो दग्धं, भो दग्धम्! 'हाय, जल गया।' इत्युच्यते।** 'जल गया' अर्थात् जलं गतम्, जलं नष्टम्। नष्टे जले किमवशिष्टम्? केवलं भस्म, जलहीनश्च तत्। जलेनैव पूर्वं तज्जीवनमासीत्। जलं गतञ्चेज्जीवनमति गतमित्यभिप्रायः। तस्माज्जलं न केवलं भेषजमपितु मूलभूतो जीवनाधार इति वेदानुशासनम्।

उषःपानद्वारा उदरशुद्धिः - यदोदरेऽजीर्णता भवति, मलग्रन्थयो वोदरनलिकास्ववरुद्धा भवन्ति तदा प्रातरि शीतोष्णजलपानेन मलनिष्कृतिर्बलाद्भवतीत्यनुभूतमेतद्। आन्त्रोष्णतां गुदामार्गोष्णतां, शारीरिकज्वालां, कण्ठशुष्कताञ्चेदं पयोऽपाकरोत्येव।

अथर्ववेदीय षष्ठकाण्डस्य चतुर्विंशतितमं सूक्तमिदमेव समर्थयति-

हिमवतः प्रस्रवन्ति सिन्धौ समह संगमः।

आपो ह मध्यं तद् देवीर्ददन् हृद्द्योतभेषजम्।। अथर्व ६.२४.१

आपः जलानि हिमवतः हिमालयात् प्रस्रवन्ति निःसरन्ति सिन्धौ समुद्रे च तानि समह समान तया संगमः संमिलन्ति। तानि दिव्यजलानि च मप हृद्द्योतभेषजं हृदयसम्बन्धितज्वलनं हृच्छूलं हृद्रोगञ्च प्रदहन्ति।

मम चक्षुषोर्यज्ज्वलनं भवति, पाष्ण्योः पादयोर्वा यत्कष्टकारिणी उष्णता वरीवर्त्ति सर्वं तन्नु भिषजां सुभिषक्तमाः चिकित्सकानामुत्तमचिकित्सका आपः जलानि निष्करन् निःसारयन्ति-

यन्मे अस्योरादिद्योत पाष्ण्योः प्रपदोश्च मत्।

आपस्तत् सर्वं निष्करन् भिषजां सुभिषक्तमाः॥ अथर्व ६.२४.२

तच्च जलं समुद्रपत्नीनां नदीनाञ्चेत्तन्नु समधिकसुखावहमुपभोग्यञ्च भवति। अत उक्तम्-

सिन्धुपत्नीः सिन्धुराज्ञीः सर्वा या नद्य स्थन।

दत्त नस्तस्य भेषजं तेन वो भुजनामहै॥ अथर्व ६.२४.३

त्रिष्वेषु मन्त्रेषु प्रवहद्भ्यो जलेभ्यः शैत्यमायाति तेन च ज्वलनस्य चिकित्सा साध्यत इत्युक्तम्। हिमवतः पर्वतान्निर्झरतो जलस्योपयोगेन तत्पानेन हृदयरोगः प्रणश्यति। उपरिष्टात् प्रपतनेनाद्रिजले वीरूधां लतानां भूरूहाणां च रसाः भवन्ति, ते च हृदयं बलं प्रददति। नेत्रयोः शीतलजलप्रसिञ्चनेन शैत्यमुद्भवति नेत्ररोगाश्चानिवार्यतयाऽपाक्रियन्ते। नदीजलावगाहनेन, जलतरणेन, जलमग्नोन्मग्नेन या शान्तिरवाप्यते तस्या अपि सूचनं मन्त्रेषु विद्यते।

अन्यत्राप्युक्तम्-

अरिप्रा आपो अप रिप्रमस्मत्।

प्रास्मदेनो दुरितं सुप्रतीकाः प्र दुःष्वप्न्यं प्र मलं वहन्तु॥ अथर्व १०.५.२४

अत्रेदं वर्णितं यद् (अरिप्राः) निर्दुष्टानि (आपः) जलानि अस्मत् रिप्रं रोगम् अपवहन्तु अपाकुर्वन्तु। (सुप्रतीकाः) शुद्धानि समुज्ज्वलजलानि अस्मद् (दुरितम् एनः) दुष्टदोषं (प्रवहन्तु) निःसारयन्तु। दुःष्वप्न्यं मलं स्वप्नदोषात्मकं मलं दूरीकुर्वन्तु।

अस्मिन् मन्त्रे जलचिकित्सया दुषितस्वप्नस्यापाकरणं निगर्दितम्। स्वप्नकाले वीर्यस्खलनं खलु भवति, तस्य चिकित्सा जलेन कर्तुं शाक्यत इति मन्त्राभिप्रायः। अपर्युषितजलेन स्नानविधिराचरणीया। नाभिस्नानेन, उपस्थेन्द्रियस्नानेन, शयनाद् वा प्राक् शीतलजलार्द्रवस्त्रास्पर्शेन कामोष्णताशान्तिरवाप्यतएवेति नात्र संशीतिः।

जलेन हि सद्योवृणचिकित्सापि निष्पाद्यते। व्रणोपरि, शस्त्राघातोपरि जलधाराप्रवाहेण रक्तानिर्झरो निरुध्यते। जलेन मर्दनं जलेन मार्जनं जलेनाभिवर्षणमित्यादिकं विविधजलकर्म सुखमेवावहति। अतः कथ्यते-

जालाषेणाभिषिञ्चत जालाषेणोपसिञ्चत।

जालाषमुग्रं भेषजं तेन नो मृड जीवसे॥ अथर्व ६.५७.२

आपः खलु निश्चयेनौषधया रोगकारणापाकर्त्र्यः वर्त्तन्ते -

आप इद्वा उ भेषजरापो आमीवचातनीः।

आपो विश्वस्य भेषजीस्तास्त्वा मुञ्चन्तु क्षत्रियात्। अथर्व ३.७.५

आपः मातृवद्रोगनियन्त्री वर्त्तते। ताः जननीवदस्मान् शुन्धयन्ति - आपो अस्मान् मातरः शुन्धयन्तु (ऋग्वेद 10.17.10)

आपः पयांसि ओषधयः एषां समेषां समानकर्म तच्च शोधनात्मकम्। अत उक्तम् -

पयस्वतीरोषधयः पयस्वन्मामकं वचः।

अपां पयस्वदित्पयस्तेन मा सह शुन्धत॥ ऋग्वेद १०.१७.१४

ऋग्वेदीयदशममण्डलस्य नवमसूक्तस्य देवता आपः वर्तन्ते। तत्र ह्युक्तम्-आपः मयोभुवः सति। जलं खलु मयः सुखमेव भावयति। यस्तत्र शिवकारका रसः सोऽस्मान् भजतां कामयमाना मातेव सेवताम्। यच्च किमपि मयि दुरितं तदापः प्रवहन्तु खलु -

आपो हिष्ठा मयोभुवः। यो वः शिवतमो रसस्तस्य भाजयतेह नः।

उशतीरिव मातरः॥ इदमापः प्रवहत यत्किंच दुरितं मयि। (ऋग्वेद १०.९)

नदीजलावगाहनेन सर्पदंशविषचिकित्सा अपि प्रसाध्यते। जलीय सर्पा भूमिसर्पापेक्षया स्वल्पविषाः भवन्ति। **''वारिविप्रहता अल्पविषाः स्मृताः'' (सुश्रुत कल्प ४.३२)** सर्पदंशे सति जलावगानं चरकेऽपि वर्णितम्-**''अवगाहनरक्तमोक्षणाः'' (चरकः विषचि०। अ०२३। ३३)** **अत्यन्तविषरोगार्तान् जलमध्ये विनिक्षिपेत्** (ऐन्द्रजालिक। कामरत्न। सर्पविषचिकित्सा। 45)

8

जैन धर्म–दर्शन में अहिंसा एवं शाकाहार

एक वैज्ञानिक विश्लेषण

डॉ॰ (श्रीमती) विनय गुप्ता

जीवन में उपयोगिता की दृष्टि से धर्म–दर्शन और विज्ञान दोनों का स्वतन्त्र महत्त्व है। ये दोनों ही सत्य तक पहुंचने के माध्यम हैं। विज्ञान जहाँ भौतिक–प्रयोगशाला में किसी वस्तु की सार्वभौमिक सत्यता को उद्घाटित करता है, वहीं धर्म व दर्शन जिज्ञासा–अनुभव के आधार पर आत्मप्रयोगशाला में सत्य को खोजता है। दोनों का मार्ग्य तो एक ही है– सत्य को खोजना। परन्तु मार्ग अलग–२ हैं।

जैन धर्म–एवं दर्शन प्रकृति के अनुरूप होने के कारण जीवनोपयोगी व व्यावहारिक धर्म–दर्शन है। इसकी मान्यताएँ वास्तविकता की सुदृढ़ नींव पर अवस्थित हैं और विज्ञान सम्मत हैं। अतएव यह एक वैज्ञानिक धर्म–दर्शन है। यह एक निर्विवाद सत्य है कि अणु, परमाणु, जीव, पुद्गल, वनस्पति आदि का विशद व सूक्ष्म विश्लेषण जैन–दर्शन में मिलता है जिसका लक्ष्य पूर्ण वीतरागता की प्राप्ति है।

जैन धर्म के दो मुख्य अंग हैं– आचार और विचार। यहां आचार का मूल आधार अहिंसा और विचार का मूलाधार अनेकान्तवाद है।

'अहिंसा परमो धर्मः' अत्यन्त प्राचीन एवं सर्वमान्य सिद्धान्त है। जिसका प्रथम रूप वैदिक परम्परा में देखने को मिलता है जिसका आरम्भ उपनिषदों से होता है। जैन धर्म सन्देश देता है कि "सभी प्राणी जीना चाहते हैं, कोई मरना नहीं चाहता। सभी सुख चाहते हैं। दुःख सभी के प्रतिकूल है[1]"। किसी को न मारना ही अहिंसा नहीं है, अपितु न मारना तो हिंसा का निषेध मात्र है। मैत्री, करुणा, प्रमोद, दया, अपरिग्रह, संयम, समता और सच्चरित्र का अनुसंधान ये अहिंसा के विधि पक्ष के प्रधान स्तम्भ हैं।[2]

मनुष्य स्वभावतः एक अहिंसक प्राणी है तथा अहिंसा ही आत्मा का स्वभाव है[3]। प्रकृति ने मानव शरीर को अहिसंक आहार के योग्य बनाया है। मानवीय शरीर का प्रारम्भिक काल मां के दूध,

---

1. सब्वेणाणा पिआऽया सुहसाया, दुह पठिकूला आधिव वहा– आचारांग सूत्र– १–३
2. (क) दया य संजमे लज्जा, दुंगछा अच्छलवादि या।
   तितिक्खा य अहिंसा या, हिरीतिएकाट्ठिया यदा। उत्तराधययन सूत्र, नियुक्ति, अध्याय–३
   (ख) भगवती सूत्र ८–७१
   (ग) उत्तराधययन सूत्र वृत्ति १–११
3. अहिंसा निउणाट्विट्ठा सव्वभूएसुसंजमो। दशवैकालिक सूत्र ८–६

गाय भैंस के दूध, अन्न, फल, वनस्पत्तियों के सहारे बीतता है और इनके प्रति मनुष्य का सहज झुकाव होता है। शारीरिक संरचना, पाचन संस्थान, पाचन–प्रक्रिया, मानसिक वृत्तियों आदि पर सम्यक् व वैज्ञानिक दृष्टि से विचार करने पर परिलक्षित होता है कि मानव एक अहिंसक प्राणी है। सात्विक निर्वाह के लिये मानव को प्रेरित करना जैन धर्म का मुख्य लक्ष्य रहा है। स्वास्थ्य रक्षा एवं आरोग्य की दृष्टि से जैन धर्म आधुनिक चिकित्सा विज्ञान के अत्यन्त निकट है। उसके अनुसार मानव शरीर को स्वस्थता प्रदान करने के लिये तथा आयुपर्यन्त शरीर की रक्षा के लिये सात्विक एवं हिंसा के निमित्तों से रहित आहार योग्य होता है।

समस्त हिंसा यत्नशून्य एव हारो युक्ताहारः[१]।

जहां प्राणिमात्र अपनी नैसर्गिक प्रवृत्ति से प्रेरित होकर अपने लिये प्रकृति–प्रदत्त प्राकृतिक भोजन से जीवन यापन करता है, वहीं आधुनिक मानव अपनी नैसर्गिक वृत्ति को न सुनता है और न मानता है। "जीवन तो भोजन के लिये है"– इसी सिद्धान्त को अपनाकर अज्ञानतावश और जीभ लोलुपता वश वह अपने नैसर्गिक आहार–शाकाहार को भूलकर अण्डाहार मांसाहार जैसे अखाद्य आहार ग्रहण कर अनेक प्रकार के नाशों को खुला निमन्त्रण दे देता है। इस समय तो मांसाहार के पीछे दो प्रमुख शक्तियां कार्य करती दिखाई देती हैं– रसलोलुपता और आधुनिक शरीर विज्ञानियों द्वारा मांसाहार पक्ष में प्रभूत प्रचार।

आहार सम्बन्धी आधुनिक वैज्ञानिक अध्ययनों से ज्ञात होता है कि मांसाहार से लाभ की अपेक्षा हानि अधिक होती है। ये तथ्य सामने आए हैं कि मांसाहार शरीर में जैव–विष (टाक्सिनस) पैदा करता है। जैव विष रक्त संचार को बाधित कर, हृदयाघात, कैन्सर आदि अनेक रोगों को आमन्त्रित करता है जो मानव का शारीरिक व मानसिक सन्तुलन बिगाड़ देता है। विश्व–विख्यात हृदय–रोग विशेषज्ञ बोस्टन के डॉ॰ पॉल उडले व्हाइट का कहना है कि मांसाहार से प्राप्त होने वाली अतिरिक्त वसा (Colestrol) की परतें हृदय की धमनियों में जमा होकर रक्त संचार को बाधित करती है जिससे हृदय पर अनावश्यक दबाव पड़ता है जो प्रायः मृत्यु का भी कारण बन जाता है। इसके साथ–साथ अमाशय व आंतों में मांस की सड़ांध से उत्पन्न विषैले कीटाणु रोगकारक होते हैं। डॉ॰ बोशई के कथनानुसार मांसाहारियों में शाकाहारियों की अपेक्षा दुगुनी विषाक्तता होती है।

अनेक शोधों से प्रत्यक्ष हुआ है कि मृत्यु डर से ग्रसित जानवर की विभिन्न स्राव–ग्रन्थियों से अनेक हारमोन्स स्रवित होते हैं। जो मानव के शरीर व मन पर गहरा विपरीत प्रभाव डालते हैं। हाल में फेले 'बर्ड–फलू' नामक रोग का खतरा मांसाहारियों को ही मूलतः है। इस समय भारत में भी इस रोग ने पांव रख दिया है। जिससे दो तरह से जीवन हानि हो रही है– एक तरफ तो संक्रमण से पक्षी मर रहे हैं और साथ ही संक्रमण से भय से निरीह मूक पक्षियों को निरन्तर बड़ी संख्या में निर्ममता से मारा जा रहा है और दूसरी ओर मानवों में पक्षियों से फैला संक्रमण भी उनकी मृत्यु का कारण बन रहा है। ये प्यारे से पशु–पक्षी इस संसार के सौन्दर्य और सुख में विशिष्ट जैवी–विधि से अपनी भूमिका का निर्वाह करते हैं जिन्हें देखभाल की व सुरक्षा की आवश्यकता है।

शाकाहारियों का शारीरिक व मानसिक स्वास्थ्य मांसाहारियों की अपेक्षा अधिक सुदृढ़ होता है। यदि हम शाकाहार व मांसाहार का आवश्यक तत्वों अर्थात् प्रोटीन, खनिज लवणों या कैलशियम

---

१. प्रवचनसार, २२४

आदि की दृष्टि से विचार करें तो ज्ञात होता है कि शाकाहार में ये सभी तत्व मांसाहार की अपेक्षा प्रचुर मात्रा में मिलते हैं। जैसे यदि हम Protins की बात करें तो एक अण्डे में जहां ये १३.३ प्रतिशत और १०० ग्राम मांस में १८.५ प्रतिशत हैं, वहीं (१०० ग्राम) मूंग में २४ प्रतिशत और सोयाबीन में ४३ प्रतिशत पाए जाते हैं। उर्जादायक कार्बोहाईड्रेट्स (energy giving) जहां (१०० ग्राम) मूंग में ५६.६ प्रतिशत, मटर में ६३.५ प्रतिशत, मैथी में ४४.१ प्रतिशत, दूध में ५१ प्रतिशत अर्थात् शाकाहारी भोजन में प्रचुर मात्रा में मिलते हैं, वहीं मांसाहारी भोजन में बिल्कुल भी नहीं होते हैं।[1] नीचे प्रस्तुत तुलनात्मक चार्ट (हेल्थ बुलेटिन नं॰ २३, भारत सरकार) से शाकाहार की महत्ता स्वतः सिद्ध होती है।

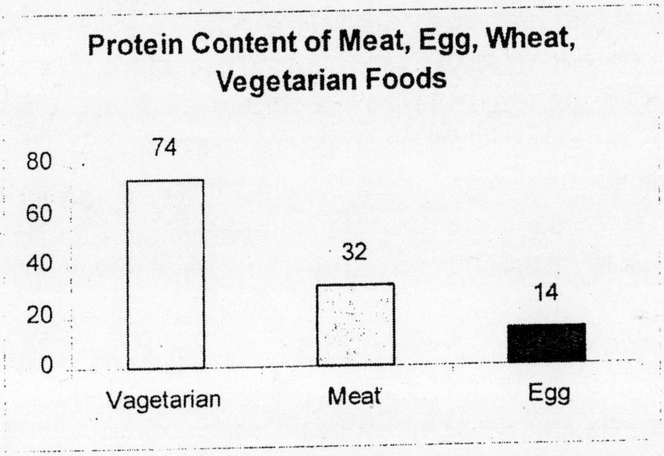

भोजन के पाचन व मल–निष्कासन में सहायक फाइबर अर्थात् रेशा भी शाकाहारी भोजन में प्रचुर मात्रा में मिलता है।

अनेक आंकड़ों से ज्ञात होता है कि मांसाहारियों की मृत्यु–दर का अनुपात शाकाहारियों की मृत्यु–दर से कहीं अधिक है।

किसी भी वनस्पति स्रोत से प्राप्त न होने कारण अण्डाहार को शाकाहार में सम्मिलित करना मानव की मात्र जिह्वा लोलुपता ही कही जाएगी। कारण कि अनेक वैज्ञानिक शोधों से ज्ञात हो चुका है कि यद्यपि कुछ अण्ड़ों में प्रजनन शक्ति कम हो सकती है परन्तु उनमें जीव–उत्पादक शक्ति नहीं है– ये तो मात्र भ्रम है।

आज विश्व के लाखों लोग पुनः शाकाहार की ओर परिवर्तित हो रहे हैं। अमेरिका, इंग्लैण्ड आदि देशों में मानव–आहार पर किये जा रहे अनेक शोधों से पता चला है शाकाहारी दीर्घायु, स्वस्थ मन मस्तिष्क वाला, उर्जा से परिपूर्ण तो रहता ही है साथ ही उसमें रोग–प्रतिरोधक शक्ति भी अधिक होती है।

शाकाहार का मानसिक स्वास्थ्य से भी गहरा सम्बन्ध है। किसी जीव की हत्या पर कोई भी भावनात्मक रूप से करूणामय, प्रेममय, सन्तुलित व संयमी नहीं रह सकता। जहां मांसाहार उत्तेजना,

---

1. ये सभी तथ्य भारत सरकार के Health Bulctin No. २३ के Food Value Chart के आधार पर है)

हिंसक प्रवृत्ति, क्रोध, भय, संवेदनशीलता, चिड़चिड़ापन आदि का संचार करता है वहीं शाकाहार शालीनता, शान्ति, प्रसन्नता, मैत्री करूणा व स्फूर्ती का संचार करता है।

मनुज प्रकृति से शाकाहारी, मांस उसके अनुकूल नहीं,

पशु–पक्षी मानव से प्राणी, वे मेवा, फल–फूल नहीं।

जैन दर्शन ने अहिंसा के सिद्धान्त को गम्भीरता से लिया है और विस्तार से इसका विवेचन किया है। प्राकृतिक सौन्दर्य व सन्तुलन और पर्यावरण संरक्षण के लिये भी अहिंसक–आहार अत्यन्त आवश्यक है। जैन दर्शन के अनुसार प्रत्येक व्यक्ति से दैनिक–जीवन में यह अपेक्षा की गई है कि वह यथासम्भव जीवों की रक्षा करें और हिंसा का त्याग करें। भूमि पर छोटे–२ कीट–कृमियों को अपना भोजन बनाने वाले जीव मेंढ़क, चूहा, सांप आदि जहां वनस्पति संरक्षक है वहीं मछली, मगरमच्छ आदि जीव जल संरक्षक है। प्रकृति की व्यवस्था में सभी जीवों का अपना–२ महत्त्वपूर्ण स्थान है। इस जैविक–चक्र में जहां हम एक कड़ी को तोड़ देते हैं, वहीं से हिंसा–चक्र आरम्भ हो जाता है और फिर हम अपनी आपदाओं पर से पार नहीं पा सकते।

पशु–पक्षी न केवल पर्यावरण–संरक्षक हैं अपितु प्राकृतिक सौन्दर्य वर्धन में भी सहायक है। साथ ही ये अपने परिवर्तित व्यवहार द्वारा पृथ्वी पर भूस्खलन, अतिवृष्टि, अनावृष्टि आदि अनेक प्राकृतिक आपदाओं की भी पूर्व सूचना दे देते हैं, जिनका विशद विवेचन हमारे प्राचीन ग्रन्थों में मिलता है।

जैनमतानुसार तो सम्पूर्ण–प्राणी–वंश की रक्षा करना मानव का परम धर्म है परन्तु हमने अपने अज्ञान, अहंकार, लिप्सा के वशीभूत होकर अनेक दुर्लभ वन्य पशु–पक्षियों का तो नामोनिशान मिटा दिया है जो मानव का प्रकृति के साथ सबसे क्रूर व्यवहार कहा जा सकता है।

जैन धर्म–दर्शन में अहिंसा को कायरता से दूर वीरत्व और विराट भावना का मार्ग माना गया है। वहां मानव को अपनी चारित्रिक उदारता से विश्वकल्याण की ओर अग्रसर होने का सन्देश दिया है।

इस प्रकार मांसाहार शारीरिक, मानसिक, नैतिक सामाजिक एवं प्राकृतिक सन्तुलन आदि सभी दृष्टियों से हानिकारक व त्याज्य है। अनेक ठोकरें खाने के बाद अन्त में मानव समाज को समग्र रूप से उस भगवती अहिंसा की शरण में आना ही पड़ेगा जो प्राणिमात्र के लिये कल्याण प्रसविनी है–

एसा भगवती अहिंसा तसथावरसवभय खेमकरी (प्रश्न व्याकरण)

हम भी अपने ओर से नतमस्तक होकर इस देवी के चरणों में अपनी श्रद्धांजलि अर्पित करते हैं और सर्वोदय के वरदान की याचना करते हैं–

सर्वे भवन्तु सुखिनः, सर्वे सन्तु निरामयाः।

सर्वे भद्राणि पश्यन्तु, मा कश्चित् पापमाचरेत्।

9

अथर्ववेदीय ऋषि ब्रह्मा की चिकित्सा-पद्धति

डॉ॰ अनुभा जैन

वेदों और उपनिषदों में मानव शरीर को ब्रह्मपुरी कहा गया है। इस पुरी में निवास करने के कारण ही ब्रह्मा को पुरुष कहा जाता है।[1] भौतिक दृष्टि से वेदों में 'अथर्ववेद' का विशेष महत्त्व है; क्योंकि इसका सीधा सम्बन्ध आयुर्विज्ञान से भी है। इसीलिए इस वेद का यही उपदेश है कि अपने क्षेत्र में निरोग बन कर रहो। यह क्षेत्र अर्थात् शरीर किसी प्रकार भी दैहिक या आध्यात्मिक व्याधि से ग्रसित न हो। शरीर को रोग रहित रखने हेतु जरा, व्याधि से विमुक्त रखने के लिए भारतीय प्राचीन वैदिक चिकित्सा पद्धति को तीन रूपों में देखा जा सकता है। शल्य क्रिया द्वारा की जाने वाली चिकित्सा आसुरी चिकित्सा कहलाती है। औषधियों द्वारा की जाने वाली चिकित्सा मानुषी तथा प्राणायाम, योग, हवन आदि के द्वारा दैवी चिकित्सा सम्पन्न की जाती है।

शरीर के विषय में चर्चा करते हुए 'अथर्ववेद' की यह मान्यता है कि शरीर का अधिष्ठाता जीव नित्य है[2] परन्तु वात, पित्त, कफ तथा रक्त, रस, मांस, मेद, अस्थि, मज्जा, शुक्र- इन 7 धातुओं से शरीर का निर्माण होता है।[3] 'ऋग्वेद'[4] और 'अथर्ववेद' में वात, पित्त, कफ के लिए 'त्रिधातु' शब्द का प्रयोग मिलता है। 'ऋग्वेद' में सायण ने त्रिधातु की व्याख्या करते हुए इन तीनों का ही नामोल्लेख किया है।[5] चरक, सुश्रुत और अष्टांगहृदय में भी वात, पित्त और कफ - तीनों को ही शरीर का आधार माना गया है ; क्योंकि इन तीनों के विकार से ही शरीर में नाना प्रकार के दोष उत्पन्न होते हैं।[6] सुश्रुत के अनुसार वात, पित्त, श्लेष्म (कफ) ये तीन शरीर की उत्पत्ति के कारण हैं।[7]

---

1. अथर्व॰, 10.2.28 ऊर्ध्वो नु सृष्टास्तिर्यङ् नु सृष्टा सर्वाः दिशः पुरुष आ बभूव
   पुरं यो ब्रह्मणो वेद यस्याः पुरुष उच्यते ।
2. वही, 18.4.28
3. वही, 18.4.29
4. ऋ॰, 1.34.6 : त्रिधातु शर्म वहतं शुभस्पती ।
5. ऋ॰, सा॰ भा॰
6. चरक, सूत्र॰, 1.57
7. सुश्रुत, सूत्र॰ 21.3

शरीर और जीव के सम्बन्ध को पुष्ट करता हुआ 'अथर्ववेद' यह स्वीकार करता है कि जीवात्मा की चेतना-शक्ति से प्रेरित प्राण हृदय में प्रविष्ट हो कर रक्तवाहिनी नाड़ियों के रक्ताशयों को पुनः-पुनः कँपाता है। उनमें भ्रमण करता हुआ रक्त जीवात्मा की स्थिरता तथा जीवन का कारण बनता है। वह प्राण रक्त को पूर्ण रक्ताशय से दूसरे रक्ताशय में असंख्य सूक्ष्म तन्तुओं वाले मार्ग के माध्यम से प्रेरित करता है। इसी प्रकार शरीर के भीतर अनेकानेक मार्गों से रक्त का संचार बराबर बना रहता है। इस प्रकार रक्ताशयों द्वारा रक्त के गमनागमन से जीवन का शरीर के साथ सम्बन्ध बना रहता है और यही शरीर का आधारस्तम्भ है।

मनुष्य के भौतिक और आध्यात्मिक विकास के लिए उसका तन-मन स्वस्थ होना अत्यावश्यक है। इस द्वि-आयामी लाभ के लिए आयुर्वैदिक चिकित्सा प्रणाली का प्रयोग किया जाता है। इस भारतीय चिकित्सा प्रणाली की विशेषता यह है कि इसमें शारीरिक स्वास्थ्य के साथ-साथ मानसिक एवम् आत्मिक स्वास्थ्य पर भी ध्यान दिया जाता है; क्योंकि इस पद्धति में शरीर, मन, बुद्धि और आत्मा को पृथक् रूप में नहीं देखा जाता जैसा कि एलोपैथी चिकित्सा प्रणाली में किया जाता है। सुश्रुत भी वात, पित्त, कफ दोष की साम्यावस्था के साथ-साथ आत्म इन्द्रियाँ और मन के प्रसन्न होने की बात कहते हैं।[1] चरक के अनुसार भी आहार एवं आचरण के साथ-साथ अपनी चेष्टाओं पर पूर्ण नियन्त्रण आवश्यक है।[2] यही कारण है कि 'यजुर्वेद' के छः मन्त्रों में प्रार्थना की गई है कि मनुष्य का मन शुभ विचारों वाला, शुभ्र एवं पवित्र हो।[3] 'ऋग्वेद' में मन की शक्ति से मात्र हस्तस्पर्श के द्वारा रोगी को निरोग करने का उल्लेख मिलता है।[4] अतः आयुर्वैदिक चिकित्सा शरीर के रोगों का उपचार किया जाता है। अतः आयुर्वैदिक चिकित्सा शरीर के रोगों को मन, चित्त और कर्मो के साथ जोड़ती है।

प्राचीन भारतीय चिकित्सा-पद्धति की एक अन्य महत्त्वपूर्ण विशेषता यह भी है कि एक व्यक्ति का रोग प्रकारान्तर से उसके सारे जीवन, कार्य-शेष और समाज के साथ जुड़ जाता है। वैदिक वाङ्मय में पर्यावरण की शुद्धता और वनस्पतियों को फलयुक्त करने हेतु की गई सम्पूर्ण प्रार्थनाएँ व्यक्ति और विश्व के स्वास्थ्य को समन्वित रूप में दृष्टिगत कर उदात्त भावनाओं के साथ स्वस्थ देह की व्याख्या प्रस्तुत करती है।'

---

1. सुश्रुत, सू॰, 15.48 : समदोषः समाग्निश्च समधातुमलक्रियः। प्रसन्नात्मेन्द्रियमनाः स्वस्थ इत्यभिधीयते ।।
2. च॰ सं॰, सूत्र॰, 7.60 : आहाराचारचेष्टासु सुखार्थी प्रेत्यचेह च। परं प्रयत्नमातिष्ठेद् बुद्धिमान् हितसाधने।।
3. द्र॰, यजु॰, 34.1-6 मन्त्र
4. ऋ॰, 10.137.7 : हस्ताभ्यां दशशाखाभ्यां जिह्वा वाचः पुरोगवी। अनामयित्नुभ्यां त्वा, ताभ्यां त्वोप स्पृशामसि।।

विश्व चिकित्सा शास्त्र के इतिहास में सम्भवतः आयुर्वैदिक चिकित्सा प्रणाली एकमात्र ऐसा उदाहरण है जिसमें रोग, स्वास्थ्य और अध्यात्म को एक साथ रख कर देखा जाता है। सम्पूर्ण स्वास्थ्य के लिए सम्यक् सन्तुलन आवश्यक है। यह सन्तुलन शरीर और मन से प्रत्यक्ष-अप्रत्यक्ष सम्बन्ध रखने वाले सभी उपादानों में होना चाहिए। स्थान तथा काल के साथ-साथ आहार सम्बन्धी आचरण भी शरीर को प्रभावित करते हैं। आयुर्वेद की यह दृढ़ धारणा है कि शरीर में रोगों का प्रवेश मन के द्वार से ही होता है। अतः यह निष्कर्ष अत्यन्त सारगर्भित है कि सभी प्रकार के रोगों से बचने के लिए चिन्तन एवम् आहार-व्यवहार में अनुशासन अनिवार्य है। यही कारण है कि भारत के सभी धार्मिक-ग्रन्थों में आहार सम्बन्धी आचरण पर विशेष बल दिया गया है। इसीलिए चरक संहिता में शरीर और मन के आश्रयभूत व्याधियों की उत्पत्ति में काल, बुद्धि और ज्ञानेन्द्रियों के विषय का मिथ्या योग, अयोग और अतियोग- ये रोगों के त्रिविध कारण स्वीकार किए गए हैं।[1]

जन्म से मृत्युपर्यन्त के अन्तराल का नाम जीवन है और इसका आधार देह है। मानव-जीवन के अधिकांश सुख-दुःख देह से ही सम्बन्धित हैं। वैदिक चिन्तन में जड़-चेतन सभी वस्तुओं के कारकरूप में पांच तत्त्वों का उल्लेख मिलता है। ये तत्त्व वायु, पृथिवी, अग्नि, जल तथा आकाश हैं। मानव देह के अस्तित्व में भी इन्हीं पांच तत्त्वों की प्रमुख भूमिका मानी गई है। यही कारण है कि वैदिक प्रार्थनाओं और रोगोपचारों में इन पाँचों तत्त्वों के स्वास्थ्य की कामना और आवश्यकता पर विशेष बल देते हुए वैदिक ऋषियों ने शरीर को स्वस्थ रखने पर अतिशय बल दिया है।[2]

भारतीय चिन्तन परम्परा में शरीर को एक महत्त्वपूर्ण स्वतन्त्र इकाई मान कर आत्मा के वाह्न के रूप में अथवा आत्मा का आधार माना गया है। आध्यात्मिक क्रम में (ब्रह्म ⇒ आत्मा ⇒ बुद्धि ⇒ चित्त ⇒ अहंकार ⇒ मन ⇒ शरीर) पंच तत्त्व निर्मित शरीर सबसे बाद की इकाई है। नश्वर तत्त्वों के संयोग से निर्मित इस शरीर का अपना महत्त्वपूर्ण स्थान है। यही शरीर वह माध्यम है जिसे वैदिक साहित्य में नौका, रथ, वाहन आदि के रूप में चित्रित किया गया है। यथा मलिन् वस्त्रों को ऊपर से एक स्वच्छ कपड़े में बांध दिया जाता है, तथैव 206 हड्डियाँ, मांस, रक्त, मल आदि ऊपर से साफ-सुथरी त्वचा में बँधे हुए हैं। अतः स्थूल शरीर, सूक्ष्म शरीर और आत्मा का वाहन है। इसीलिए भी इसका स्वच्छ रहना अत्यावश्यक है। 'मनुस्मृति' में

---

1. च० सं०, सूत्रस्थानम्, दीर्घञ्जीवितीयाध्याय:, 1.45 : कालबुद्धीन्द्रियार्थानां योगो मिथ्या न चाति च ।
द्वयाश्रयाणां व्याधीनां त्रिविधो हेतु संग्रहः ।।
2. अथर्व०, 19.1.14 : पृथिवी शान्तिरन्तरिक्षं

देहधारियों को शुद्ध करने वाले पदार्थों का उल्लेख करते हुए ज्ञान, तप, अग्नि, आहार, मिट्टी, मन-विचार, जल, वायु, कर्म, सूर्य आदि को प्राणियों का शोधक पदार्थ बताया गया है।[1]

अथर्ववेदीय ऋषि ब्रह्मा-दृष्ट अनेक मन्त्रों में रोगों का परिचय तथा रोग-लक्षण देते हुए उन्होंने भौतिक तथा आध्यात्मिक दोनों दृष्टियों से उन रोगों का उपचार भी बताया है। भौतिक रूप से ये मन्त्र जहाँ रोग और उनके निदान की ओर संकेत करते हैं; वहीं आध्यात्मिक छवि से कर्म, सच्चरित्र और आत्मिक विकास का भी मार्ग प्रशस्त करते हैं।

अथर्ववेदीय ऋषि ब्रह्मा ने शरीर में होने वाले कतिपय विचारों के उपचारार्थ कुछ महत्त्वपूर्ण सर्वसुलभ प्राकृतिक उपायों का निर्देश किया गया है, जिनसे तत्तत् दोषों को दूर किया जा सकता है।

रक्त स्राव :

'चरक-संहिता' के अनुसार उत्तम विधि से भक्षण किए हुए आहार रस से जो रक्त की उत्पत्ति होती है; वह 'शुक्ल रक्त' होता है।[2] ऋषि ब्रह्मा ने भी शरीर में प्रवाहित होने वाले रक्त-चक्र की ओर संकेत किया है।[3] उनके अनुसार लाल-वर्ण वाले रक्त को वहन करने वाली धमनियाँ मानव को जीवन प्रदान करती हैं। यदि ये धमनियाँ रक्त को प्रवाहित करना बन्द कर दें[4] तो सम्पूर्ण जीवन ही अस्त-व्यस्त हो जाता है। अत: जीवन को सुचारू रूप से चलाने के लिए यह अत्यावश्यक है कि धमनियाँ जीवन-दायक लाल वर्ण वाले रक्त का उचित प्रकार से वहन या संचार करती रहें।

ऋषि ब्रह्मा ने रक्त स्राव के उपचारार्थ उपाय संकेतित किए हैं। उनके अनुसार इन धमनियों के कट जाने से और इनमें रक्तस्राव होने पर कटे हुए भाग पर धमनियों के सिरों को बाँध देना चाहिए।[5] रक्तस्राव को रोकने का दूसरे उपाय का भी उल्लेख ऋषि ब्रह्मा के ही मन्त्रों में मिलता है कि रेत या शर्करा बाँधने से रक्त का बाहर की ओर होने वाला प्रवाह रुक जाता है।[6] उनके अनुसार बारीक मिश्री के चूर्ण से भी रक्तस्राव को रोका जा सकता है।[7]

---

1. मनु॰, 5.105 : ज्ञानं तपोऽग्निराहारो मृन्मनो वायुर्पाञ्जनम्। वायुः कर्मार्ककालौ च शुद्धेः कर्तृणि देहिनाम् ॥
2. च॰ सं॰ सूत्रस्थानम्, विधिशोणितीयाध्यायः, 24.1-2:
   विधिना शोषितं जातं शुद्धं भवति देहिनाम् । देशकालौकसात्म्यानां विधिर्यः संप्रकाशितः ॥
3. अथर्व॰, 1.17
4. वही, 1.17.1: जामयस्तिष्ठन्तु हतवर्चसः।
5. अथर्व॰, 1.17 : तिष्ठावरे तिष्ठ पर उत त्वं तिष्ठ मध्यमे। कनिष्ठिका च तिष्ठति तिष्ठादिद्धमनिर्यही।
6. वही, 1.17.4 : परि वः सिकतावती धनूर्बहत्मक्रयीत् ।
7. वही, 1.17.2 : शतस्य धमनीयां सहस्रस्य हिराणाम्। अस्थुरिन्मध्यमा इमाः साकमन्ता अरंसत् ॥

'अथर्ववेद' में रक्तस्राव रोकने के लिए जल-चिकित्सा का भी मुख्य रूप से उल्लेख किया गया है।[1] 'पैप्पलाद-संहिता' में कुश को आस्रावभेषज कहा है।[2] 'अथर्ववेद' में अन्यत्र रक्तस्राव को रोकने के लिए विषाणका नामक ओषधि का उल्लेख मिलता है।[3] 'भावप्रकाश निघण्टु' में इसी ओषधि का अपर नाम 'मेषशृंगी' वर्णित है।[4]

कुष्ठ रोग

कुष्ठ रोग त्वचा का रोग होता है; जिसमें शरीर का रंग विकृत हो जाता है। कुष्ठ रोग की उत्पत्ति में सर्वप्रथम त्वचा; फिर लसीका, रक्त, मांस आदि क्रम से दूषित होते हैं। परन्तु वैदिक ऋषियों के मतानुसार वनस्पतियाँ सूर्य की किरणों से प्रकाश, तेल व ऊर्जा ग्रहण करके हरे रंग को धारण करती हैं तथा रोग नाशक बनती हैं जिनके उपयुक्त प्रयोग के द्वारा ही रोग का उपचार किया जा सकता है।

'अथर्ववेद' में अन्यत्र भी कुष्ठ रोग का वर्णन प्राप्त होता है।[5] 'अष्टांगहृदय' में कुष्ठ की उत्पत्ति के विशेष कारणों का उल्लेख मिलता है यथा मिथ्याएवम् अनियमित आहार-व्यवहार, सज्जन आदि का वध, पूर्वजन्मकृत-कर्मफल के कारण वातादि दोष आदि।[6]

ऋषि ब्रह्मा ने इस रोग के उपचार के लिए 'आसुरी वनस्पति' का संकेत किया है। यह ओषधि मात्र कुष्ठ रोग को ही नहीं, अपितु त्वचा के रंग को भी ठीक कर देती है।[7] यह वनस्पति श्याम वर्ण की होती है तथा ऋषि के अनुसार इसकी जड़ें महत्त्वपूर्ण हैं।[8]

कुष्ठरोगी के लिए व्रत, उपवास, जप आदि के अतिरिक्त सूर्यकिरण-चिकित्सा को बहुत उपयोगी बताया गया है।[9] इसके अतिरिक्त ऋषि कुष्ठ रोग के लिए रामा तथा कृष्णा ओषधि का संकेत करते हैं।[10] सायण के मत में इन्द्रवारुणी नामक ओषधि कुष्ठ (कोढ़) कास, श्वास, प्रमेह, गंडमाला और विष रोगों की नाशक है। सुश्रुत में कुष्ठनाशन के लिए घी में नील डाल कर उसे

---

1. अथर्व॰, 2.3.1: अदो यदवधावति - अवत्कमणि पर्वतात् । तत् ते कृणोमि भेषजम्।।
2. पै॰ सं॰, 20.54.3 : असृक्स्थानम् अथो आस्रावभेषजम् ।
3. अथर्व॰, 6.44.1
4. भाव॰, गुडू॰, 257-259, पृ॰ 256-257
5. द्र॰ अथर्व॰, 1.23, 1.24
6. अष्टांग, निदानस्थानम्, 14.1-3
7. अथर्व॰, 1.24.2
8. वही, 1.24.4 : श्यामा सरूपकरणी पृथिव्यसा अध्युद्भुता।
9. अथर्व॰ : भास्करा
10. वहीं, 1.23.1 : रामे कृष्णे असिक्नि च

व्रण पर लगाने का वर्णन मिलता है।[1] ऋषि ब्रह्मा द्वारा उल्लिखित आसुरी वनस्पति को भावप्रकाश में 'राजिका' नाम से भी इंगित किया गया है, जो कुष्ठ और खुजली की नाशक है।[2]

निद्रा रोग :

जब कार्य करते-करते मन थक जाता है एवम् इन्द्रियाँ भी थकने के कारण अपने-अपने विषयों से निवृत्त हो जाती हैं, तब मनुष्य को शयन करने की आवश्यकता का अनुभव होता है।[3] इस प्रकार मनोयुक्त इन्द्रियों का विषय से युक्त होना ही निद्रा की स्थिति है। आधुनिक चिकित्सा विज्ञान में इस निद्रा के न आने के मुख्य कारणों में एक कारण तनाव भी बताया जाता है और इसके इलाज के लिए ELDEP 75, VALIUM जैसी नशीली दवाइयां प्रयोग की जाती हैं। इससे इन्द्रियों की चेतना शक्ति नष्ट हो जाती है और गाढ़ निद्रा प्राप्त होती है।

परन्तु ऋषि ब्रह्मा ने इस रोग के उपचार के लिए प्राकृतिक उपाय बताया है। उनके अनुसार ध्यान व दृढ़ संकल्प से निद्रा को प्राप्त किया जा सकता है। रात्रि के समय जब चन्द्रमा की चाँदनी इस धरातल पर अपने शुभ्र व उज्ज्वल रूप में फैली होती है[4] तथा वायु अनुकूल गति से संचार करता है[5] तो ऐसी स्थिति में मनुष्य किञ्चित् ध्यान करके, मन शान्त करके निद्रा को प्राप्त कर सकता है।[6] सम्भवतः वैदिक ऋषि पहले से ही यह बात जानते थे कि आने वाला समय Stress, Tension, Pollution और Pressure से भरपूर होगा इसलिए अपने मन्त्रों में उन्होंने इसके उपचार का संकेत किया होगा।

बन्ध्यात्व रोग :

वैदिक काल से ही पुत्र रूप में सन्तानोत्पत्ति श्रेयस्कर मानी जाती थी। ऋषि ब्रह्मा ने एक ऐसी ओषधियाँ का संकेत किया है जो ऋषभक के नाम से जानी जाती है। जिसका सेवन करने से बंध्यात्व दोष दूर हो जाता है। यह ओषधि बल व शक्ति को बढ़ाती है, शरीर को पुष्ट करती है तथा गर्भाशय के दोष को दूर करती है। ऋषि ब्रह्मा ने मात्र इस ओषधि का नाम ही नहीं बल्कि इसके प्रयोग करने की विधि एवं स्थान का भी वर्णन किया है। उनके अनुसार इसके बीज कल्याणकारी हैं।[7] ये ओषधियाँ गहरे समुद्र भूमि में उत्पन्न होती हैं।[1]

---

1. सुश्रुत॰, चि॰, 9.37
2. भाव॰, धान्य॰, 72-74, पृ॰ 398
3. द्र॰, अथर्व, 4.5
4. अथर्व॰, 4.5.1: सहस्रशुङ्गो वृषभो यः समुद्रादुदारचरत् ।
5. अथर्व॰, 4.5.2: न भूमिं वातो अति वाति नाति पश्यति कश्चन।
6. वही, 4.5.1: तेना सहस्येना वयं नि जनान्त्स्वापयामसि ।
7. वही, 3.23.5 : यानि भद्राणि बीजान्मृषभा जनयन्ति च। तैस्त्वं पुत्रं विन्दस्व मा प्रसूर्धेनुका भव।।

प्राचीन रोग-निदान पद्धति ने प्राय: शारीरिक रोगों का जनक भी मन को ही माना है। यही कारण है कि सभी ऋषियों ने एकमत होकर मन की शुद्धि, स्थिरता और शान्ति पर बल दिया है। वेद में आत्मा को भी औषधि के रूप में स्वीकार किया गया है। चरक के अनुसार ईर्ष्या, शोक, भय, क्रोध, अहंकार और द्वेष आदि मन के विकार अर्थात् मानसिक रोग प्रज्ञापराध से ही उत्पन्न होते हैं।[2] ऋषि ब्रह्मा के सूक्तों का चिकित्सा-परक व्याख्या करते हुए एक जिज्ञासा उठती है कि आखिर शुद्ध-ज्ञान में डूबे हुए ऋषियों को सामान्य चिकित्सात्मक नुस्खे लिखने की क्या आवश्यकता थी। यदि उन्हें रोगों निदान के लिए ज्ञात जड़ी-बूटियों को लिपिबद्ध करना था तो चरक-संहिता जैसे शुद्ध चिकित्सा परक ग्रन्थ लिख दिए होते; परन्तु वस्तु-स्थिति ऐसी नहीं है। इन मन्त्रों में मात्र रोग या शरीर की चिकित्सा को प्रमुखता नहीं दी जाती। यहाँ रोग के माध्यम से शरीर, शरीर के माध्यम से आचरण, आचरण के माध्यम से मन और मन के माध्यम से आत्मा की एक सुनिश्चित भव्य यात्रा को रेखांकित किया गया है और इस चर्मोत्कर्ष की स्थिति में भौतिक जगत् से ऊपर उठ कर वह सात्त्विकता में समाहित हो जाता है। ऋषि ब्रह्मा प्राकृतिक तत्त्वों के महत्त्व तथा उनसे किए जाने वाले रोग-निदान से परिचित थे। मानव शरीर भी इन्हीं पंचतत्त्वों से मिल कर बना है। इसीलिए इन तत्त्वों की शुद्धता पर बल देते हुए ऋषि ब्रह्मा कामना करते हैं कि वायु समान रूप से मनुष्य के लिए पवित्र हों; अमृत रूपी जल सदा पृथ्वी पर विद्यमान रहे। वनस्पतियों तथा मानव को सदा सूर्य का प्रकाश ग्रहण होता रहे।[3] वैदिक ऋषि सम्भवत: इस बात से पूर्ण परिचित थे कि आने वाले समय में चाहे कितनी भी आधुनिक से आधुनिक पद्धति क्यों न विकसित हो जाए परन्तु सूर्य, जल, अग्नि, वायु, रक्त आदि प्राकृतिक तत्त्वों का कोई भी समतत्त्व नहीं हो सकता। अत: उन्होंने रोगों का निदान भी इन्हीं प्राकृतिक तत्त्वों के माध्यम से ही बताया है।

सूर्य :

ऋषि ब्रह्मा सूर्य की असीम ऊर्जा एवं जीवन के लिए उसकी अनिवार्यता से भलीभाँति परिचित थे। इसीलिए उन्होंने अपने सूर्यपरक मन्त्रों में ऐसा संकेत किया है कि सूर्य का प्रकाश प्राणि-जगत् के लिए अमृत है।[4] जीवन के लिए ऊर्जा का मिलना अत्यन्त आवश्यक है, जो हमें सूर्याग्नि से प्राप्त होती है। अत: सूर्य दर्शन एवं सूर्य की किरणों का प्रत्यक्षरूप में सेवन करने से

---

1. वही, 3.23.6: यासां द्यौ: पिता पृथिवी माता समुद्रो मूलं वीरुधां बभूव। तास्त्वा पुत्रविधाय दैवी: प्रावन्त्वोषधय:।।
2. च॰ सं॰, सूत्र॰, 7.52: ईर्ष्या-शोक-भय-क्रोध-मान-द्वेषादयश्च ये। मनोविकारास्तेऽप्युक्ता: सर्वे प्रज्ञापराधजा:।।
3. अथर्व॰, 8.1.5 : तुभ्यं वात: पवतां मातुरिश्वा तुभ्यं वर्षन्त्वामृतान्याद:।
   सूर्यस्ते तन्वेर शं तपति त्वां मृत्युर्दयतां मा प्र येष्ठा।
4. वही, 8.1.1 : इहायमस्तु पुरुष: सहासुना सूर्यस्य भागे अमृतस्य लोके।

मनुष्य स्वस्थ रहता है।[1] सूर्य जब उदय होता है, तो उसके प्रकाश से वनस्पति और मानव में कल्याण का संचार होता है और कई रोग उत्पन्न करने वाले कीटाणुओं का नाश होता है।[2] सूर्य में अनन्त शक्ति है जिसे वनस्पतियाँ सूर्यकिरणों के माध्यम से ग्रहण करती हैं। इससे ही वनस्पति में रोग नाश करने की क्षमता तथा बल-बुद्धि का गुण आता है।[3]

वेदों में सूर्य-किरणों से चिकित्सा का सविस्तर वर्णन प्राप्त होता है। सूर्य को इस चराचर जगत् की आत्मा कहा गया है।[4] उपनिषद् में भी सूर्य को मानव जगत् का प्राण कहा है[5] क्योंकि सूर्य रोगों को दूर करता है।[6] रोगों को दूर करने के साथ ही यह बुद्धि को भी शुद्ध करता है और ज्ञान की वृद्धि करता है।[7] सूर्य की किरणों में जीवन-शक्ति होती है और रोगों को नष्ट करने की क्षमता होती है। इसीलिए 'ऋग्वेद' में कहा गया है कि उदय होता हुआ सूर्य हृदय के सभी रोगों को, पीलिया और रक्त की न्यूनता को दूर करता है।[8] इसी बात को 'अथर्ववेद' में ऋषि भी संकेत करते हैं कि उदय होते हुए सूर्य की अवरक्त किरणें हृदय की बीमारियों और खून की कमी को दूर करती हैं।[9]

'अथर्ववेद' में एक अन्य स्थल पर सूर्य-किरण-चिकित्सा से ठीक होने वाले अनेक रोगों का वर्णन प्राप्त होता है। यथा सिरदर्द, रक्त की कमी, ज्वर, पीलिया, जलोदर, यक्ष्मा, नेत्ररोग आदि[10] एक अन्य सूक्त में 'सूर्यः कृणोतु भेषजम्' सूर्य-चिकित्सा से कुष्ठ रोग ठीक होने का वर्णन है।[11] इससे ज्ञात होता है कि वेदों के अनुसार प्रायः सभी सामान्य और असामान्य बीमारियाँ सूर्य-किरण-चिकित्सा से ठीक हो सकती है।

जीवन के लिए ऊर्जा का मिलना अत्यन्त आवश्यक है जो हमें सूर्याग्नि से प्राप्त होती है, जिसे आधुनिक चिकित्सा पद्धति विटामिन डी तथा के के रूप में मानती है। सूर्य की ऊर्जा को

---

1. वही, 8.1.4 : मा च्छित्था अस्माल्लोकादग्ने सूर्यस्य सदृशः ।
2. अथर्व॰, 8.2.11: वैवस्वतेन प्रहितान् यमदूतांश्चरतोऽप सेधामि सर्वान्।
3. वही, 1.24.1 : सुपर्णो जातः प्रथमस्तस्य त्वं पित्तमासिथ। तदासुरी युधा जिता रूपं चक्रे वनस्पतीन्।।
4. ऋ॰, 1.115.2 : सूर्य आत्मा जगतस्तस्थुषश्च ।
5. प्रश्नो॰, 1.8: प्राणः प्रजानामुदयत्येषु सूर्यः ।
6. ऋ॰ 10.100.8 : अपामीवां सविता साविषत्।
7. वही, 8.18.10 : अप सेधत दुर्मतिम्, आदित्यासः।
8. वही, 1.50.11: उद्यन्नद्य मित्रमह आरोहन् उत्तरां दिवम्। हृद्रोगं मम सूर्य हरिमाणं च नाशय।।
9. अथर्व॰, 1.22.1: अनु सूर्यम् उदयतां हृद्द्योतो हरिमा च ते। गो रोहितस्य वर्णेन तेन त्वा परिदध्मसि ।।
10. अथर्व॰, 9.8.1-22
11. वही, 9.83.1-4

एकत्रित करके शक्ति/ऊर्जा (सौर ऊर्जा) से बड़े से बड़े प्रयोग किए जा रहे हैं। स्पष्टतः इतनी अधिक शक्ति/ऊर्जा देने वाला सूर्य वेदों में इसीलिए पूजनीय है।

जल :

जल जहाँ एक ओर स्वस्थ जीवन के लिए प्रकृति का सर्वश्रेष्ठ उपहार है, वहीं चिकित्सा के साधन के रूप में जीवनदायी तत्त्वों से युक्त भी है। ऋषि ब्रह्मा ने अपने अथर्ववेदीय मन्त्रों में जल से होने वाले उपचारों का संकेत किया है। उनके अनुसार हृदय की जलन को दूर करने[1], आँखों, एड़ियों व पंजों की जलन को शान्त करने[2] सद्योव्रण चिकित्सा तथा रक्तस्राव को रोकने आदि रोगों में जल उपयोगी है। यही नहीं, जल अप्रत्यक्ष रूप से भी ओषधियों और वनस्पतियों को संवर्धित करके उनमें ओषधीय गुणों का संचार करता है।[3]

'अथर्ववेद' के अतिरिक्त 'ऋग्वेद' में भी जल-चिकित्सा को बहुत अधिक महत्त्व दिया गया है। यह वेद इस मत का स्पष्ट रूप से समर्थन करता है कि जल में सभी औषधियों के गुण विद्यमान हैं। इसीलिए जल को अमृत मान कर यह वर्णन किया गया है कि इससे सभी रोगों की चिकित्सा सम्भव है।[4] चरक-संहिता, अष्टांग-हृदय तथा सुश्रुत-संहिता में भी ज़ल के विविध रूपों तथा उनके उपयोग तथा उनसे होने वाली चिकित्सा-पद्धति के विषय में विस्तृत वर्णन मिलता है।[5]

वायु :

मानव जीवन का आधार प्राण है। प्राणों की सत्ता से ही जीवन की गतिविधियाँ होती हैं। इसी से शरीर में बल, स्फूर्ति, उद्यम, उत्साह और ओजस्विता है। यदि प्राणशक्ति का संरक्षण, पोषण और संवर्धन किया जाए तो शरीर को व्याधियों से मुक्त किया जा सकता है।

वैदिक ऋषि भली-भाँति इस तथ्य से परिचित थे कि वायु ही जीवन है।[6] शरीर के अन्दर जाने वाली वायु समस्त धमनियों, फेफड़ों से होकर जब बाहर निकलती है तो अनेक रोगों से हमारा निदान करती है। ऋषि ब्रह्मा ने स्वयं दृष्ट अथर्ववेदीय मन्त्रों में इन्हीं दोनों वायु-श्वास और

---

1. अथर्व०, 6.24.1 : हिमवतः प्रस्रवन्ति सिन्धौ समह सङ्गमः। आपो ह मह्यं तद् देवीर्ददन् हृद्योत-भेषजम्।।
2. वही, 6.24.2: यन्त्रे अक्ष्योरादिद्योत पार्ष्ण्योः प्रपदोश्चत् यत्। आपस्तत् सर्व निष्करन् भिषजां सीभषत्तमाः।।
3. वही, 6.57.2: सिन्धुपत्नीः सिन्धुराज्ञी सर्वा या नद्य स्थन । दत्त नस्तस्य भेषजं तेना वो भुनजामहै ।।
4. ऋ० 1.23.20 : अप्सु मे सोमो अब्रवीद् अन्तर्विश्वानि भेषजा आपश्च विश्वभेषजीः।
5. च० सं०, सूत्र०, अ० 27, जलवर्ग, च० सं०, सूत्र०, अ० 45, जलवर्ग, अष्टांग, सूत्र०, अ० 5, तोयवर्ग।
6. अथर्व०, 8.2.3 : वातात् ते प्राणमविदं।

उच्छ्वास को 'प्राण' तथा 'अपान' कहा है।[1] उन्हें ज्ञात था कि शरीर में स्वच्छ वायु के निरन्तर संचार से ही दीर्घायु प्राप्त होती है, क्योंकि जीवन-दायिनी वायु शारीरिक शक्ति में वृद्धि करती है, धैर्य की क्षमता का विकास करती है। साथ ही मानव शरीर की इन्द्रियों के कार्य करने की क्षमता को सुव्यवस्थित करती है।[2]

ऋषि ब्रह्मा के अतिरिक्त अन्य ऋषि भी वायु की इस परम शक्ति से अवगत थे तभी तो वे वर्तमान, भूत और भविष्य तीनों कारणों की निर्भरता का प्राण पर ही स्वीकार करते हैं।[3] 'ऋग्वेद' में भी अनेक मन्त्रों में वायु को भेषज, विश्वभेषज, अमृत आदि विशेषणों से युक्त किया गया है। यह वेद वायु को अमृत की निधि स्वीकार करता हुआ इसे दीर्घायु का साधन मानता है।[4] ऋग्वैदिक ऋषि------ के अनुसार वायु हृदय को शक्ति तथा मनुष्य को दीर्घायु प्रदान करने वाला भेषज है।[5] 'मनुस्मृति' में भी यह स्पष्ट उल्लेख मिलता है कि प्राणायाम के द्वारा शरीर के समस्त दोषों को नष्ट किया जा सकता है।[6]

इस प्रकार ज्ञात होता है कि प्राणायाम करना, शुद्ध वायु का सेवन और घर में शुद्ध वायु का अबाध प्रवेश मनुष्य के शरीर और हृदय को स्वच्छ रखते हैं, पुष्ट करते हैं और दीर्घायु प्रदान करते हैं।

वनस्पति :

प्राकृतिक तत्त्वों में प्राणी-जीवन के लिए वनस्पति का भी अपना महत्त्वपूर्ण योगदान है। शरीर पंचतत्त्वों से मिल कर बना है और उन्हीं पंचतत्त्वों से निर्मित इन वनस्पतियों में वह तत्त्व शक्ति है जिससे रोगों का नाश होता है।[7] ऋषि ने अथर्ववेदीय मन्त्रों में इन्हीं परम शक्तियों से युक्त तथा रोगों का नाश करने वाली ऋषभक, आसुरी आदि वनस्पतियों का संकेत किया है। मात्र रोग-नाशक ही नहीं अपितु आधुनिक युग में भी इन विभिन्न वनस्पतियों का विभिन्न प्रकार से विधिवत् उपायों के द्वारा दैनिक जीवन में प्रयोग किए जाने वाले टूथपेस्ट, साबुन, सौंदर्य प्रसाधन भी बनाए जाते हैं जिनका मूल इन्हीं वेदों में मिलता है।

---

1. वही, 2.16.1 : प्राणापानौ मृत्योर्मा पातं स्वाहा ।
2. अथर्व०, 7.53(54).2 : सं क्रामतं मा जहीतं प्राणापानौ ते सयुजाविह स्ताम्।
3. वही, 11.4.15 : प्राणे ह भूतं भव्यं च प्राणे सर्वं प्रतिष्ठितम् ।
4. ऋ०, 10.186.3 : यददो वात ते गृहे, अमृतस्य निधिर्हितः।
5. वही, 10.186.1 : वात आ वातु भेषजम्।
6. मनु, 6.71: दह्यन्ते ध्मायमानानां धातूनां हि यथा मलाः। तथेन्द्रियाणां दह्यन्ते दोषाः प्राणस्य निग्रहात् ।।
7. अथर्व०, 8.2.28: अग्नेः शरीरगसि पारयिष्णु रक्षोहासि सपत्नहा अथो अमीवचातनः पृतुद्रुर्नाम भेषजम् ।।

स्पष्टत: ऋषि ब्रह्मा प्राकृतिक तत्वों के महत्त्व तथा उनसे किए जाने वले रोग-निदान से भली-भाँति परिचित थे। मानव शरीर भी इन्हीं पंचतत्त्वों से मिलकर बना है और इसी में विलीन हो जाता है। ये पंचतत्त्व हैं- पृथ्वी, अग्नि, जल, आकाश तथा वायु। इसीलिए इन तत्त्वों की शुद्धता पर बल देते हुए ऋषि कामना करते हैं कि वायु समान रूप से मनुष्य के लिए पवित्र हो, अमृत रूपी जल सदा पृथ्वी पर विद्यमान रहे और वनस्पतियों और मानव को सदैव सूर्य का प्रकाश ग्रहण होता है।

भारतीय सभ्यता की आधारशिला चतुर्वर्ग-धर्म, अर्थ, काम तथा मोक्ष हैं। शरीर के अन्तर्गत शरीर, मन, बुद्धि और आत्मा ये ही 4 अङ्ग अनन्त कामनाओं एवम् आवश्यकताओं के केन्द्र माने जाते हैं। ज्ञान में धर्म का जितना अंश रहता है, उसी के अनुसार बुद्धि में स्थिरता आती है। जिस प्रकार बुद्धि धर्म से सम्बन्धित है, उसी प्रकार शरीर का अर्थ से, मन का काम से और आत्मा का मोक्ष से सम्बन्ध है।

वेद में आत्मा को ओषधि के रूप में भी लिया गया है; क्योंकि ऋषि ब्रह्मा के सूक्तों के चिन्तन एवं मनन करने से ऐसा प्रतीत होता है कि वे लोग आत्मिक शुद्धि पर विशेष बल देते थे। यदि व्यक्ति का मन एवं बुद्धि, शुद्ध तथा विषय रहित है तो शरीर स्वयं स्वस्थ रहेगा। अत: यह उनकी दृढ़ धारणा थी कि मनुष्य को मन के विकारों से दूर रहना चाहिए तभी आत्मिक बल संवृद्ध होगा।

निष्कर्षत: वैदिक ऋषियों की यही धारणा है कि प्राय: शारीरिक रोगों का जनक भी मन ही है। यही कारण है कि सभी ऋषियों ने एकमत होकर मन की शुद्धि, स्थिरता और शान्ति पर बल दिया है। मन शान्ति का एकमात्र उपाय सत्य-वचन और तदनुरूप आचरण है। सत्य ही मनसा-वाचा-कर्मणा मन के स्थिर हो जाने पर ही क्रिया फलीभूत होती है[1]; क्योंकि किसी भी कार्य को पूर्ण करने तथा रोगों का उपचार करने के लिए मन की संकल्प शक्ति का होना परमावश्यक है।[2]

1. पातञ्जलयोगप्रदीप, साधनपाद, 36 : सत्यप्रतिष्ठायां क्रियाफलाश्रयत्वम् ।

2. द्र० का०, 27

Section VI

AGRICULTURAL SCIENCE

1

Agriculture in Sanskrit Sastras

Deepak Deshpande

The first sign of change of civilisation of man from a wandering hunter who sustained on natural fruits, hunting and fishing to a stable resident occurred with the development of agriculture. The history of agriculture is the history of "civilised" man and goes back to remote unknown past.

World's oldest literature 'Vedas' refer to agriculture many times indicating a highly developed civilised society. 9000 yr. Old (7000 B.C.) cottonseed has been discovered in Mehergarh, now in Pakistan.[1]

This shows the antiquity of well developed agriculture in ancient India. No wonder Indian ancient literature is replete with description of various aspects of Agriculture.

Source Material :

Ṛgveda advises a reckless gambler deprived of his belongings to resort to agriculture, a noble profession that gives man wealth.[2] Increase in food production through Agriculture is advised in Atharvaveda.[3] Epics like Mahabharat & Ramayan have ample references to Agriculture. Manusmruti and Kautilya's Arthshastra Patanjali's Mahabhasya, Ayurvedic Texts of Carak and Sushrut deal with agriculture extensively. There are many more texts on agriculture and we will briefly summarise the contents of these **Texts.**[4]

In another chapter 'Vaishyavarg' in Amarkosh, Amarasimha describes different kinds of soils and their suitability for cultivation of specific crops. He had also surveyed the land of Northern India and described their characteristics and suitability of growing various crops. Amarkosh has been translated in major languages of the world. Amarasimha divided land into 3 classes namely, Jangala, Anupa and Sadharana. In fact, Amarasimha is the father of soil science.

Varahmihir (505-587 AD) is considered as great expert on Astronomy and Astrology. In his Bruhat Samhita he has discussed many things about Astrology, Geology, Geography, Ayurved, Botany, Zoology and Agriculture. He had described the methods of propagation (grafting) of fruit trees, as well as several signs of locating underground water for digging wells. In 'Visva-Vallava' Chakrapani has dealt in detail as to how one can have an approximate idea regarding water below the surface of different kinds of lands, based on certain characteristics on the land. In fact, he had composed this work after seeing the works of that type written by Varahan Sarasvata, Gargeya, Laghna, Samvasyu, Hemadri and from his own experience.

Varahmihira in Bruhat Samhita gives the following etiology and diagnosis of the diseased condition of a plant. "Cold climate (low temperature) wind (dryness) and Sun (high temperature) are the causes of diseases. When the plant is diseased, the leaves become yellow, buds do not develop or their growth is arrested, branches become dry and the sap exudes" Kashyap says, "those plants that have yellow leaves, that are fruitless and denuded of leavesș and these are caused by coldness, excessive heat, too much rain, dry wind and by intermingling of roots of different plants are to be known as diseased and are to be treated accordingly." He then prescribes the following remedies. "To cure the plant, first scrape off or otherwise remove the parts offected with a knife, then apply mud kneaded with ghee and viḍanga (Embelia ribes), to these parts after which milk diluted with water should be poured at the roots." Agnipuran has almost similar prescription. Surapala gives a very detailed account of growing various types of plants in his book 'Vrksayurveda'. Sowing of seeds was a highly developed art in ancient India.

Probably, "Krish Parasara" is the best-known text available and published. It is attributed to Parasara. G.P. Majumdar, who translated it in English, places the work in 950-1100 AD. In 243 verses in the text, Parasara describes : (1) Forecast of rainfall and the influence of planets and stars on agricultural operations (2) Practical details of agricultural operations (3) Rice cultivation, ploughing, seed collection, transplantation of seedlings, thinning, weeding, draining of water from fields and conservation of water in fields. Nala Ropana i.e. planting a kind of seed in one corner of the field to protect the crop; finally harvest and threshing and storing of paddy, (4) Proper care of draught animals. Number of bulls used is eight and measurements. In the concluding sentence Parasara describes agriculture as the eternal duty of all four Varnas, only author to recommend it as an occupation without stigma, He describes cultivation as ideal profession, better than cattle rearing, trade and state service.

Vrkshayurved Parasara (1st Century BC or AD) gives advanced botanical knowledge in six parts : (1) plant morphology; (2) nature, properties and classification of soils; (3) description and distribution of forests; (4) flowers ș parts, functions and classification (5) definition, function and classification of fruits (6) plant limbs- root, stem, bark, heart wood, sap, excretions, oleaginous products, spines and prickles, seeds and embryonic plants.

"Krishi-Sukti" by Kashyap is the treatise on rice cultivation. It has been translated in Dutch language by G.Y. Wojtilla. We still follow the method of rice cultivation described by Kashyap 2000 years ago except chemical fertilizers and pesticides.

'Upavan-Vinod' is a part of "Sārṅgdhara-paddhati" written by Sārṅgadhara (1283-1301 AD). Chapters in this text describe : (1) Glory of trees, (2) Good and evil omens relating to residence near trees, (3) Selection of soil for planting trees, (4) Classification of plants, (5) Sowing of seeds, (6) Methods of planting, (7) Watering of plants, (8) Rules for protection of plants, (9) Construction of a Garden House, (10) Examination of soil, where wells are to be dug, (11) Rules for nourishment of plants, (12) Kunapa water i.e. nutrient solution, (13) Treatment of plants, (14) Botanical marvels, (15) Assessment of price of things.

Amarkosh may be the first Encyclopaedia of the world prepared by Amar in the first century B.C. In one chapter titled 'Bhumivarg' he describes 12 types of land depending upon fertility of soil, irrigation and physical characteristics. Each type of land has been given a Sanskrit name like Urvara (fertile land), Usar (barren land), Maru (desert land), Aprahat (fallow land), Sadavala (grassy land), Pankila (water-logged land), Jalpraymanupa (watery), Kaccha (land contiguous to water), Savkara (land full of pebbles and lime kankar), Sarkaravati (sandy), Nadimatruka (land watered from river) and Devmatruka entirely on rain. Ancient Indians extensively employed irrigational methods of cultivation is further corroborated by Megasthenes. He writes, "India has many huge mountains, which abound in fruit trees of every kind and many vast planes of great fertility- more or less beautiful, but all alike intersected by multitude of rivers. The greater part of the soil is moreover under irrigation and consequently bears two crops in the course of the year." Arthaśāstra mentions a special Government Officer called Superintendent of Agriculture, who assessed land at rates varying according to different methods of irrigation. These statements indicate that agriculture was quite advanced in ancient India. In the Yajur Veda, distinct references to the rotation of crops are found. The Taittiriya Samhita distinctly mentions that in the course of a year, two crops were harvested from the same field. It also mentions different seasons for ripening of different crops and the proper times for harvesting them. Kauṭilya gives directions for seasonal cultivation and harvesting.

Botanical studies are found in some of the earliest texts. The classification of plants is given in Atharva Veda on the basis of structure and uses. The physiology and life of plants is described in Brhadaranyakopnishad. The plants are compared with the human body. The life of plants is also described in

Nyayadarshan. In the opinion of Udayan the tree has to pass through several stages- birth, death, disease, sleep, wakefulness, activity, ceasing (resting), etc. We find the description of several plants and their nutrition in Vishnupuran, Bhagawatpuran, Matsyapuran, etc. The concept of the flow of nutritional juices in a tree is also described. One can find the detailed study of plants in Sushrut, Charak Samhita, Agnipuran, etc. For the germination of seed the favourable climate, soil and water are necessary. The information on the conditions suitable for growing various plants, Characteristics of good soil, application of manure, how to diagnose the disease of plants and their control measures, different methods of propagation, etc. is given in Brnhat Samhita and Agnipuran. It is stated in Shadarshansamuchay that the plants take their food from the soil. As the food is suitable or unsuitable the plant body becomes plump or dry. Plant is a living body and some of its movements are like the living animal. This has been stated in Mahabharata. The experts could distinguish the gender of flowers and knew that pollination was necessary for the development of fruit seeds (Harit Samhita and Charak Samhita). In Mahabharata and Bhagwat puran the plants are classified as (1) those which bear fruit without flowering etc. Umber, Vad and Pipal (2) Annual plants ṣ once they yield fruit and die e.g. banana grain crops (3) Plants which need support of a tree e.g. vines (4) Twaksar ṣ do not need support e.g. Bet (6) Drum- the tree first flowers and then yields fruit e.g. mango. Varahmihir says that is it possible to get off-season fruits.

There are several texts in Sanskrit which are entirely devoted to the science of agriculture- i.e. Krishi Sashatra.

Having briefly summarized the contents of some of the books, we shall now discuss the Science & Agriculture as it appears in different ancient texts.

Classification of land :

Carak and Susruta samhitas mention different types of land. To reap the best yield from agricultural operations, the farmer should have a working knowledge of agricultural properties of different types of land. To arouse an increased awareness of the agricultural properties of land, according to ecological conditions and biological potentials, our ancient agricultural scientists have classified land, taking into consideration various factors like geographical and climatic conditions, fertility, etc. Lands, according to ecological conditions, were distinguished as agneyi, vayavi, Mahendri and varuni. Among these, Varuni is described as that land where water is available in plenty, is beautiful, fertile, often resorted to by cattle and best suited for growing corns in plenty. Lands are classified as jangala (arid), anupa (marshy), and sadharani (common). Sadharani is described as the best suited for all types of cultivation.[5]

Method of cultivation :

The word 'agriculture' means tillage of soil leading to production of crops. The Vedic word krsi also denotes agriculture and krsi means the act of ploughing. Hence, selection of suitable soil and preparing the soil by ploughing before sowing are very essential and our ancients have given great attention to these aspects. Agricultural operations carried out in those days are summed up in ancient texts as ploughing (कृषन्तः), sowing (वपन्तः), Reaping (लुनन्तः), and threshing (मृणन्तः), The bull and the plough were the most necessary accessories for cultivation. In some places horses also appear to have been used to plough the field. Though, generally two bullocks were yoked to the wooden plough to make furrows in the soil, on ceremonial occasions like sacrifice etc., more than two bulls, were yoked.

Sometimes six, eight or twelve or even twentyfour bulls were used. Before tilling and ploughing, the ground was worshipped for a good yield.

The plough was considered the most sacred and essential implement in agricultural operations and was known by different names like hala, sira, Iangala, sita etc. We have heard of kings like Janaka who ploughed the fields themselves, Balarama always carrying the plough and known by the name Haladhara. It is said that phala is a special ploughing instrument with sharp iron edge fixed in the front and this destroyed all insects under the earth.

The stumps were weeded out from the ground by a hoe known as stambaghna. Since it is necessary to remove weeds, thorn, stones etc., before actual sowing, the special plough known as Langala was used extensively drawn by more than two oxen. Repeated ploughing also was done to enable the soil to get the required quality. Ancient texts refer to the preparation of the field by ploughing three times in heavy rain and often by deep ploughing. The custom to plough the field over and over again from one end to other in the reverse direction was in vogue to make cultivation more fruitful and intensive.

Manure :

After ploughing the field deeply, the soil was allowed to dry up. There was a general belief that if the soil was allowed to dry up well there was no necessity of adding even manure. Krsiparasara and other works prescribe the method of preparing manure from cattle dung, dry leaves etc. Generally, dung manure was used for all types of crops. Soil mixed with honey and ghee was also considered good.[6] It is also said that the drain waters flowing from the kitchens of farmers made the crops grow well. Ancient farmers knew several methods of keeping the land fertile. They knew that the crops drew sustenance from the earth. Elaborate rules for the preparation of different kinds of manure are given in texts like Agnipurana, Brhatsamhita etc.[7] Sesame, cowdung, barley powder, meat, fish and water mixed in fixed proportions were regarded as effective manure. Methods of preparing kunapajala, a fertilizer liquid, are also prescribed for growing crops, flowers, plants etc. Most of the ingredients used were natural organic sources. The knowledge of manure etc., in those days was apparently the result of extensive practical observations. Physical effects of manure upon the texture and water-holding power of the soil were known to the farmers. They had then knowledge of the requirements of different crops and the different harvesting seasons. This might have led to the rudimentary form of crop rotation. This rotation of the crops helped the farmers to improve the fertility of the land. Many a time farmers kept their land fallow a common practice of keeping the land free of all vegetation for a specified period by occasional ploughing of soil during hot and dry weather. This helped in the removal of infested root and in controlling the breeding up of pests and nematos.

Sowing :

Sowing of seeds was carried on after ploughing the field deeply. Ancient texts like krsiparasara prescribe rules for collection and preservation of seeds. Seeds were collected in the months of Magha and Phalguna. Good, matured, pulpy seeds of uniform character were collected to get good crops. Farmers were prohibited from keeping the seeds on ghee, oil, salt, buttermilk, lamp etc. We have evidences from ancient texts that seeds were treated in a special manner[8] to get special results. Many botanical marvels were also effected by treating the seeds in special manner. Sometimes, ploughing was done in the field while the seeds were already scattered in it. Sometimes two or three different kinds of seeds were sown in the same field. This was done when one was the principal and the other secondary like masa and tila.[9] Auspicious days were selected for sowing seeds like the fullmoon day. Generally special rituals were performed at the time of sowing and these rituals are fully described in the Grhyasutras.

It is very interesting to note that there are many proverbs in regional languages in vogue form the past several centuries, which throw light on agricultural operations of ancient India. They have been serving as guidelines to farmers even now. Through these popular sayings farmers are advised to sow seeds apart from each other.

Weeding and Supervision :

After sowing seeds and supplying adequate water to the fields, the farmers were advised to remove the weeds from the fields very carefully and protect the crops properly from birds, insects and rats. The importance of weeding has been stressed in all texts dealing with agriculture. While laying down the duties of a king, Manu in Sanskrit and Tiruvalluvar in Tamil have expressed identical ideas. As a husbandman plucks up weeds and preserves his corn, let a king destroy his opponents and protect his people.[10]

Agriculture without supervision is fruitless and our ancients made arrangements to strictly supervise the operations related to agriculture. It is advised that the owner of the field must look after the field himself. It is justifiably said that if the owner does not look after the field, it will fail him like a wrathful wife. If the owner fails to supervise the agricultural operations, Goddess of prosperity will desert him and in her place adversity will enter his field. Such sayings are popular all over India which emphasize and inculcate the importance of the supervision of cultivation by the owner himself.

Irrigation :

Even though cultivation was undoubtedly dependent on rains to a great extent, and that people knew how to collect and conserve rainwater, the system of artificial irrigation was also prevalent from very ancient days. In the Sindhu valley civilization river water was stored by building huge dams and was allowed to flow through channels. In Vedic times, wells were dug and the water was lifted by means of wheels and irrigation was done through long pipes. Frequent references to rivers indicate that people were living mostly near rivers and rivers played an important role in the social, economic and cultural history of a particular land and its people. Irrigation through canals also was existent in the Vedic period and the word kulya is used to mean a canal.[11] A river is described as a cow and the canal as a calf. Rituals were performed at the time of directing the course of a river and this suggests the opening ceremony of letting river water into a canal. In the epicperiod, there are evidences of huge reservoirs, for irrigation and that they were constructed at higher levels easily.[12] Panini mentions the methods of irrigation with water drawn from the wells.[13] Large leather buckets were used for lifting water. Yoked bullocks with ropes were used to lift water from the deep wells for irrigation purposes. Poet Kamban while describing the beauty of river Sarayu, has mentioned about the canals through which the river reached the peasants.[14] Periodic supply of water was done through channels and the Naradasmrti mentions two classifications of dyke, viz., kheya (dug into the soil to drain excess water) and bandhya (constructed to prevent water from flowing out. It also states that erection of dykes in the middle of another man's field was not prohibited as it was advantageous for irrigation purposes while the loss was negligible.[15]

Protection of crops :

Crops were generally classified as कृष्टपच्य (grown by cultivation) and अकृष्टपच्य (naturally growing in the jungle like the grains Nivara etc.)[16] They were further classified according to the time of sowing. Those sown on the fullmoon day was called Asvayujaka (autumnal), Vasantaka and graisma crops are mentioned (vernal and summer crops). Kautilya mentions rainy crops (varsika sasya) followed by haimanamusti (autumnal harvest) and haimanasasya followed by vasantikamusti (vernal harvest) and vernal crops follwed by rainly harvests.

A part from the usual crops like yava, vrihi, etc., fibrous plants like hemp, dye-stuffs like manjista (madder), nili (indigo) etc., were also grown. Large varieties of fruits and vegetables were also grown. Ancient texts like Susrutasamhita and Carakasamlinata provide long lists of various crops which were cultivated in those days. Great care was taken to protect the standing crops from animals, birds and robbers. Observers called cavakas were appointed to protect barley crop from deers.[17] Shadow figures made of straw called cancabhirupa were placed in the fields to frighten birds. Care was taken to protect the crops from locusts (salabha), hawks (syena) and rats.

Harvesting operations :

During the reaping operaions the ripe grain was cut by a sickle and then mowing was done. Then the crops were gathered on the threshing floor called khala where grain was separated from chaff. [18] A special kind of harvesting was done in ancient India as implied by the word अपस्किरनी mentioned by Panini. In some parts of the country reaping was done, not from one end of the field to the other, but in an unsystematic manner. Some cereals were harvested by uprooting the whole plant.

Measurement of land and boundaries :

The cultivated lands were divided into different holdings known as ksetra. Some kind of survey of the areas and boundaries must have been done for the distributions of plots among individual peasants. Rules for expressing the area of a field in terms of measures are known as kanda, kula etc.[19] The sutra literature refers to the existence of separate fields or ksetras which were enclosed.[20] It also becomes clear that the ownership of land vested in private imdividuals who had every right to gift, sell or lease it. If the lands were not enclosed by fences, the boundary was determined by trees, water embankments, anthills, heaps of stones etc. Naradasmrti mentions five types of boundaries viz., dhvajini (trees), matsyini (rivers and tanks), naidhani (hidden signs), bhayavarjita (settled by argument of parties) and rajasasanatita (laid down by the order of the king). Manu and Kautilya have prescribed two kinds of boundaries for land viz., visible and invisible.48 Smrti works have laid down elaborate rules of settling boundary disputes between neighbouring landowners.

Having considered the texts and different agricultural operations. We now see some of the interesting stanzas in Sanskrit texts and their meaning-

मूलाग्रादार्जवं वृत्तं शाखानेक समन्वितं।।

तत्त्तुपुंस्त्वंभवेन्मूले स्थूलं स्त्रीत्वं कृशाग्रकं।।

स्थूलाग्रं कृशमूलंतु षंढमेतदुदीरित्तं।। मयमत, अ० १५

"A tree that goes straight from bottom to top and has many branches is called 'Male' tree. A tree that is broad at the bottom and which is tapering and becomes narrow at top is called ' Female' tree and the tree which is narrow at bottom and becomes broad at top is called 'Eunuch' tree. For all auspicious 'Muhurt', only 'Male' tree is to be used.

That the science of textiles & clothes was also well developed in the ancient times, is best illustrated by following description in 'Parashriy krishi'

त्वस्त्रं चतुर्विधं प्रोक्तं कौषेयं लोमजं तथा।

कार्पासकंच वार्क्ष्यच पंचमं चर्म गद्यते।।

कृमि कोष समुद्धूतं कौषेयामि तिगद्यते।

सूक्ष्मा सूक्ष्मौ मृदु स्थूलौ तंतवः स्यूर्यथाक्रमं।।

जन्तु लोमोद्भवं वस्त्रं लोमजं नाम कीर्तितं।

वैंतानत्वच गुरुता कौषेयस्य गुणग्रहः।।

लोमजं वसनं श्रेष्ठं सर्वेषामुपपद्यते।
अंशीतताचा लघुता लोमजस्य गुणग्रहः।।
बहु कार्पास सूत्रोत्थं कार्पासमिति गद्यते।
रूप सौक्ष्म्यादि भेदेन तच्चानेक विधं भवेत्।।
तत्सर्वं सर्व योग्यंहि सर्वेषामपि संमतं।
श्वेतं वाप्यथवा चित्रं शुभ कार्ये प्रशस्यते।।
वृक्षत्वक् संभवं वार्क्ष्यं तज्ज्ञेयंतु चतुर्विधं।
छल्नी चतुर्विधा प्रोक्ता स्थूला तनु शरा मृदु।।
लघुता गुरुता चैव वार्क्ष्यस्य गुणसंग्रहः।।
कौषेय मथ कार्पासं वार्क्ष्यं वस्त्रं शुभं भवेद्।
एक जाति भवं वस्त्रमुत्तमं संप्रचक्षते।।
दिजातिं संभवं मध्यं बहुजातंतु गर्हितं।
सूक्ष्म रोमा सुखस्पर्शी कोमलांगस्तु यो भवेद्।
ब्रह्म जाति सविज्ञेय श्चर्मचास्य गुणोत्तरं।।
एक जाति भवं द्वैजातं सुखदायकं।
त्रि चतुर्जाति संभूतं लोमजं न सुखावहं।। पराशरीय कृषिः

Clothes are of four types. First is Kaushey (silk), second is Lomaj (woolen) third is Karpas (cotton) and the fourth is Varkshya (treebark). fifth type which is useful like cloth is leather Kaushey is made from the nest of silkworm. Thin and Thick, soft and hard are the characteristics of string of silk. Silk cloth is considered good if it is long broad and thick. Woolen clothes are made from the hair of the animals and are called Lomaj. This cloth is good and is available to all. Woolen cloth is light and protects us from cold. Cloth made from cotton is karpas. It is of different types due to various colours and mixing thin and thick threads. Since this is available to all, this is good for all purpose. White and different coloured clothes are considered auspicious. Clothes made from tree bark are called Varkshya. Its threads are of four types viz thick, thin, rough and soft. Light and long cloth of tree bark is good. Silk,woolen and tree bark clothes are always pure. It means that cotton clothes are sometimes impure. Cloth made from only one type of thread is good Cloth made from two types of threads is considered of medium quality and cloth made form different types of threads is considered bad. Since this cloth is not uniform, it wears out fast. Leather of thin hair which is soft in touch is considered good leather and the animal whose leather is as above is called 'Brahman' animal.

Amongst the wollen, cloth made from the hair of only one animal is considered good, that made from two animals is pleasant but cloth made from the hair of many animals is not longer lasting.

The above shows a great development of Agriculture and related activities in ancient India.

References :

1. www.baluchonline.com/culture & www.hvk.org/hvk/articles/0502/66
2. Rgveda X 34.13
3. Atharvaveda 6.142
4. Raychowdhari S.P. (1964). Agriculture in Ancient India ICAR, NewDelhi.
5. Susruta Samhita, svatra sthana 35, Caraka Samhita, Kalpa I
6. Atharvaveda 3.17.91
7. Brhatsamhita 55-17,18. Agni purana 281 - 8, 9
8. Sivatattvaratnakar VI 10. 55 - 60
9. Astadhyayi V. 5-58
10. Manusmruti III. 10
11. Rgveda III 45.3, X 43.7 Atharvaveda III. 131
12. Mahabharat Sabha Parva 5.80
13. Astadhyayi II 3.123
14. Ramayana Balkanda
15. Narad smruti II 17.20
16. Astadhyayi III 1.114
17. Mahabhasya I 1.3
18. Gobhila Grhya Sutra IV 4.29
19. Manu smruti, VII 19
20. Baudhayan Dharmsutra III 2.2.1, 2 Gautam Dharma Sutra XII 18.

2

वेदों व भारतीय–ग्रंथों में वर्णित ऋतु–विज्ञान व कृषि–विज्ञान

डॉ॰ चित्तरंजन दयाल सिंह कौशल,

विदेशी विद्वानों में ही नहीं, बल्कि शिक्षित भारतीयों में भी यह एक प्रचलित धारणा है कि वेद, उपनिषद् व दर्शन आदि प्राचीन भारतीय ग्रंथों में वर्णित हमारी प्राचीन धरोहर मुख्यत: धर्म और दार्शनिकता से ही सम्बन्धित है। प्राचीन ग्रंथों में विज्ञान एवं प्रौद्योगिकी का उल्लेख नहीं है। अपनी उज्ज्वल वैज्ञानिक परम्परा के गौरवपूर्ण इतिहास को संसार के सामने प्रस्तुत करने की आज नितान्त आवश्यकता है। संस्कृत केवल संस्कार भाषा ही नहीं, अपितु विज्ञान व प्रौद्योगिकी के क्षेत्र में भी अपना बहुमूल्य योगदान देने की क्षमता रखती है।

शून्य की परिकल्पना, अंकों की स्थानीय मानपद्धति, वर्गमूल, घनमूल व घातांक आदि के बारे में प्राचीन भारतीयों को जानकारी थी। दो प्राचलों (Variables) वाली द्वितीय डिग्री के बीजगणितीय समीकरणों का हल व चक्रीय चतुर्भुजों का वर्णन ब्रह्मगुप्त ने सदियों पहले प्रस्तुत किया था। प्राचीन भारत में तारामण्डलों की उपस्थिति, ग्रहों तथा चन्द्रमा की गति तथा चन्द्रग्रहण व सूर्यग्रहण के समय व अवधि के बारे में पूरी जानकारी थी। पृथ्वी सूर्य के चारों ओर घूमती है, यह तथ्य सदियों पहले यहां ज्ञात था।

भारतीय चिकित्सक आज भी चिकित्सा के लिए चरक संहिता व अष्टांग–हृदय तथा शल्य क्रिया के लिए सुश्रुत–संहिता आदि आयुर्वेद के ग्रंथों की सहायता लेते हैं। आयुर्वेद पद्धति भारत तथा कुछ अन्य एशियाई देशों में काफी प्रचलित है। अब पश्चिमी देशों में भी इस क्षेत्र में गहन रुचि उत्पन्न हो रही है।

लौह, जस्ता, ताम्बा व इनकी मिश्र धातुओं की धातुकी में प्राचीन भारत के योगदान को सम्पूर्ण विश्व स्वीकार करता है। पीतल व कांसे की भारतीय कला–कृतियों में सुन्दरता व तकनीकी कौशल का मिश्रण स्पष्ट परिलक्षित होता है। मन्दिरों की अत्युत्तम वास्तुकला, निकास–तंत्रों सहित नगर–योजना तथा सड़कों, पुलों, किलों तथा बंदरगाहों के निर्माण की वास्तुकला भी अत्यन्त विकसित थी। हमारे प्राचीन ग्रंथों में सिंचाई व खेती से सम्बन्धित उचित विधियां स्पष्ट रूप से वर्णित है। जलपोत निर्माण व नौचालन में हमें विशेषज्ञता प्राप्त थी। भारत से पश्चिम में स्थित व दक्षिण पूर्व एशिया के देशों से हमारा समुद्री व्यापार भी होता था। रेशमी व सूती वस्त्रों में भारत की उत्कृष्टता सारे विश्व को ज्ञात थी।

इस उज्ज्वल वैज्ञानिक परम्परा से सम्बन्धित जानकारी का प्रसार समाज में वैज्ञानिक चेतना जगाने व विज्ञानमयी संस्कृत–भाषा को समझाने में एक प्रबल प्रेरणास्रोत सिद्ध होगा। इससे सभी देशवासियों में आत्मसम्मान की अनुभूति और राष्ट्रीय गौरव की एक झलक भी परिलक्षित होगी।

प्रस्तुत पत्र में भारतीय ऋतु—विज्ञान व कृषि—विज्ञान से सम्बन्धित महत्त्वपूर्ण विषयों का समावेश किया गया है।

ऋतु—विज्ञान—

प्राचीन खगोलिकी में काल—चक्र, प्रमुख विक्रम—संवत्, राशिचक्र व नक्षत्र, मास व ऋतु—विज्ञान का वर्णन मिलता है। सभी प्राचीन खगोलिकियों की भाँति प्राचीन भारतीय खगोलिकी भी टेलिस्कोप की खोज से पहले के काल से सम्बंधित है। उस समय के प्रायोगिक आंकड़े लम्बे समय तक किये गये आकाशीय प्रेक्षणों पर आधरित थे। इस पत्र में प्राचीन भारतीय साहित्य में उपलब्ध विशाल समय खण्डों की गणना युग, भारतीय संवतों, राशिचक्र, चंद्रमा की गति पर आधारित मास प्रणाली से सम्बन्धित ऋतु विज्ञान का उल्लेख वर्णित है। अधिक मास सम्मिलित करने के कारण चान्द्रमासों तथा मौसम में एकरूपता बनी।[3] प्राचीन भारत के लब्धप्रतिष्ठ खगोल शास्त्रियों में लगध (१५०० ईस्वी पूर्ण) का वेदांग—ज्योतिष, पटना के आर्यभट्ट प्रथम (४७६ ईस्वी) का आर्यभटीय व आर्यभट्ट—सिद्धांत, उज्जैन के वराहमिहिर (५०५ ईस्वी.) की पंचसिद्धान्तिका व बृहत्संहिता, बल्लभी (गुजरात) के भास्कर प्रथम (६०० ईस्वी) का महाभास्करीयम् व लघुभास्करीयम्, भिल्लमाला राजस्थान के ब्रह्मगुप्त (६०० ईस्वी) का ब्रह्मस्फुट—सिद्धान्त प्रसिद्ध है। दासपुर (मध्यप्रदेश) के लल्ला (७०० ईस्वी) सुप्रसिद्ध खगोलविद् रहे हैं। इनकी शिष्यधी—वृद्धिदा पुस्तक सुप्रसिद्ध है। इन्होंने सूर्यग्रहण की वैज्ञानिक व्याख्या प्रस्तुत की।

सवितुश्च यदन्यथोन्यथा प्रतिदेशं सकलं समीक्ष्यते।
न च कुत्रचिदित्यवेत्य कः कुरुते राहुकृते ग्रहे ग्रहम्।।

शिष्यधीवृद्धिदा, २०—२६

अर्थात् विश्व के विभिन्न भागों में लोग सूर्य के भिन्न—भिन्न भागों का ग्रहण देखते हैं। कुछ स्थानों पर ग्रहण बिल्कुल भी दिखाई नहीं देता है। इससे यह स्पष्ट हो जाता है कि ग्रहण का कारण राहु नहीं है।

वटनगरा के वटेश्वर (८८० ईस्वी) का वटेश्वरसिद्धान्त, प्रकाशपट्टन के मंजुल (९३२ ईस्वी) का लघुमानस व बृहन्मानस, आर्यभट्ट द्वितीय (९५३ ईस्वी) का महासिद्धान्त, श्रीपति (९९९ ईस्वी) का सिद्धान्तशेखर व ज्योतिषरत्नमाला आदि का भी उल्लेख मिलता है। उज्जैन के राजा भोज (१००० ईस्वी) विरचित राजमार्तण्ड व विद्वज्जनवल्लभ खगोलविद्या के सुप्रसिद्ध ग्रन्थ हैं। भास्कर द्वितीय (१११४ ईस्वी) विरचित सिद्धान्तशिरोमणि, लीलावती बीजगणितम् व कर्णकौतूहलम्, श्रीधर द्वितीय (१२२७ ईस्वी) की लघुखेचरसिद्धि व केरल के परमेश्वर (१३७६ ईस्वी) का दृग्गणित, गोलदीपिका व ग्रहण मंदरा भारतीय खगोल—विद्या के प्रमाणभूत ग्रन्थ रहे।

भारतीय विचारधारा के अनुसार काल, जो सभी जीवों के अस्तित्व को मिटा देता है, एक चक्रीय घटना है। कालः कलयति सर्वाणि भूतानि इति। समय का चक्र ब्रह्मा के एक दिन अर्थात् सृष्टि के जीवन—काल और ब्रह्मा की एक रात अर्थात् पुनरुत्पत्ति के काल के बीच घूमता है। इस उत्पत्ति, लय, पुनरुत्पत्ति और लय को काल—चक्र के नाम से जाना जाता है।

भारतीय खगोलिकी में युग की संकल्पना अद्वितीय है। कलि, द्वापर, त्रेता व कृत– ये चार मूलभूत युग हैं। कलियुग से दोगुणा द्वापर, तीन गुणा त्रेता व चार गुणा कृतयुग का काल है।

१. कलियुग	४,३२००० वर्ष	२. द्वापर	८,६४००० वर्ष
३. त्रेता	१२,९६००० वर्ष	४. कृत	१७,२८००० वर्ष

इन चार युगों का कुल समय अर्थात् ४३,२०,००० वर्ष का एक महायुग होता है। एक मन्वन्तर में ७१ महायुग और एक कृतयुग मिला कर ३०,८४,४८००० वर्ष होते हैं। आकाशगंगा के चारों ओर चक्कर लगाने में सूर्य जितना समय लेता है, एक मन्वन्तर का समय लगभग इसी के बराबर होता है। एक कल्प में १४ मन्वन्तर और एक कृतयुग मिलाकर ४,३२,००,००,००० वर्ष अर्थात् १००० महायुग होते हैं। यह समय लगभग सृष्टि की आयु के बराबर है। ब्रह्मा का एक दिन २ कल्प के बराबर होता है

भारतीयों में अधिकतर कलि, विक्रम, शक व शालिवाहन संवतों का उपयोग किया जाता है। द्वापर युग के बाद के काल को कलियुग कहा जाता है। महाभारत युद्ध की समाप्ति और श्री कृष्ण की मृत्यु के बाद से इसका आरम्भ माना जाता है। ३१०२ ईस्वी पूर्व कलियुग संवत् का प्रारम्भ हुआ। विक्रमसंवत् शकों पर विजय के स्मरणोत्सव के रूप में मनाने के लिए उज्जयिनी के राजा विक्रमादित्य द्वारा आरम्भ किया गया। ५७ ईस्वी पूर्व दीपावली के बाद कार्तिकमास की शुक्ल प्रतिपदा को शकसंवत् का आरम्भ हुआ। गुजरात में आज भी यह प्रचलित है। उत्तरी भारत के कुछ भागों में आज भी थोड़े से परिवर्तन के साथ विक्रमसंवत् का प्रचलन है। नव वर्ष का आरम्भ चैत्रमास की शुक्ल प्रतिपदा से माना जाता है। आकाशमान में दृश्यमान सूर्यपथ को १२ समान (प्रत्येक ३० अंश) भागों में बांटा गया है, जिन्हें राशि कहा जाता है। नक्षत्रों की सहायता से चंद्रमा के पथ को अंकित करने की पद्धति का प्रचलन है। प्रत्येक नक्षत्र १३ डिग्री, २० मिनट के बराबर है। इसे फिर चार समान भागों में विभाजित किया गया है। प्रत्येक भाग (पाद) का मान ३ डिग्री, २० मिनट है। प्रत्येक राशि २.२५ नक्षत्रों अथवा ९ पादों से बनी है।

मास पक्ष और तिथि भारतीय कैलेण्डर के महत्त्वपूर्ण अंग है। चन्द्रमा पूर्णिमा के दिन जिस नक्षत्र में होता है, चान्द्र मास का नाम उसी नक्षत्र के नाम पर रखा गया है। मास का नाम पता लग जाने पर आकाश में तारों के स्थान व ऋतुओं का ज्ञान स्वतः हो जाता है। मासों को छः ऋतुओं में बाँटा गया है।

चैत्र–वैशाख	मार्च–अप्रैल	वसन्त–गर्मी
ज्येष्ठ–आषाढ़	मई–जून	ग्रीष्म–गर्मी
श्रावण–भाद्रपद	जुलाई–अगस्त	वर्षा–बरसात
आश्विन–कार्तिक	सितम्बर–अक्तूबर	शरद्–बरसात
मार्गशीर्ष–पौष	नवम्बर–दिसम्बर	हेमन्त–सर्दी
माघ–फाल्गुन	जनवरी–फरवरी	शिशिर–सर्दी

अधिकमास अथवा पुरुषोत्तम मास – ३६५.२५६ दिन के सौर वर्ष और ३५४.३७२ (२९.५३१×१२) दिन के चान्द्र वर्ष का अन्तर १०.८८४ दिन का है। यदि यह अंतर संचित हो जाए तो चान्द्र मास और ऋतुओं का आपसी मेल समाप्त हो जाएगा। अतः जब सौर–मासों व चान्द्र–मासों का अन्तर कुल मिला कर एक चान्द्रमास के बराबर हो जाता है, तब वर्ष के बीच में एक अधिक मास जोड़ दिया जाता है। ऐसा ३३ सौर मासों में एक बार होता है।

पन्द्रह निमेष की एक काष्ठा होती है। तीस काष्ठा की एक कला और तीस कला का एक मुहूर्त्त होता हैं। तीस मुहूर्त्त के एक 'दिन–रात' होते हैं। एक दिन में तीन–तीन मुहूर्त्त वाले पाँच काल होते हैं। प्रातःकाल, संगवकाल, मूयकाल, अपराह्ण काल व सायम् काल। इनमें पन्द्रह मुहूर्त्त व्यतीत होते हैं। पन्द्रह दिन–रात का एक पक्ष कहलाता है। दो पक्ष का एक मास कहा गया है। दो सौरमास की एक ऋतु होती है। तीन ऋतुओं का एक अयन होता है। दो अयनों का एक वर्ष माना जाता है। अथर्ववेद के एक मंत्र में संवत्सर को निरूपित करते हुए ऋतु, आर्तव, हायन, समा, मास का उल्लेख किया गया है।

ऋतून् यज ....आर्तवान् उत हायनान्।
समाः संवत्सरान् मासान्.......।।

अथर्ववेद, ३.१०.९

कौटिल्य के अर्थशास्त्र में काल को तुट या त्रुटि, लव, निमेष, काष्ठा, कला, नालिका, मुहूर्त्त, पूर्वभाग, अपरभाग, दिवस, राशि, पक्ष, मास, ऋतु, अयन, संवत्सर, युग आदि इकाइयों में विभक्त किया गया है।

कालमानम् तुटो, लवो, निमेषः, काष्ठा, कला, नालिका, मुहूर्त्तः, पूर्वापरभागो, दिवसो, रात्रिः, पक्षो, मासः, ऋतुरयनं, संवत्सरो युगमिति।

अर्थशास्त्रम् , २२०.३०

यजुर्वेद में १२ मासों के वैदिक नाम और ६ ऋतुनाम दिए गए हैं। मासों के नाम चैत्र आदि के स्थान पर मधु–माधव आदि दिए गए हैं।

मधु–माधव	चैत्र–वैशाख	वसन्त (Spring)
शुक्र–शुचि	ज्येष्ठ–आषाढ़	ग्रीष्म (Summer)
नभस्–नभस्य	श्रावण–भाद्रपद	वर्षा (Raining)
इष–ऊर्ज	आश्विन–कार्तिक	शरद् (Autum)
सहस्–सहस्य	मार्गशीर्ष–पौष	हेमन्त (Winter)
तपस्–तपस्य	माघ–फाल्गुन	शिशिर (Cold)

मधुश्च माधवश्च वासन्तिकावृतू। —**यजुर्वेद, १३.२५**
शुक्रश्च शुचिश्च ग्रैष्मावृतू —**यजुर्वेद, १४.६**

नभश्च नभस्यश्च वार्षिकावृतू। —यजुर्वेद, १४.१५

इषश्चोर्जश्च शारदावृतू । —यजुर्वेद, १४.१६

सहश्च सहस्यश्च हैमान्तिकावृतू । —यजुर्वेद, १४.२७

तपश्च तपस्यश्च शैशिरावृतू । —यजुर्वेद, १४.५७

कृषि—विज्ञान

प्राचीन भारतीय कृषि से सम्बन्धित पुरातात्त्विक तथ्य, वैदिक, बौद्ध व जैन काल में भारतीय कृषि और प्राचीन भारतीय कृषि प्रणाली की विशेषता आदि को आधुनिक वैज्ञानिक दृष्टिकोण पर कसने का प्रयास किया है।

भोजन जीवन की मूलभूत आवश्यकताओं में से एक है। प्रारम्भ से ही भोजन की प्राप्ति मनुष्य की एक मुख्य गतिविधि रही है। कृषि प्रत्येक सभ्यता का अभिन्न अंग रही है। प्राचीन साहित्य से उपलब्ध तथ्यों द्वारा यह प्रामाणिक रूप से कहा जा सकता है कि वैदिक काल में भारत में मनुष्यों तथा पालतू पशुओं के लिए खाद्य पदार्थ उत्पन्न करने तथा उनका भंडारण करने हेतु कृषि का विज्ञान के रूप में विकास हुआ। प्राचीन काल में कृषि एक बहु आयामी गतिविधि थी जिसमें अनाज तथा दालों का उत्पादन, वानिकी, चिकित्सीय एवं सजावटी पौधे, पशु—पालन तथा अन्य सम्बन्धित पहलू शामिल थे। इस प्रगति की जानकारी हमें पुरातात्विक खोजों तथा पराशर व सुरपाल जैसे विद्वान् वैज्ञानिकों द्वारा कृषि पर लिखे गये ग्रंथों से मिलती है। सातवाहन, मगध व विजयनगर के राजाओं ने उपज बढ़ाने व उत्कृष्ट खाद्यान्न पैदा करने के लिए कई योजनाओं का कार्यान्वयन किया था।

मानव जीवन अन्न पर निर्भर है। अन्न की प्राप्ति कृषि से होती है। कृषि—कार्य गौरव का कार्य माना जाता था। कृषि और अन्न से ही मनुष्यों का जीवन चलता था। अथर्ववेद के अनुसार जो कृषि विद्या में निपुण होते थे, उन्हें 'कृष्टराधि' और 'उपजीवनीय' अर्थात् सफल आजीविका वाला कहा जाता था।

ते कृषिं च सस्यं च मनुष्या उप जीवन्ति।

कृष्टराधिरुपजीवनीयो भवति।

अथर्ववेद, ८.१०.२४

कृषि—विशेषज्ञों को 'अन्नविद्' नाम देते हुए कहा गया है कि सर्वप्रथम उन्होंने कृषि के नियम बनाये। यजुर्वेद में राजा का प्रमुख कर्त्तव्य बताया गया है कि वह कृषि की उन्नति करे, जन—कल्याण करे और धन—धान्य की वृद्धि करे।

कृष्यै त्वा क्षेमाय त्वा रय्यै त्वा पोषाय त्वा।

यजुर्वेद, ९.२२

शतपथब्राह्मण में पूरे कृषिकार्य को चार शब्दों में उल्लेखित किया गया है।

कृषन्तः, वपन्तः, लुनन्तः, मृणन्तः।

शतपथब्राह्मणम्, १.६.१.३

१. कर्षण – खेत की जुताई करना

२. वपन – बीज बोना,

३. लवन – पके खेत की कटाई करना

४. मर्दन – मड़ाई करके स्वच्छ अन्न को प्राप्त करना।[3]

भारतवर्ष चावल, जूट, कपास, दालों, आम, नींबू, काली मिर्च, गेहूं, राई, अलसी, सेब, नाशपाती तथा अखरोट आदि के उत्पादन के प्राचीनतम केन्द्रों में से एक था। गेहूं, जौ, कपास आदि का उत्पादन तथा हल, बीज–ड्रिल, पहियेवाली गाड़ियों आदि उपकरणों का उपयोग हड़प्पा सभ्यता (लाहौर से लगभग १६० किलोमीटर दूर) की कुछ मुख्य विशेषताएं हैं। प्राचीनतम कृषि में केला, गन्ना, साबूदाना, पाम आदि पौधों के वानस्पतिक वर्धन की जानकारी उपलब्ध थी। खेती तथा अन्य उपयोगी कार्यों में पशुओं को पालतू बनाकर सहयोग लिया जाता था। अनाज के व्यापार में परिवहन के लिए नौकाओं का उपयोग किया जाता था। कपास की सफाई, कताई व बुनाई में प्रयुक्त उपकरण भी प्राप्त हुए हैं।

ऋग्वेद की ऋचाओं में पर्यावरण, पशुपालन वानिकी, कृषि–संसाधनों व कार्यप्रणाली के विभिन्न पहलुओं का वर्णन है। इनमें मानव जाति के लिए सूर्य, वर्षा, वृक्ष तथा पशुपक्षियों की उपयोगिता पर भी प्रकाश डाला गया है। कृषि–सूक्त और अक्ष–सूक्त में खेती के महत्त्व को दर्शाया गया है। अथर्ववेद की ऋचाओं में पादप–सुरक्षा, पवित्र उपवनों व अशोक, पीपल, बरगद आदि पवित्र वृक्षों का वर्णन है।[4]

कृषि पराशर (४०० ईस्वी पूर्व) पुस्तक पूर्णत: कृषि से सम्बन्धित प्राचीनतम पुस्तक मानी जाती है। इसमें कृषि–प्रबन्धन, कृषि–औजार, पशुविज्ञान, बुआई, प्रत्यारोपण, पादप–संरक्षण, निराई, जलधारण, निकासी, कटाई, उपज मापन, अनाज–भण्डारण आदि विषयों का समावेश है।

कौटिल्य के अर्थशास्त्र (३०० ईस्वी पूर्व) में कृषि से सम्बन्धित अलग प्रखण्ड है। इसमें कृषि–प्रक्रियाओं, मौसमी फसलों और नदी–तलों के उपयोग का वर्णन है।

वराहमिहिर–कृत बृहत्संहिता (५०० ईस्वी पूर्व) के कृषि प्रखण्ड में मिट्टी के वर्गीकरण, सिंचाई प्रणालियों व कृषि औजार आदि विषयों पर प्रकाश डाला गया है।

सुरपाल कृत वृक्षायुर्वेद (१००० ईस्वी) पुस्तक में पादप विज्ञान से सम्बन्धित क्रियाओं व जानकारियों का विशाल संग्रह है। इसमें वृक्षारोपण के विभिन्न पहलुओं, पादप–वर्गीकरण, पौधों के गुणन, पोषण व संरक्षण, बीज व बुवाई, भू–जल, मिट्टी–वर्गीकरण व चयन, पशुओं व फसलों को बढ़ाने के लिए जमीन–उपयोग के वनस्पति–संसूचक, विहार–उद्यानों तथा बागवानी के आश्चर्यों का समावेश है।

निष्कर्ष रूप में कहा जाता है कि वेदों में व भारतीय शास्त्रों में विद्याओं के सूत्र विद्यमान है। वेदों में जहां धर्म, आचार शिक्षा, नीतिशिक्षा, सामाजिक जीवन, राजनीतिक शास्त्र, अर्थशास्त्र, आयुर्वेद आदि से सम्बन्धित पर्याप्त सामग्री उपलब्ध होती है, वहीं विज्ञान के विविध अंगों से सम्बद्ध सामग्री भी प्रचुर मात्रा में उपलब्ध है। वेदों में भौतिकी, रसायन–विज्ञान, वनस्पति–शास्त्र, जन्तु–विज्ञान, प्रौद्योगिकी, ऋतु–विज्ञान व

कृषि–शास्त्र, गणित, ज्योतिष, पर्यावरण, वृष्टि एवं भूगर्भ से सम्बन्धित सामग्री बहुलता से प्राप्य है। यह सामग्री बहुत अधिक बिखरी हुई है।

'यच्छब्द आह तदस्माकं प्रमाणम्' अर्थात् जो शब्द कहता है, वह हमारे लिए प्रमाण है। महर्षि पतंजलि के इस वैज्ञानिक मन्तव्य को अपना कर यहां ऋतु–विज्ञान व कृषि–विज्ञान से सम्बद्ध कुछ चिन्तन प्रस्तुत किया गया। यह चिन्तन निस्संदेह वैदिक व भारतीय मनीषा की वैज्ञानिकता को प्रमाणित कर सर्वत्र स्थापित करने में सहायक होगा।

सन्दर्भ ग्रन्थ सूची

१. भारतीय संस्कृति कोष (खण्ड १ से १० तक)
: भारतीय संस्कृति कोष मण्डल, पुणे, १९७०

२. इंडियन एस्ट्रोनामी — ए सोर्स बुक
: बी.वी. सुब्बरायप्पा व के. वी. शर्मा
नेहरू सेंटर, मुम्बई, १९८५

३. वृक्षायुर्वेद — सुरपाल
: एशियन एग्रीहिस्ट्री फाउडेशन, सिकन्दराबाद १९९९

४. कृषि पराशर —
: जी.पी. मजूमदार एवं एस. सी. बनर्जी
दि एशियाटिक सोसायटी, कोलकाता, १९६०

3

प्राचीन भारत में कृषि की वैज्ञानिक प्रगति

श्रीमती (डॉ०) इन्दु शर्मा

भारत में मनुष्य के उभयपक्षीय पूर्णत्व के उन सभी साधनों का प्रावधान उपलब्ध होता है जिनका सम्बन्ध मनुष्य के अभ्युदय और निःश्रेयस से है। मानव सभ्यता के विकास में कृषिविज्ञान का बहुत महत्त्व है क्योंकि अन्न के बिना जीवन सम्भव नहीं। अन्न प्राप्ति का एकमात्र साधन कृषि है। इसका आविष्कार सृष्टि की उत्पत्ति के साथ ही हुआ था। अथर्ववेद में इन्द्र द्वारा हल की रेषा को पकड़ने, पूषा द्वारा उसकी रक्षा करने और हल की रेषा द्वारा रसयुक्त होकर मनुष्यों को रस प्रदान करने का उल्लेख मिलता है।[1] अन्यत्र कहा गया है कि यह कार्य विद्वानों और कवियों द्वारा भी सम्पन्न किया जाता था।[2] वस्तुतः कृषि के बिना समाज के जीवन–पोषण की कल्पना भी नहीं की जा सकती। यह मानवीय कल्याण का साधन है। यजुर्वेद में कृषि की उन्नति को, जनकल्याण को तथा धन–धान्य की वृद्धि को राजा का प्रथम कर्तव्य स्वीकार किया गया है।[3] अथर्ववेद में विराट् ब्रह्म के मनुष्यों के पास पहुँचने का उल्लेख है। इसमें उसे इरावती (अन्नसमृद्धि) के नाम से अभिहित किया गया है और उसको दुहकर कृषि और सस्य की प्राप्ति का उल्लेख है– **ते कृषिं च सस्यं च मनुष्या उप जीवन्ति।**[4] जब मनुष्य की क्षुधा शान्त नहीं होती, वह न तो शारीरिक स्वास्थ्य का लाभ कर सकता है और न ही बौद्धिक विकास से सम्पन्न हो सकता है। यही कारण है कि अन्नसमृद्धि का दोहन मानवों के लिए परम आवश्यक घोषित किया गया है। आज भी अन्न को मानव जीवन की तीन मुख्य आवश्यकताओं में प्रथम स्वीकार किया जाता है। जब भी चर्चा होती है– रोटी, कपड़ा और मकान की होती है। मकान, कपड़ा और रोटी की नहीं होती। अथर्ववेद में यह बात सूत्र रूप में प्रतिपादित है। इसके अनुसार बुद्धियों को भरण–पोषण हेतु अन्न में रखने का परामर्श दिया गया है – **अन्ने समस्य यदसन् मनीषाः।**[5] ऋग्वेद में कृषि विद्या के विकास की चर्चा करते हुए कहा गया है– सर्वप्रथम देवगण आगे आये। उनके पास अपनी–अपनी कुल्हाड़ियाँ थीं। उन्होंने जंगलों को काटकर साफ किया। प्रजाजन भी उनके साथ थे। उन्होंने उपयोगी लकड़ियों को नदियों के किनारे रख दिया जहाँ कहीं घास फूस था उसे जला दिया।[6] वास्तव में भूमि को कृषि योग्य बनाना कृषि की सर्वप्रथम अपेक्षा है। ऋग्वेद और अथर्ववेद के अनुसार कृषि का आविष्कार वेनपुत्र पृथु के द्वारा किया गया।[7] उसका जन्म वैवस्वत मनु की परम्परा में हुआ था।[8] अन्यत्र इन्द्र को कृषि कर्म का अधिष्ठाता माना गया है और मरुतों को कृषिकर्मी।[9] पृथु के कृषि के आविष्कारक होने का समर्थन महाभारत और भागवतपुराण में भी उपलब्ध होता है।[10] उन उल्लेखों से यह सिद्ध हो जाता है कि कृषियोग्य भूमि तैयार करने के लिए वनों का काटा जाना, चट्टानों और कंकरों आदि का हटाना अनिवार्य है।

यजुर्वेद में प्रश्नोत्तर शैली में भूमि को बीज बोने के लिए सर्वोत्तम स्थान माना गया है।[11] अन्यत्र कहा गया है कि कृषि उत्तम अन्न को देने वाली होनी चाहिए।[12] आरम्भ से ही हमारे देश में कृषि को सर्वाधिक

महत्त्व प्राप्त होता रहा है। कृषि के आधुनिकीकरण के पश्चात् भी पारम्परिक कृषि आज भी विद्यमान है। अथर्ववेद में दर्शाया गया है कि कृषि के लिए उत्तम भूमि के साथ–साथ उत्तम बीज, हल, बैल और किसान की भी आवश्यकता होती है। इसी आशय के मन्त्र ऋग्वेद और यजुर्वेद में भी मिलते हैं। अथर्ववेद के अनुसार सुख प्राप्ति के लिए हलों को जोतना और जुओं को पृथक्–पृथक् बाँधना आवश्यक है।[13] यही कारण है कि वैदिक–वाङ्मय में हल को तैयार करने, तैयार की हुई भूमि में बीज बोने, हल के आकार–प्रकार, किसानों के बैलों के पीछे चलने, बैलों और मनुष्यों की कल्याण कामना, जुती हुई भूमि का अभिनन्दन, जुती हुई भूमि का घी और शहद से सिंचन आदि के विविध सन्दर्भ वेदों में विद्यमान हैं।[14]

इन सन्दर्भों से पता चलता है कि उस युग के लोग कृषि कर्म से भली भांति परिचित थे। इसके लिए जिन यन्त्रों की आवश्यकता होती थी, उनकी गुणवत्ता का वे ध्यान रखते थे। जलवायु की अनुकूलता, भूमि की उर्वरता के लिए आवश्यक खाद के महत्त्व से भी वे भली भांति परिचित थे। 'कृषि से सुखप्राप्ति' सूक्त यही सिद्ध करता है कि कृषि से भाग्य की वृद्धि होती है। जो लोग भूमि की अवस्था, वृष्टि की परिस्थिति, जलवायु की अनुकूलता को जानते हैं वे कृषि से लाभ उठाकर सुखी हो सकते हैं। यजुर्वेद में अन्न के लिए 'उर्वर्य' शब्द का प्रयोग हुआ है।[15] उर्वर्य से अभिप्राय उर्वरा भूमि में उत्पन्न होने वाला अन्न है। उर्वरता कृषियोग्य भूमि की उपजाऊ शक्ति को कहा जाता है। जिस भूमि में अन्न नहीं बोया जाता उसे 'खल' कहा जाता है। यह वह भूमि है जिसमें कृषि से उत्पन्न अन्न को साफ करके उसे भण्डारण के योग्य बनाया जाता है।[16] जो भूमि कृषि योग्य नहीं होती उसे ऊसर कहा जाता है। इसका कारण उसमें खारी मिट्टी का होना है। खारी मिट्टी के कणों के लिए शतपथब्राह्मण में ऊषरसिकता शब्द का प्रयोग किया गया है।[17] ये सन्दर्भ यह सिद्ध करते हैं कि वैदिक काल में भी लोग क्षारमृत्तिका के दोषों से परिचित थे। अथर्ववेद में इसके लिए 'इरिण' शब्द का प्रयोग हुआ है।[18] एक अन्य मन्त्र के अनुसार अन्न के लिए उत्तम भूमि और वर्षा आवश्यक हैं। वर्षा से भूमि की उर्वराशक्ति बढ़ती है तथा अन्न ही सारी प्रजा का आधार है। भूमि को पर्जन्य की पत्नी कहा गया है।[19] इसका उद्देश्य वर्षा को भूमि की उर्वरा शक्ति में वृद्धि का साधन सिद्ध करना है। यजुर्वेद और अथर्ववेद दोनों में ही खाद के प्रयोग का परामर्श दिया गया है।[20] इस बात का भी उल्लेख है कि बोने से पहले बीज को पानी में भिगोया जाए। उसमें ओषधियाँ भी डाली जाएँ ताकि उसकी गुणवत्ता बढ़ सके।[21] हमारे मनीषी अन्न की गुणवत्ता में वृद्धि करने वाले स्रोतों से भली–भांति परिचित थे। आजकल भी बीज को भिगोकर बोने की परम्परा प्रचलित है।

कौटिल्य अर्थशास्त्र के 'सीताध्यक्ष:' प्रकरण में कृषि–सम्बन्धी अनेक सन्दर्भ मिलते हैं।[22] जिसके अनुसार धान के बीज को सात दिन रात की ओस और दिन की धूप में रखने का परामर्श दिया गया है। जबकि मूंग, उड़द आदि के बीजों को तीन या पाँच दिन। इसके अनुसार बोने से पहले प्रत्येक बीज को सोने से स्पृष्ट पानी में भिगोना चाहिए। कौटिल्य अर्थशास्त्र में बीज को वर्षा के अनुपात तथा ऋतु के अनुसार बोने का परामर्श दिया गया है. इतना ही नहीं, उन्होंने बीज के अंकुरित होने पर छोटी मछलियों की खाद देने का परामर्श भी दिया है। उनका कहना है कि यदि साँप की केंचुली और बिनौलों को एकसाथ रखकर जला दिया

जाए तो जहाँ तक उसका धुआँ फैलेगा वहाँ तक कोई साँप नहीं आ सकेगा। कृषि सुरक्षा सम्बन्धी अन्य साधनों की चर्चा करते हुए कौटिल्य ने कहा है कि खलिहान में पुआल, भूसा आदि न छोड़ा जाए। आग को खलिहान से दूर रखा जाए और खलिहान में जल का प्रबन्ध अवश्य हो।

संहिताओं में यज्ञ को संसार की नाभि माना गया है– **यज्ञो वै भुवनस्य नाभिः।**[23] यजुर्वेद के यज्ञ से कृषि का होने वाले लाभों का विस्तृत विवरण उपलब्ध होता है। इनमें मुख्य हैं – यज्ञ द्वारा अन्न, फूल और फल की वृद्धि, अच्छी फसल के बाधक तत्वों का विनाश, उत्तम वृष्टि और कृषि की सम्भावना, अक्षय धन और धान्य की प्राप्ति, यज्ञ द्वारा अग्नि की ऊष्मा, सूर्य की ऊष्मा, सूर्य की किरणों का प्रकाश और उत्तम भूमि का प्राप्त होना।[24] इसके अनुसार यज्ञ से बादल बनते हैं। बादलों से वृष्टि और वृष्टि से अन्न उत्पन्न होता है और इसी अन्न से प्राणिमात्र की उत्पत्ति होती है–

अन्नाद्भवन्ति भूतानि पर्जन्यादन्नसंभवः।
यज्ञाद्भवति पर्जन्यो यज्ञः कर्मसमुद्भवः।।[25]

गीता में अन्न शब्द का प्रयोग व्यापक अर्थों में हुआ है। यह मात्र अनाज का पर्याय नहीं अपितु उन सभी स्थूल और सूक्ष्म पदार्थों का द्योतक है जिनके खाने से प्राणियों के शरीर आदि की पुष्टि होती है। भावतः समस्त प्राणियों के शरीर में रज और वीर्य की उत्पत्ति खाद्य पदार्थों से होती है और इनके संयोग से विभिन्न प्राणियों की उत्पत्ति होती है। उत्पत्ति के बाद उनका पोषण भी अन्न के द्वारा होता है। इससे पूर्व यही बात तैत्तिरीयोपनिषद् में कही गई है कि सब प्राणी अन्न से ही पैदा होते हैं और उत्पन्न होकर अन्न से ही जीते हैं – **अन्नाद्ध्येव खल्विमानि भूतानि जायन्ते अन्नेन जातानि जीवन्ति।**[26] वेदों में कृषि के लिए आवश्यक पदार्थों में उर्वरा भूमि, उत्तम बीज, वायु, धूप, जल, खेत की सुरक्षा तथा खादविषयक स्फुट निर्देश उपलब्ध होते हैं। यजुर्वेद के एक मन्त्र में उर्वरा भूमि को नमस्कार किया गया है। **नमः उर्वर्याय।**[27] इससे अभिप्राय है कि जिस भूमि की उर्वरा शक्ति बोए हुए बीज को ठीक ढंग से उगाने में समर्थ हो वह सचमुच नमस्करणीय है। जहाँ तक बीज का सम्बन्ध है तद्विषयक निर्देश ऋग्वेद और यजुर्वेद में संयुक्त रूप से मिलते हैं। इनके अनुसार बीज को भूमि के परिष्कार के पश्चात् ही बोना चाहिए।[28] अच्छी फसल के लिए धूप अनिवार्य है, क्योंकि सूर्य की किरणों से पौधों में ऊर्जा ही नहीं आती, अपितु ग्लूकोज के रूप में उन्हें भोजन भी मिलता है। इसकी पुष्टि करते हुए कहा गया है कि कृषि को सभी अग्नियों का सहयोग प्राप्त हो – **विश्वे भवन्त्वग्नयः समिद्धाः।**[29] कृषि के लिए वायु का महत्त्व भी उतना ही अधिक है जितना कि धूप का। इसके परिणामस्वरूप फसलों को रक्षक तत्त्व अथवा अवि–तत्त्व प्राप्त होता है। इसका उल्लेख करते हुए अथर्ववेद में कहा गया है–

अविर्वै नाम देवता ऋतेनास्ते परीवृता।
तस्य रूपेणेमे वृक्षा हरिता हरितस्रजः।।[30]

यजुर्वेद में उपलब्ध कृषि के साथ वृष्टि का उल्लेख उन्हें परस्पर सापेक्ष सिद्ध करता है. तदनुसार ही उसमें कृषि और उपज के लिए जल और ओषधियों के उपयोग का निर्देश है।[31] वर्षा से पृथ्वी को जीवनी शक्ति मिलती

है। तदनुसार ही यह प्रार्थना की गई है कि वर्षा समय पर होती रहे।[32] कृषि के लिए खाद उतना ही आवश्यक है जितना कि जीवन के लिए अन्न। वेदों में खाद के लिए करीष, शकन् और शकृत शब्दों का प्रयोग हुआ है।[33] इसके अतिरिक्त घी, दूध, शहद और ओषधियों के रसमिश्रण से भी खाद डालने का विधान है।[34] कौटिल्य के अर्थशास्त्र में स्नुही के दूध का भी खाद के रूप मे प्रयोग किए जाने का उल्लेख है। कृषि के लिए अन्य आवश्यक पदार्थ हैं – सुरक्षा एवं कृमिनाशन। यजुर्वेद में सू, प्रसू, सीर और लय शब्दों का प्रयोग कृषि की प्रक्रिया को स्पष्ट करने के लिए किया गया प्रतीत होता है। सू उत्पन्न करने की क्षमता का द्योतक है अर्थात् उत्तम बीज, प्रसू उत्तम उत्पादन का, सीर हल के प्रयोग से उत्पन्न धान्य का वाचक तथा लय कृषि नाशक तत्त्वों का पर्याय।[35]

कृषिनाशक तत्त्वों को **'ईति'** कहा गया है। एक मन्त्र में तर्द (कृषि के नाशक कीट पतंगे) को नष्ट करने, बिल में रहने वाले चूहों को नष्ट करने, इनका सिर और इनकी पीठ तोड़ने और इनके मुँह बाँधने का परामर्श दिया गया है जिससे जौ की फसल की रक्षा हो सके।[36] इसी सूक्त में अन्य हानिकारक कीट–पतंगे, टिड्डी आदि से फसल की रक्षा करने का परामर्श दिया गया है। ऐसे भी सन्दर्भ मिलते हैं जिनमें प्राकृतिक शक्तियों से आग्रह किया गया है कि वे कृषि को प्राकृतिक आपदाओं से बचाएँ। तदनुसार ही अतिवृष्टि के निषेध के साथ–साथ अनावृष्टि से बचने के लिए भी प्रार्थना की गई है और फसल को तीव्र धूप और हिमपात से बचाए रखने के लिए भी परामर्श दिए गए हैं।[37] इससे यह सिद्ध हो जाता है कि वैदिक मनीषियों का कृषि विज्ञान पर्याप्त मात्रा में व्यापक था। वे उत्तम कृषि में सहायक सभी तत्त्वों से भली भांति परिचित थे। उत्पादों की गुणवत्ता की महत्ता को भली भांति समझते थे। अन्न को जीवनदायिनी शक्ति स्वीकार करते थे। इसीलिए इसकी पौष्टिकता और आरोग्यता पर पूरा ध्यान दिया जाता था।

वैदिक संहिताओं में कृषि के लिए सिंचाई के लिए जिन साधनों का उल्लेख मिलता है, वे हैं – वर्षा, नहर, नदी, तालाब और कूप इत्यादि।[38] ऋग्वेद में सिंचाई के इन साधनों को दिव्याः (वर्षा का जल) खनित्रिमाः (कुओं का जल), स्वयंजाः (स्रोत का जल), समुद्रार्थाः (समुद्र में मिलने वाली नदियों का जल) कहा गया है।[39] यजुर्वेद और तैत्तिरीयसंहिता में इन्हीं साधनों का उल्लेख मिलता है।[40] इससे यह तथ्य स्पष्ट हो जाता है कि उन दिनों में भी कृषि मात्र वर्षा पर आधारित नहीं होती थी, सिंचाई के अन्य साधन भी प्रयोग में लाए जाते थे। आज भी बहुधा क्षेत्र में इन्हीं परम्परागत साधनों का प्रयोग किया जाता है।

वैदिक संहिताओं में मुख्य अनाजों के नामों का उल्लेख प्राप्त होता है, वे हैं – धान, जौ, उड़द, तिल, मूंग, चना, कंगुनी, अणु, सांवा, नीवार, गेहूँ, मसूर इत्यादि।[41] इसके अतिरिक्त कटाई के हिसाब से चार मुख्य फसलों का उल्लेख उपलब्ध होता है। वे हैं – ग्रीष्म में कटने वाली जौ तथा गेहूँ, वर्षा में कटने वाली धान, शरद में कटने वाली धान, हेमन्त और शिशिर में कटने वाली उड़द और तिल मुख्य हैं।[42] वर्तमान में भी इन्हीं चार ऋतुओं में फसलों की कटाई होती है। परन्तु अब ग्रीष्म ऋतु में कटने वाली फसल रबी और शरद् ऋतु में कटने वाली फसल खरीफ के नाम से प्रसिद्ध हो गई है। ऐसा प्रतीत होता है कि रबी और खरीफ का वर्गीकरण तैत्तिरीयसंहिता के आधार पर किया गया है।[43]

निष्कर्षत: यह कहा जा सकता है कि प्राचीन भारत में कृषि को भारतीय अर्थव्यवस्था में महत्त्वपूर्ण स्थान प्राप्त था। इसकी उन्नति एवं विकास के हर सम्भव प्रयत्न किए जाते थे। कृषिविदों को सम्मान से देखा जाता था। कृषिकर्म एक सम्मानजनक कर्म समझा जाता था। भूमि की जुताई से लेकर अनाज के भण्डारण तक वे सभी उपाय किए जाते थे जो उत्पादन की वृद्धि तथा सुरक्षित भण्डारण के लिए अपेक्षित माने जाते हैं। कृषि की सुचारूता को व्यवस्थित बनाए रखना राजा का दायित्व माना जाता था। खाद के रूप में केवल पौष्टिक खादों का ही प्रयोग किया जाता था। प्रयोग में लाए जाने वाले कीटनाशक भी विषाक्त नहीं होते थे। आज भी परम्परागत कृषि–साधन लुप्त नहीं हुए हैं। वैदिक संहिताओं में कृषि सम्बन्धी सन्दर्भ उस समय में उपलब्ध साधनों के भरसक प्रयोग के साक्षी सिद्ध होते हैं। उन दिनों में अन्न समृद्धि को पर्याप्त महत्त्व प्राप्त था। जो कृषिविद्या में निपुण होते थे उन्हें कृष्टराधि और उपजीवनीय कहा जाता था। कृषिविशेषज्ञ अन्नविद् के नाम से प्रसिद्ध थे। कृषि मानवीय कल्याण का साधन मानी जाती थी। तदनुसार ही यजुर्वेद में **'कृष्यै त्वा क्षेमाय त्वा रय्यै त्वा पोषाय त्वा'** (यजु० ९–२२) के माध्यम से कृषि की सुव्यवस्था का दायित्व राजा का स्वीकार किया गया है।

सन्दर्भ :

१. इन्द्रः सीतां नि गृह्णातु तां पूषाभिरक्षतु।
सा नः पयस्वती दुहामुत्तरामुत्तरां समाम्।। अथर्ववेद, ३.१७.४

२. वही, ३.१७.१

३. कृष्यै त्वा क्षेमाय त्वा रय्यै त्वा पोषाय त्वा। यजुर्वेद, ९.२२

४. अथर्ववेद, ८.१०.२४

५. वही, २०.७६.४

६. ऋग्वेद, १०.२८.८

७. वही, ८.९.१०

८. अथर्ववेद, ८.१०.(४) १०–११

९. देवा इमं मधुना संयुतं यवं सरस्वत्यामधि मणावचर्कृषुः।
इन्द्र आसीत् सीरपति शतक्रतुः कीनाशा आसन् मरुतः सुदानवः।। अथर्ववेद, ६.३०.१

१०. शान्तिपर्व, ५९.११३–११५; भागवतपुराण ५.१६.२३

११. यजुर्वेद, २३.४५–४६

१२. वही, ४.१०

१३. अथर्ववेद, ३.१७.१; ऋग्वेद १०.१०१.४; यजुर्वेद १२.६७

१४. अथर्ववेद ३.१७.१–९

१५. यजुर्वेद, १६.३३

१६. ऋग्वेद, १०.४८.७

१७. शतपथब्राह्मण, काण्ड ६

१८. अथर्ववेद, ४.१५.१२

१९. यस्यामन्नं व्रीहियवौ यस्य इमाः पञ्च कृष्टयः।
भूम्यै पर्जन्यपत्न्यै नमोऽस्तु वर्षमेदसे। १२.१.४२

२०. यजुर्वेद, १२.७०: अथर्ववेद, ३.१७.९

२१. यजुर्वेद, १.२१

२२. कौटिल्य अर्थशास्त्र, पृ० २३८–२४४
२३. तैत्तिरीय ब्राह्मण, ३.९.५.५
२४. यजुर्वेद, १८.१–२७
२५. गीता, ३.१४
२६. तैत्तिरीयोपनिषद्, ३
२७. यजुर्वेद, १६.३३
२८. (क) कृते योनौ वपतेह बीजम्। यजुर्वेद, १२.६८
(ख) वपन्तो बीजमिव धान्याकृतः। ऋग्वेद, १०.९४.१३
२९. यजुर्वेद, १८.३१
३०. अथर्ववेद, १०.८.३१
३१. यजुर्वेद, १८.३५
३२. वही, २२.२२
३३. अथर्ववेद, १.१४.३; ऋग्वेद, १.१६१,१०; १.१६४.४३
३४. यजुर्वेद, १२.७०
३५. वही, १८.७
३६. अथर्ववेद, ६.५०.१
३७. वही, ७.१८.१–२
३८. अथर्ववेद क्रमशः, ४.१५.१; २०.१७.७; १.४.३; १.६.४; १.६.४
३९. ऋग्वेद, ७.४९.२
४०. यजुर्वेद, २२.२५; तैत्तिरीयसंहिता, ४.५.७.१–२
४१. यजुर्वेद, १८.१२
४२. तैत्तिरीय संहिता ७.२.१०.२
४३. द्विः संवत्सरस्य सस्यं पच्यते। तैत्तिरीयसंहिता, ५.१.७.२

Section VII

MISCELLANEOUS

1

Buddhist Concept of Basic Material Element: A Scientific Critique

Dr. Narasingha Charan Panda

The structural exposition of the beings has been made with two reals, viz. *nāma* and *rūpa*. The term *nāma* (*namati iti nāmam, cintanam vā*) stands for the inner personality, expressing itself through the *citta* (consciousness) and *cetasika* (psychic factors). *Rūpa* (matter) refers to the physical personality, within and without, gross in nature and material in character with all its varieties of manifestations. Matter is that which appears or manifests itself to the senses (*rūpati iti rupam*) without ever suggesting either reality or stability, substance or substratum. So matter is what materializes. Different meanings are given of this materializing— as pressure, pain, disappearance or change. Thus matter is something that disappears. In Modern science matter is anything that has mass (heaviness) and occupies space. To understand this material world knowledge of matter is most essential. An element consists of only one kind of atoms.

The elaborate discussion of the *Dhammasa Ūgaṇi* begins by defining *rūpa* or matter as "चत्वारो च महाभूता चतुन्नञ्च महाभूतानाम् उपादाय रूपम्[1]" "the four *mahābhūtas* or great basic elements that proceeding from the grasping of that is called *rūpa*. Buddhaghośa explains it by saying that *rūpa* means the four *mahābhūtas* and those which arise depending on them as a modification of them. In the *rūpa* the six senses including their affections are also included. In the *Samyutta Nikāya* the Buddha says "O *bhikkhus* it is called *rūpam*, because it manifests (*rūpyati*); how does it manifests? It manifests as cold, and as heat, as hunger and as thirst, it manifests as the touch of gnats, mosquitoes, wind, the sun and the snake; it manifests, therefore it is called *rūpa*"[2]. This *rūpa* or matter is atomic.

Matter is regarded as the collective organism or collocation, consisting of the fourfold substratum of colour, smell, taste and contact. The unit possessing this fourfold substratum is known as *paramāṇu*, which is the subtlest form of *rūpa*. It cannot be pierced through or picked up or thrown away. It is indivisible, unanalysable, invisible, inaudible, untastable and intangible. It is not permanent, but is like a momentary flash into being. The simple atoms are called *dravya-paramāṇu* and the compound atoms *samghāta-paramāṇu*.[3] A *dravya-paramāṇu* never arises or exists in isolation. It always arises and exists in combination with other *dravya-paramāṇus*. A collection of them, forming a unity and having a simultaneous origination and a simultaneous cessation, is called *samghāta-paramāṇu*, 'aggregate-atom', i.e. molecule.[4] (*na vai paramāṇurūpamekaṁ pṛthagbhūtam asti / saṁghātasthaṁ tu tad rūpyate eva*[5]). The smallest *saṁghāta-paramāṇu* is an octet consisting of the four primary elements and four of the secondary elements, viz. *rūpa, rasa, gandha* and *sparśa*. That the four primary elements always arise simultaneously and that the secondary elements cannot arise independently of the primary, are the two fundamental principles involved in the conception of the *samghāta-paramāṇu*. This is a brief note of the two kinds of *paramāṇu* postulated by the Vaibhāṣikas. According to Prof. T.A. Stcherbatsky "the combined atoms (*saṁghāta-paramāṇu)* alone appear in phenomenal reality, the simple ones, or infra-atomic elements, presumably, were relegated to transcendental reality, in accordance with the general character of a Buddhist element."[6] The organs of sense are also regarded as modifications of atomic matter. Seven such *paramāṇus* combine together to form an *aṇu*, and it is in this combined form only that they become perceptible. The combination takes place in the form of a cluster having one atom at the centre and others around it. The point which must be remembered in connection with the conception of matter is this, that the qualities of all the *mahābhūtas* or primary elements are inherent in the *paramāṇus*.

The ultimate units of matter elements are known as *paramāṇu* or the properties. These properties are of four different kinds, viz. *pṛthivī* (earth particles), *ap* (water particles), *vāyu* (air particles), and *tejas* (fire particles). These are known as four *mahābhūtas* or great elements.[7] Most of the schools of Indian thought, notably the Sāṁkhya, the Vedānta and the medical tradition as represented by Caraka and Suśruta, recognizes five great elements, such as: *pṛthavī, ap, tejas, vāyu* and *ākāśa*. That *ākāśa* (ether or space) is the fifth, is admitted by the Nyāya-Vaiśeṣika too.

In Buddhism, although hypothetically these *mahābhūtas* are said to be of different types, they are seldom found in isolation[8]; each type remains in a mixed up condition with the remaining types. Each particular class of *mahābhūtas* contains its one characteristic feature, e.g. the earth particle (*pṛthavī paramāṇu*) as got hardness; water particle has got humidity; fire has got heat and air particle has got motion. It is said that these four types remain in each constituted object in relative proportion. All these four primary elements are characterized with colour (*varṇa*) and shape (*saṁsthāna*).

Although Buddhism does not ignore the concept of space (*ākāśa*), but enters it in a list of six *dhātus* with *vijñāna*. In many respects it differs from the other four: it is a non-corporeal (*amūrta*) substance devoid of tactility (*sparśa*) and characterized by ubiquity (*vibhu*), absolute continuity and infinite magnitude. As such, unlike the other four substances, it is not, in the ultimate analysis, composed of atoms. Thus, although *ākāśa* is introduced as a *bhūta* (element), in view of its peculiar characteristics, it has to be distinguished from the other four and is, in a way, on a par with such intangible substances as *kāla*, time.

The four principle characteristics of matter are referred to and enumerated in many canonical texts.[9] This has also been elucidated in the commentarial works, viz. Aṭṭhasālinī and Visuddhimagga. They are called *mahābhūtas* or great primary elements, because of manifestation of greatness. It means they manifest with immense vastness with respect to phenomena 'grasped at' (*upādinnasantati)* and those of un-grasped at (*ana-upādinna-santati*). In case of *upādinna-santati*, they appear as the material forms, as the bodies of various kinds of beings, like human beings, animals, birds, etc. But in case of *ana-upādinna-santati*, they appear into a world system, tremendously vast, traditionally calculated as 1203450 *yojanas* in circumference containing big waters, mountains, as well as the great famous trees like Jambu tree, Kappa tree, etc.[10]

These four primary elements are being elaborated as under:

Pṛthavī-dhātu (earth element):

The word *pṛthavī* (*paṭhavī)* is derived from the root *prath-* (*puttha-*), meaning to expand or extend and directly refers to the vastness of the earth. It is the feminine form of the noun *pṛthu* which means vast, broad, expansive, etc. *Dhūtu* means that which bears its own characteristic marks (*attano sabhāvaṁ dhāretiti dhātu*). So, the literal as well as nearer meaning of the term is 'the element of extension'. It does not stand only for geometrical extensions of length, width and height, but also for solidity and hardness. Whatever is hard and solid is the element of earth.

यं अज्झत्तं कक्खळं खरिगतं उपादिन्नं, से" यथीदं - केसा लोमा नखा दन्ता तचो मंसं ङ्ख यं वा पनङ्‌ पि चिञ्चि अज्झतं पच्चत्तं कक्खळं खरिगतं उपादिन्नं। अयं वुच्यतावुसो, अज्झत्तिका पठवी धातु ।[11]

So, the earth element is that which is hard (*kakkhaiam*), rough (*khara-gatam*), hardness (*kakkhaiattam*), rigidity (*kakkhai abhavo*) both internal and external.

For the Abhidharma, too, kakkhaia and khara which mean hard and rigid respectively bring out the essential nature of earth element.12 The first is said to represent its characteristic (*lakkhaṇa*) and the second, its mode (*ākāra*).[13] Similar definitions are met with in other schools of Buddhist thought. The Mahāvastu says that when the living beings who lived at a new evolution of the world began to eat whole mouthfuls of the essence of this earth as food, their bodies came to possess the characteristics of gurutva (heaviness), kharatva (roughness), kakkhatatva (hardness).[14] The Abhidharmakośa and its commentary, too, use the latter two terms in defining the pathavī-dhātu. In the Abhidharma Samuccaya it is defined as: 'kaīhiṇatā', a term which could be interpreted as meaning 'solidity'. As such this interpretation is almost the same as that explained by the Theravādins. Thus, there is a common outlook among the Buddhist scholars in maintaining that what is called primary element of earth, which stands for the phenomenon of hardness, solidity or compactness in matter. It is a repellent energy which gives to matter the characteristics of dimension, expansion, weight and pressure.

***Apo-dhātu* (water element):**

The term *ap* is derived from the root *āp-*, which means 'to pervade' or 'to obtain' (*āpnoti*). *Āpo-dhātu* represents the fact of viscidity (*sineha*) and cohesion or binding together in matter-'*rūpassa bandhanattam*'.[15] It is the binding quality of matter, as it acts as a binding force of the material elements. In the Sanskrit Buddhism, i.e. in the Abhidharmakośa, *āpo-dhātu* is defined in a similar manner. It stands for *dravatva* (liquidity) and *snehatva* (viscidity) in material things[16]. The Vaiśeṣika School also accepts '*ap*' substance with the same two characteristics as: '*rūparasasparśavatya āpo dravāt snigdhat*'[17]. But according to the Vaiśeṣikas liquidity and viscidity are qualities inherent in the '*ap*' substance. No such dichotomy is recognized by the Buddhists.

However, cohesion is that which diffuses itself throughout its co-existent qualities or that which increases the bulk of them.[18] Although fluidity (*paggharaṇa*) is its distinctive characteristic, yet cohesion is its particular function. It is interesting to note that this quality of cohesion is not perceptible to the sense of touch or sense contact, for it is an entirely 'internal' quality which does not exert its energy in a way which can be apprehended by the sense organs.

***Tejo-dhātu* (fire element):**

The word *tejas* (*tejo*) is derived from the root *tija-niśāne* 'to be sharp' (cf. Greek *stigma*, Latin *in-stig-are*, English *sting*, German *stechen*, etc.). In Sanskrit, starting from Vedic times, *tejas* first expressed the meaning of a sharp ray of light, point of a flame and then fire and light. In Buddhism also *tejo-dhātu* signifies the phenomenon of heat, the term being used is *usmā* or *usumā*.[19] *Tejo* (*tejas*) is the essential material quality of thermal energy or temperature in its various aspects and degrees of cold (*sita-tejo*) and heat (*usma-tejo*). This, too, is a primary quality inherent in all matter. Both *paîhavī* and *āpo* are essentially affected by changes in temperature; for heat can expand and even melt solids, while liquids are congealed by cold; evaporation and condensation are produced by changes in temperature and variations of caloricity.

It was only in the Nineteenth Century that **Science found** it possible to agree with Buddhism, when it was discovered that temperature was due to motion: the kinetic theory. *Tejas*, then, is that essential material quality which gives 'life' to matter, without which the other primary qualities could not function. It stands for the maturation of diseased matter, for digestion and assimilation of food (*paripacana*)[20], for evolution as well as involution. For this is the element which heats, matures, sharpens and imparts heat to all other material elements.

***Vāyo-dhātu* (air element):**

The word *vāyu* (*vāyo*) is derived from the root *vā-* 'to blow', 'to move',etc. The meaning of this word is also known as motion, vibration, oscillation and pressure. It is a material energy to keep the things in a particular position by generating pressure all around. It is defined as the air (*vāyu*), that which belongs to air (*vāyogatam*), fluctuation; and the inflation of form (*thambhitattaṁ rūpassa*[21]). The standard term used in the Pāli commentarial works to describe the *vāyo-dhātu* is '*samudīraṇa*', which means mobility or motion.[22] There is general agreement among the scholiasts that *vāyo-dhātu* is representative of mobility or motion (*īraaṇ, samudīraṇa*).

This (air-element) is not to be understood merely as the movement of air, or of an object in space, but far more as the continuous adjustment between the two opposite forces of extension (earth), which repels and cohesion (*āp*) which attracts. This oscillation between attraction and repulsion causes a natural friction which is the generator of heat. Thus *vāyu* is as much a condition for the arising of *tejas*, as *tejas* is a condition for the origination of movement. It is through vibration that extension is felt as pressure; through vibration also cohesion becomes magnetic and centripetal. In a way, vibration is the most important of the four great elements as it is mainly through this essential quality that matter is known to us.

The general position assigned to the elements may now be examined. If some one will examine how the elements are interpreted in other systems of Indian Philosophy one would notice that in Buddhist thought they are assigned a comparatively primary position. What the Sāṁkhya considers as elements are not the ultimate irreducible constituents of matter, for they are evolved immediately from the *tanmātras* and ultimately from the Prakṛti. According to the Vedāntins the elements are produced from the *sūkṣma-bhūtas*. For the Jains the ultimate constituents of *puggala* (matter) are not four elements (*dhātū-catukka)* but the homogeneous atoms. The latter are recognized as the essential causes of the former.[23] The Nyāya-Vaiśeṣikas postulate four kinds of atoms corresponding to the four elemental substances, viz. *pṛthavī, ap, tejas* and *vāyu*. This may be described as an attempt to reconcile the older theory of the elements with the later atomic theory.

In Buddhism, unlike in many other systems of Indian thought, the elements (*mahābhūtas*) are assigned a primary position in the sense that they are recognized as the ultimate irreducible data of matter. It is of course true that a given instance of matter consists of not only the four elements but also a set of *upadā-rūpas*, such as colour, smell, etc. But these so called *upadā-rūpas*, as conceived by the Buddhists, are always depended on, therefore secondary to, the elements.

The Buddhist concept of *bhūtas* (atoms) may be compared with the atoms of ancient system of Greek Philosophy, Leibnitz's philosophy of spiritual atomism and logical atomism of Vaiśeṣika philosophy with the physical properties of modern science. The Buddhist conception of the atoms lying always in group seems to remove an ancient difficulty of Nyāya-Vaiśeṣika idea about a combination of two atoms (*dvyaṇuka*) and perceptible manifestation of the combination of three atoms (*trasareṇu*). Thus, these thinkers take recourse to *ārambhavāda* (theory of new origin) and *īśvara kāraṇatā-vāda* (theory of causal agency of God) in order to explain that combination. These two ideas on the other hand are criticized by the Buddhist thinkers. Buddhists view seems to be more rational when they hold that the atoms always are found together. Unlike the Monads of Leibnitz, the atoms, according to Buddhists, are not spiritual but physical. These atoms again are momentary and dynamic. This idea, however, is not generally found in Greek school of atomism; evidently, a scientific approach is seen here and the approach is very close to that of in modern physics. These atoms in their composite state are subject to change from moment to moment.

Thus, although the modern conception of Physics, like field of forces, seat of energy, or dynamism or space time units are not clearly found in Buddhist thought, it is evident that these ancient Buddhist thinkers grasped the correct nature of the universe as multiversity, evanescent and changing at every moment. Though appearing as a continuous and a single one, the universe is a composition of momentary units just flowing on. These, abstract and hypothetical units whether they are thought as *dharmas* or atoms are only comparable with the views of modern scientists like John Eddington or like Sir James Jeans.

From the foregoing description of the four primary elements it would appear that *pṛthivī* stands for solidity, heaviness and extension, *ap* for viscidity and cohesion, *tejas* for the temperature of cold and heat and *vāyu* for motion. These four elements are coordinate and represent four distinct forces or phenomena in the realm of matter. Besides, another fundamental feature of these elements is that they always remain together (*sahajāta*). No element can exist independently of the other three.[24]

Modern Scientific concept of material elements was latent, since the very out set of creation. Hence, in Buddhism the basic four material elements are discussed from the scientific outlook. In the Science three forms of matter existṭ solid, liquid and gas. Solid is compared with earth element as it has mass (weight or heaviness), which also occupies space. Water element is a liquid form of matter, whereas heat (*tejas*) and air (*vāyu*) are considered as gaseous form of matter. This way, the basic material elements are analyzed in Buddhist thought bearing the idea of Modern Science.

References

1. Dhammasaūgaṇi, pp. 124-179.
2. Saṁyutta Nikāya, III.86.
3. Abhidhammakośa Bhāṣya, I.13.
4. *Ibid.*, II.22.
5. *Ibid.*, I.13.
6. *The Central Conception of Buddhism*, chapter IV, p.14 (2005 Edition, Motilal Banarsidass, Delhi)..
7. तत्थ भूतरूपं चतुब्बिधं - पृथवीधातु, आपोधातु, तेजोधातु, वायोधातू ति। विसुद्धिमग्ग, निद्देस 14.35; (cf. bhūtāni pṛthavīdhāturaptejovāyudhātavaî- Abhidharmakośa, I.12).
8. *Saṁcitā daśarūpiṇaī* Abhidharmakośa I.35 comm. and *na vai paramāṇurūpamekaṁ pāthagbhūtamasti*, *ibid.*, I.13.
9. *Sāmaūüaphala Sutta of Dīgha Nikāya*, I.76 (*cātumahābhūtiko*); Ānanda Vagga's fifth sutta of Aūguttara Nikāya, I.222; *Upādāna parivaṭṭa sutta of* Saṁyutta Nikāya, III.56, *sanka sutta of* Majjhima Nikāya, I.185, etc. (*cattāro mahābhūta ṣ pṛthavī-dhātu, āpo-dhātu, tejo-dhātu, vāyo-dhātu*).
10. *Pātali-simbali-jambu, devānaṁ päricchattako /*

Kadambo kapparukkhoca, sirisena bhavati sattamo// Aṭṭhasālinī, 242.

11. Majjhima Nikāya, *Mahāhatthipadopama Suttam* (I.185).

12. DhammasaŪgaṇi, p. 177; cf. VibhaŪga Pāli, p. 82 (P.T.S.).

13. *Tattha paÎhamaṁ lakkhaṇa-vacanaṁ dutiyam ākāra-vacanam* ṣ Visuddhimagga, p. 286 (P.T.S.).

14. *yato ca bhikṣavaste satvā taṁ pṛthavīrasamālopakāra* ṣ *kamāhāram* ṣ *ihārensuî atha teṣāṁ kāye gurutvaṁ ca kharatvaṁ ca kakkhaÎatvaṁ ca upa.ıīpate* ṣ Mahāvastu, I.339 (P.T.S.).

15. DhammasaŪgaṇi, p. 226; VibhaŪga Pāli, p. 83 (P.T.S.).

16. Abhidharmakośa, (comm.) I.

17. Vaiśeṣika Sūtra, II.1.2

18. *Âpeti sahajātarūpāṇi pattharati; appāyati vā brūheti ´vaṄÑhetīti āpo* ṣ Abhidhammāttha-vibhāvani, p. 167.

19. DhammasaŪgaṇi, p. 177;VibhaŪga, p. 83 (*usmā usmārtatam usumam usumāgatam ajjhattika tejodhātu'*).

20. Aṭṭhasālinī, p.332.

21. Vibhanga Pāli, p. 84 (3.1.5).

22. Aṭṭhasālinī, p. 332, Abhidhammātthavikāsinī, p. 251; cf. *samudīraṇā vāyudhātu*-Abhidharmakośa Bhāṣya (sphuṭārthā) II.22.

23. cf. *Adeśamattamutto dhātucatukkassa kāraṇaṁ jo du* ṣ *so neo paramāṇo pariṇāmaguṇo sayamasaddo*, Paücastikāyasāra, p.28.

24. Abhidharmakośa, Ch. II, p. 248.

2

The Śukla Yajurveda and Science

Prof. M.L. Wadekar

The Ṛgveda Yajurveda, Sāmaveda and Atharvaveda - these are four Vedas. Each Veda has several recensions. There is a vast Vedic literature in the form of the Saṁhitās. Brāhmaṇas, Āraṇyakas and Upaniṣads. Moreover, the Śrauta, Gṛhya, Dharma and Śulva Sūtras are also available severally in accordance with different recensions. The Yajurveda has two recensions - the Śukla and Kṛṣṇa. The Śukla Yajurveda has two subrecensions - Kāṇva and Mādhyandia. The sage Yajñyavalkya is the promulgator of the Mādhyandina recession of the Śukla-Yajurveda. The Vedas incorporate hymns in praise of different gods and goddesses - the different powers of nature-spontaneously proclaimed by our ancient sages. Though this is true in case of most of the Mantras, there are several other topics, which are found dealt with in them. The Mantras contain description of the Vedic rituals - the Yajñas also. The social and cultural condition of that period is also reflected in the Vedas. The supreme and unparalleled speculations of the Vedic sages regarding the unknown secrets of nature are also met within the Vedas. Beside all these aspects of knowledge, the Vedas contain information relating to different branches of science. In this article, we shall deal with only the Śukla-Yajurveda, hence the truths of science found in the Ṛgveda, Sāmaveda and Atharvaveda are not noted here. Though there was no invention of new and modern equipments like telescope, microscope etc., they have discovered many scientific truths - this is quite wonderful. The earth moves round the Sun - this theory of Copernicus against that of Tolamy has been indicated in the Mantra - āyaṁ gauḥ (Yaj. 3.6).[1] Thousands of years ago, the sages knew it. The moon is not self-luminous and is illumined by the rays of the Sun – this is clearly stated in the Mantras - suṣumṇaḥ sūryaraśmiḥ. (Yaj. 18.40)[2] and candramasam yataujāḥ (Yaj. 23.99)[3]. The Śukla-Yajurveda mentions 27 constelations, six seasons, 12 months of the year and also the intercalary month (adhikamāsa)[4] - this shows that the sages

---

1. आयं गौः पृश्निरक्रमीदन्मातरंपुरः। पितरं च प्रयन्त्स्वः।। ३.६
2. सुषुम्णः सूर्यरश्मिश्चन्द्रमागन्धर्वः.....१८.४०
3. को वेद चन्द्रमसं यतोजाः। २३.४७

   वेदसूर्यस्य बृहतोजनित्रमथो वेदचन्द्रमसं यतोजाः।। २३.९०
4. २७ नक्षत्र, १२ महीने, ६ ऋतु, अहंस्पति, मलिम्लुच वातो वा मनो वा गन्धर्वाः सप्तविंशतिः। ९.७

had minute and accurate knowledge about the movement of the earth. The year was not of 360 days. The three sources of energy– the Sun, Agni, and water are clearly referred to in the Mantra - divaspari (Yaj. 12.18-20)[1]. The concept of hydro-energy is as old as Vedas. The Vedic sages knew the power that supported the earth and the sky. This power has upheld the sky with both of its hands. (sa dādhara pṛthivīm dyāmutemām (Yaj. 23.1)[2]. yena daurugrā pṛthivī ca dṛḍhā (Yaj. 32.6)[3]. Is this not the suggestion of the gravitational force, invented by Newton? The Śukla-Yajurveda (17.59) even mentions aero-plane. (Vimāna-eṣa divo Madhya āsta).[4] The physics and astronomy have intrinsic relation with mathematics. Without minute and accurate calculations, astronomical theories are not possible. In the Mantra– ekā ca me (Yaj. 18.24)[5], there is mention of odd numbers from one to twentythree in ascending order. The next Mantra – catasraśca me – refers to even numbers from four to twenty-eight in ascending order. The Mantra– tryaviśca me etc. following it alludes of half numbers like 1½, 2½, 3½, 4½. The concept of zero

---

इषश्चोर्जश्चशारदावृत्, सहश्चसहस्यश्च हैमन्तिकावृत्, तपश्च तपस्यश्च शैशिरावृत्। १५.५७

मधुश्च मविवश्च वासन्तिका वृत १३.२४

शुक्रशुचिश्च ग्रैष्मावृतू १४.९

नभश्च नभस्यश्च वार्षिकावृत (श्रावण) (भाद्रपद) १४.१५

1. दिवस्परि प्रथम जज्ञेऽग्निरस्मद्वितीयं परिजातवेदा:।
तृतीयमप्सु नृमणाऽजस्रमिन्धानऽनं जरते स्वाधी:।।
विद्या ते अग्ने त्रेधा त्रयाणि विद्या ते धाम विभृता पुरुत्रा।
विद्या ते नाम परमंगुहा यद्विद्मा तमुत्सं यतऽआजगन्य।।१७।।
समुद्रे त्वा नृमणाऽप्स्वन्तर्नृचक्षाऽईधे दिवो अग्नऽऊधन।
तृतीये त्वा रजसि तस्थिवाऽसमपामुपस्थे महिषाऽअवर्धन्।।२०।। १२.८.२०

2. हिरण्यगर्भ: समवर्तताग्रे भूतस्य जात: पतिरेकऽआसीत्।
स दाधार पृथिवीं द्यामुतेमां कस्मै देवाय हविषा विधेम।।२३.१

3. येन द्यौरुग्रा पृथिवी दृढा येनस्व: स्तभितं येननाक:।
योऽन्तरिक्षे रजसो विमान: कस्मै देवाय हविषा विधेम।।३२.९

4. विमानऽएष दिवो मध्य आस्तआपब्रिवान्रोदसीऽअन्तरिक्षम्।
स विश्वाचीरभिचष्टे घृताचीरन्तरा पूर्वमपरंच केतुम्।।१७.५७

5. एका च मे तिस्रश्च मेल्प पंञ्च च मे - यज्ञेन कल्पन्ताम्। १८.२४
चतस्रश्च मे....१८.२५
त्र्यविश्च ये त्र्यवी च मे। १८.२९
षष्ठवाट् च मे पष्ठौही च मे १८.२७, इमा मे अग्न इष्टका धेनव: दश च दश च....

is completely Vedic and genuinely Indian and this is our most extra-ordinary discovery for the world. The Mantra – ekā ca daśa ca śatam ca– enumerates all tens – not ever thought by any mathematics in the world. They are – śata, sahasra, ayuta, niyuta, prayuta, arbuda, nyarbuda, samudra, Madhya, anta, parārdha. The Śulva-sūtras knew the theory of Pythagoras – the square of the hypotenuse is equals to the squares of the two sides of a rectangular triangle. The sages knew the minutest parts of the body - the internal organs like the heart (hṛde tvā - Yaj. 6.25; 37.19)[1], Mind (manase tvā - Yaj. 37.19), Head (agnirmūrdhā - Yaj. 13.14;[2] 15.20)[3], Foetus, its outer cover (Yaj. 19.76[4]; 8.28)[5], anus (Yaj. 39.10) etc. - without the help of any external instruments like microscope. They also knew the formative elements of the body like bodily hair, skin, blood, fat, flesh, muscles, bones, marrow, semen. These are referred to in the Mantra - lomabhyaḥ svāhā (Yaj. 39.10)[6]. The foetus, water around it, its outer cover, birth of a child after ten month - all these are clearly mentioned in the Mantra - ejatu daśa māsyaḥ (Yaj. 8.28)[7]. From the point of view of the Sexology, the Mantras 20 to 31 of the Adhyāya 23 should be considered.[8] The Śatapatha Brāhmaṇa[9] gives etymology of the word hṛdaya (heart), as being constituted of three letters - hṛ to draw, da to give, ya to control - which refer to the function of the heart. The Atharvaveda has enumerated several diseases, the minute germs that cause them and their varieties -

---

1. हृदे त्वा मनसे त्वा दिवे त्वा सूर्याय त्वा। ऊर्ध्वमिममध्वं दिवि देवेषु होत्रा यच्छ।।९.२५
हृदे त्वा मनसे त्वा दिवे त्वा सूर्याय त्वा। ऊर्ध्वोध्वरं दिवि देवेषुधेहि।।३७.१७
2. अग्निर्मूर्धा दिव: ककुत्पति: पृथित्याऽअयम्।
अपाꣳरेता सि जिन्वति....१३.१४
3. १३.१४ - १५.२०
4. रेतो मूत्रं रिजहाति योनिं प्रविशदिन्द्रियम्। गर्भो जरायुणावृत उज्वं जहाति जन्मना।। १७.७९
5. एजतु दशमास्यो गर्भो जरायुणा सह। यथायं वायुरेजति यथा समुद्रऽएजति।
एवायं दशमास्योऽअस्रज्जरायुणा सह। यस्यै ते य ज्ञियो गर्भो यस्मै योनिह्रियण्ययी।। ८.२८
अङ्गान्यह्रुता यस्य तं मात्रा समजीगमꣳस्याहा।। ८.२७
6. लोमभ्य: स्वाहा लोमभ्य: स्वाहा त्वचे स्वाहा त्वचे स्वाहा लोहिताय स्वाहा लोहिताय स्वाहा मेदोश्य: स्वाहा मेदोभ्य: स्वाहा। माꣳसेभ्य: स्वाहा माꣳसेभ्य: स्वाहा स्नावभ्य: स्वाहा स्नावभ्य: स्वाहास्श्नभ्य: स्वाहास्थभ्य: स्वाहा मज्ञभ्य: स्वाहा मज्ञभ्य: स्वाहा रेजसे स्वाहा पायवे स्वाहा।। ३७.१०
7. तत्रैव
8. ताऽउभौ..... उत्सवक्थ्याऽअवगुदं धेहि समञ्जिचारया वृषन्। स स्त्रीणां जींवभोजन:। यकासकौ शकुन्तिका हलगिति वञ्चति। आहन्ति गभे पसो निगलगलीतिथारका।। ऊर्ध्वामेनामुच्छ्रापय...२३.२०-३१
9. शतपथ १४.८.४.१ तदेतत्र्यक्षरꣳहृदयमिति। हृ अभिहरन्ति, द ददन्यस्मै यमित्येकमक्षरंमेति स्वर्गंल्लोकं य एवं वेद।

all these scientific details of Pathology. the medicinal and surgical remedies for various diseases are documented. The Śukla-Yajurveda, however, refers only to the diseases and the medicinal plants. The very first Mantra[1] has alluded to the cows that are Anamīvā (without any diseases) and Ayakṣmā (without T.B). In the Mantras (75-101)[2] of the twelfty Adhyāya of the Śukla-Yajurveda, there is very scinctific and practical description of the medicinal plants. The sages knew many plants that could destroy the diseases like T.B., Hernia, Swellings and Wounds. One Mantra has noted that the sages had knowledge of 107 kinds of the medicinal plants, their places of origin and their practical utility for diseases. The theory that the plants, their places of origin and their practical utility for diseases. The theory that the plants are living, invented by the modern scientists like Jagadishcandra Bose was not only known to the sages but they had so much confidence about practical applicability of the herbs for the diseases that they declared that even by taking the medicinal plants in hand only the root cause of the disease itself would vanish. Now a days, the use of water only as medicine is being advocated. This hydro-therapy is referred to in the Śukla-Yajurveda. The sages says, "The water is nectar! The water is medicine!" (Yaj. 9.6)[3] The water removes all kinds of diseases and impurities – physical and mental. (Yaj. 6.17)[4] The medicines are not only for the human beings but are also for animals – the cow, the horse, the goat, the sheep (Yaj. 3.59)[5]. Thus the Veterinary medical science was also known. Like the plants, the use of metals as remedy of diseases was also known. How much the science of metallurgy had made advancement can be glimpsed from the Mantra (Yaj. 18.13)[6],

---

1. इन्द्रायभागं प्रजावतीरनमीवायक्ष्मा।
2. या ओषधी: पूर्वा जाता देवेभ्यस्त्रियुगं पुरा। मनै नु बभ्रूणामहꣳशतं धामानि सप्त च।। १२.७५
   शतं वोऽम्ब धामानि सहस्रमुत वो रुह:। अधाशतक्रत्वो यूयमिमं मेऽअगदं कृत।।
   इष्कृति नाय...... यदिमा वाजयन्नाहमोषधींहस्तऽआदधे। आत्मायक्ष्मस्य नश्यति पुरा जीवगृभो यथा।।१२.८५
   या फलिनीर्याऽफलाऽपुष्पा याश्च पुष्पिणी:। १२.८७
   यं जीवमश्नावहै न स रिष्यति पूरुष:।। १२.९१
   ..... कामाय पहृदे। १२.९२
   द्विपाच्चतुष्पादस्माकꣳसर्वमस्त्वनातुरम्। १२.९५
   नाशयित्री बलासस्याशंसिऽउपचितामसि। अथो शतस्य यक्ष्माणां पाकारोरसि नाशनी।। १२.९७
3. अपस्वन्तरमृतमप्सु भेषजमपामुत प्रशस्तिष्वश्वा भवत वाजिन:। ७.९
4. इदमाप: प्रवहतावद्यं च मलं च यत्। यच्चाभिदुद्रोहानृतं यच्च शेपेऽभीरुषम्। ९.१७
5. भेषजमसि भेषजं गवेऽश्वाय पुरुषाय भेषजम्। सुखं मेषाय भेष्यै।। ३.५७
6. हिरण्यं च मेऽयश्च मे श्यामं च मे लोहं च मे सीसंस च मे त्रपु च मे।। १८.१३

in which all kinds of metals like the gold, silver, copper, iron, lead, tin etc. There is mention of varma (armour) for the protection of the body (marmāṇi te varmaṇā – Yaj. 1939)[1]. The use of purified gold for the longevity, health and vitality is to be found in the Mantras (Yaj. 34.50 – 51.52)[2]. The patent of the medicinal plants like Apāmārga and dūrvā (Yaj. 35.11 and 13.20)[3] should be made by Indians before the Foreigners understand their medicinal properties and try for their patents. There is mention of the golden, the silver and the copper needles (Yaj. 23.37)[4], during the process of the Aśvamedha Yajña. This should be properly viewed from the point of view of the science of surgery and even post–mortem. There is description of science of Agriculture, along with different kinds of instruments of it, in the Mantras – sīra yuñjanti etc. (12.67–73)[5]. The Mantra – vrīhayaśca me (Yaj. 18.12)[6] has clearly referred to all kinds of food grains like – rice, wheat, Java, uḍada, masura, sesame, moong etc. Like the science of the botany the seers had accurate knowledge of the zoology as well. The Mantras 20–40 of the Adhyāya 24[7] have enumerated more than 108 animals, birds and aquatics, whose identification nowadays is really difficult. Besides this, the sages have exhibited their acquaintance with the science of Psychology in the Śivasaṁkalpa sūkta[8] with the refrain–tanme manaḥ śivasaṁkalpamastu – in which the exposition of the extremely subtle nature of mind is found. The philosophical speculations, regarding the inscrutable nature and the origin of the world of reflected in the

---

1. मर्माणि वर्मणा च्छादयामि, सोमस्त्वा राजामृतेनानुवस्ताम्। १७.४९
2. आयुष्यं वर्चस्यं रायस्पोषमौद्भिदम्। इदं हिरण्यं वर्चस्य जनैत्राया विशतादुमाम्।५०।।
   न तद्रेक्षाथ्ंसि न पिशाचास्तरन्ति देवानाभोजः प्रथमजथ्ं ह्येतत्।
   यो बिभर्ति दाक्षायण हिरण्य स देवेषु कृणुते दीर्घमायुः स मनुष्येषु कृणुते दीर्घमायुः।।५१।।
   यदाबध्नन्दाक्षायणा हिरण्य शतानीकाय सुमनस्यमानः।
   तन्मऽआबध्नामि शतशारदायायुष्मान्जरदष्टिर्यथासत्।।५२।। ३४.४०-४२
3. अपाघमप किल्विषमप कृत्यामपो रपः। अपामार्ग त्वमस्मदप दुःष्वन्यथ्ंसुव।। ३४.११
   काण्डात्काण्डात्प्रोरहन्ती परुषः परुषस्परि। एवा नो दूर्वे प्रतनु सहस्रेण शतेन च।।१३.२०
4. रजता हरिणीः सीसा युजो युज्यन्ते कर्मभिः। अश्वस्य वाजिनस्त्वचि सिमाः त्र्यम्यन्तु शम्यन्तीः।। २३.३७
5. सीरा युञ्जन्ति कवयो... १२.९७-७३
   सीरा, सुम्नया, युगा, शुन, कीनाशा, शुनासीरा, लाङ्गलं, इत्यादि।
6. व्रीहयश्च मे सबाश्च मे माषाश्च मे तिलाश्च मे मुद्गाश्च मे खत्वाश्च मे प्रयंगवश्च मेऽणवश्च मे श्यामाकाश्च मे नीपाराश्च मे गोधूमाश्च मे मसूराश्चमे यज्ञेन कल्पन्ताम्।। १८.१२
7. वसन्ताय कपिञ्जलान्।। २४.२०-४०
8. यज्जाग्रतो....तन्मे मनः शिवसंकल्पमस्तु। ३४.१-९

Sūkta ṣKasmai devāya[1] and the Puruṣa Sūkta. The first Mantra refers to the śreṣṭ hatama[2] – The Best of the best Action, it is pointed out at the end that there is no other way but doing the action only – selflessly, one should desire to live for hundred years in this world. Not indulging in selfish motives, one should work for the welfare of the society at large.[3] The welfare of all includes that of oneself.

Kurvanneveha karmāṇi jijīviṣet śatam samāḥ

Evem tvayi nānyatheto'sti na karma lipyate nare.

* * *

Bernal, J.D	The Origin of life, Weidenfield and Nicholson, London, 967
Bloomfield	Religion of the Veda.
Bloomfield	Athrvaveda and Gopath Brahmain, Chaukhambha Vidya Bhavan. Kashi
Chawla	The Rigvedic Deities and their Iconic form, Jyotsna Chawla, MRML
Devitt, W	Biology of the cells, Evolution Approach, W. B. Saunders, company London.
Fox, S.W (ed)	The origin of life prebiological system and of their Molecular structure. Academic press, Newyork, 1965.
Griffith	Hymns of Rigveda, Motilal Banarsidas, New Delhi.
Gaud, P.N.	The exact Sciences in Veda, IQJM 42 IV 1974-75.
Haldane J.B.S	The Origin of life, The Rationalist Annual, 1929.
Irvin	Origins of Form and Structures
Jha V.N.	Metaphors in Vedic Literature/Editor V.N. Jha, CAS in University of Pune, Pune.
Keith	The Development and History of Sanskrit Literature, Sanjaya prakashan.
Lokesh Chandra	Dictionary of Buddhist Iconography in 2 Vols. International Academy of Indain Culture and Āditya Prakashan, Delhi.
Maxmullar	History of Ancient Sanskrit Literature, Panini Office, Allahabad.
Maxmullar and Keith	Vedic Index

1. तदेवाग्निस्तदादित्यस्तद्वायुस्तदु चन्द्रमाः।
.....कस्मै देवाय हविषा विधेम।। अध्याय ३२
2. इषे त्वोर्जे त्वा वायवस्थ देवो वः सविता प्रार्पयतु श्रेष्ठतमाय कर्मणे।।१.१
3. कुर्वन्नेवेह कर्माणि जिजीविषेच्छतꣳसमाः। एवं त्वयि नान्यथेतोऽस्ति न कर्म लिप्यते नरे।। ४०.२

Machonell and Keith Vedic Index on Names and Subjects, Chaukhamba Vidya Bhavan, Kashi

Machonell, A.A. Vedic Reader for Student, Oxford University press, London.

Machonell, A.A. History of Sanskrit Literature, MRML. Nai Sarak, Delhi

Machonell, A.A. Vedic Mythology

Nason, A The biological world, John Wiley & Sons Inc. New York.

Oparin, A.I. Modern Concepts of the Origin of life on the earth, Recent Development in plant Science. Today and Tomorrow Printers and publishers, New Delhi.

Parkinson. W.D. Introduction to Geomagnetism, Scottish, Academic press, Edinbrgh.

Parker, E.N. Cosmical magnetic fields, their Origin and activity, Clarendon Press. Oxford. 1979.

Rao Kusumanjali New Interpretation of Indian Art and Culture, Editor, M.S. Nagaraja Rao.

Robertis E.D.P Essentials of cell and Molecular Biolgoy. Halt Saunders, Japan.

Thatte, A.P. The mighty little atoms, Basic Science, IBH, Publishing company.

Tewari, S.P Hindu Iconography. Agam Kala Prakashan. Delhi.

Upadhya. B. Vedic Sahita and Sanskrati, Sharda Mandir Varanasi.

Volland, H. Atmospheric Electrodynamics, Springer, New York, 1984.

Whitney. W.D. Atharvaveda Saṁhitā Translation Vol.1 & 2, Motilal Banarsidas. Delhi.

Weber,Albrecht The History of Indian Literature, Chaukhamba Sanskrit Series Office, Varanasi.

Wilson Rgveda Saṁhitā English Transl

3

Ethics and Religion – Scientific Perspective

Ashok Aggarwal
Vibha Aggarwal*
Vipin Parkash and Sapana Sharma

INTRODUCTION

Bioethical Issues:

Man's relationship with the environment is as old as time immemorial. One can easily find reflections on ecology in the Hindu and Buddhist scripture and movements. People of that time had made plants, microorganisms and animals as the theme of their scriptures. Similarly, painters have also brushed animals and plants in their paintings. At the same time, the Rishis and poets have mentioned in each of verse some kinds of relationship with environment. Vedic rishis were aware of the importance of trees, shrubs and plants. So they have remembered them as their mothers or protectors or the relievers of pain.

Osadhirti matarah, veerudhah parayisnavah, vanaspatih Samita and so on. Large trees were considered to be life-giving and were called as *'prano vai vanaspathih'*. They give stgrength to the earth and make the clouds rain *"vanaspathin Devamindramavardhayat"*. Therefore, all people used to pray *"Osadhayah te mulam, ma hinsisam"*. It means that medicinal plants or herbs are of immense value, do not destroy them.

In Manusmriti also, it has been said that the person who cuts fruits trees, will be punished with highest degree of penalty; while the one who cuts flowering trees, shrubs and bushes will be penalized accordingly.

Phala-druma-chede,uttamasahasam dandyah

Pushpa drumachedi madhyamam

Valligulma-lata-chedi, Karsa pana satam

History is the witness that ancient man used to clean the atmosphere by performing *'yajna'*.

The ancient Indian scriptures including that of the Hindus and the Buddhist accord different tales of different animals and plants. For example, snake is regarded as favourite of Lord Shiva, tiger as carrier of Durga, Garuda as that of Lord Vishnu, rat of Ganesha, Swan of Saraswati etc. What is derived from all this relationship is the bioethical or ecoethical awareness to conserve environment.

Bioethics is defined as the branch of ethics, philosophy and social interaction with the living beings. It takes into account the potential affect of biotechnology on social, political and economical behaviour of human kind.

Ethics mythology and religion have protected certain plants. Thus, they are concerned. Some examples are:

- Lord Krishna in 'Bhavishya Puran' addressing Yudhisthra says ' one who plants one peepal, one neem, one fig tree, ten tamarind, three each of wood apple, bael and five mango trees will never see narak or hell.

* The Aswatha (peepal tree- *Ficus religiosa*) is held in high esteem of human kind. In twenty six stanza of ten chapter of Shrimadbhagwadgita, Lord Krishna enumerating all his specialities (Vibhuti) say of all the trees. *I am Aswattha* 'Further in fifteenth chapter- shloka 1-3, Lord Krishna says, "they speak of the imperishable ashvattham as having its roots above the ground (hanging roots) and branches below. Its branches extend below and above, nourish by the nodes, with sense for its twigs and below in the world of men stretch fourth the roots resulting in action. In 15:3, he says, "its real form is not thus perceived here, nor its end and nor beginning, nor its foundation."

- Similarly, vata tree (*Ficus benghalensis*) is associated with Hindu Gods or demons. Such deities are believed to perch on its branches. Hence, the tree is not cut for fuel. Like this, there are many plants like kela, haldi, amla, tulsi, rice, til seeds. Nariyal, chandan wood, putran jiva, neem, barley and kamal, which were considered sacred in our Hindu mythology(Yava).

Lord Buddha has also drawn attention towards nature and its components e.g. Water, fire, birds and plants, which also work like us, have the same feelings which we have. So to purify our conduct, we should learn from them to contribute for the protection of environment. Lord Krishna further states, "Therefore, the scriptures be your authority in deciding what act to be done and what not to be done". Having known what is said in the ordinance of the scriptures you should act here (Dua, 1999).

The primary concern of bioethics is now on ethical, legal and sociological issue.. Even advanced countries like USA and UK have started legislation both for plants and animal welfare and on threatened and endangered animal and plant species. It is now a moot question as to how far biotechnology should be encouraged in the light of Indian Ethical structure. Some of the issues should be taken for consideration are:

1. The validity of preparing human models for human diseases such as the transgenic model of cancer.
2. The condition under which recombinant organisms can be released into the environment.

Biotechnological Ethics:

It is true that newly developed biotechnological knowledge has already made very important landmark to the health and welfare of humankind. But on the other side, these biotechnological advances have some drawbacks or ethical concerns relating to the certain transgenic organisms used as food.

1. Transfer of human genes to food animals e.g. Transfer of human genes to animals like sheep.
2. Transfer of genes from animals whose flesh is forbidden for human consumption by certain religious groups to animals that they usually eat.
3. Transfer of animal genes into food plants that may be of particular concern to some vegetarians.

According to Ignacimuthu (1988) that biotechnology is one of the most used abused word in modern biology with definitions to suit a wide range of views. Hence, the use of biotechnology should be realized well in time with the Indian welfare thought.

Ecological Ethics:

In Bhagwad Gita, Lord Krishna unfolds the interrelationship of humankind with nature as. "Earth, fire, water, air, ether, mind and also ego; these constitute my nature; eight fold devided. This indeed is my lower (material) nature; the other than this, by which the whole universe is sustained known it to be my higher (spiritual) nature in the form of *jīva* (the life principal), Arjuna" (Dua, 1999).

Gautama Buddha favoured the optimum use of natural resources. This aspect may be seen in the *Kasibhardvaja sutta* of the *suttani patta* wherein the importance of farming is stated. He was born under a tree. He got enlightenment under the banyan tree. He also achieved the *Mahapannirvana* under a tree

The Rio conference 1992 and similar other pursuits in this regard have certainly added importance to ecological ethics. At present there is global realization about impending dangers associated with serious environmental deterioration. Some of the consequences as a result of population explosion and industrialization are:

- Degradation of ecosystems
- Over exploitation of natural resources
- Deforestation
- Depletion of wild life due to poaching
- Depletion of biodiversity and genetic resources
- Pollution
- Green house effect

Finally, it can be said that today the world is well aware of the advances made in the field of science and technology and also could not ignore the threats of biological and chemical weapons and the ways the applications of science and technology are being perceived in the development of ammunition, missiles and the deadly biological and chemical agents affecting the mankind, animals, birds and plants. The scientific and technological pursuits for the attainment of welfare of the society as a whole should necessarily care for the ethics and morality and that can be possible when all of us think that all things in this world are pervaded by the almighty and whatever he had bestowed upon us due to his kindness. We should enjoy them with renuniation so that a healthy balance of physical and spiritual conditions could be maintained i.e. ecological, biological and ethical.

Last but not the least, it can be said that all government agencies, NGO'S and private organizations should come forward for the protection of environment. Extension activities on behalf of every research institution should be must and people should be made aware of hazardous effects of ecological imbalance.

References :

Ignacimuthu, S. 1988. Basic Biotechnology, New Delhi, Tata McGraw Hill, pp. 298.

Dua, K.K. 1999. Bhagavad Gita and environment, New Delhi, Koshal Book Depot, pp. 17.

4

Growth of Science as evidenced by archaeological excavations in India

Rajesh Purohit

The genesis of science in India dates back to the prehistoric period when man started interacting with his environment. Initially it evolved as stone and bone tool making dexterity and then slowly it developed into a rudimentary science and eventually became a perfect science based on principles. The evolution of scientific maneuvering was common to all pre historic men in the world and India was no exception to this. However, it is believed migration and transmigration too played a significant role in the growth of various traits of culture including the technique of tool making, hunting, fishing, food gathering and later on farming, transportation so on and so forth in certain geographical pockets. Originally the science was in its most rudimentary state used as a necessity for obtaining utilitarian objects essentially employed for his subsistence. In early stages of culture, the study of the material sciences was neither formalised nor separated from other branches of knowledge. Conscious attempts to study the theory of science independently was perhaps never attempted so as to know its scientific implication based on scientific principles. In most cases, technological discoveries took place without any knowledge of the underlying scientific principles, through hit and trial, and by experience. It was predominantly focused on the utilitarian aspects only.

Man was a nomad and used to move from place to place in search of games and plants just as any other animal. However unlike animals this extraordinary animal, 'man' was not equipped with natural tools as a corporeal part of his body that could help him for his subsistence. Man's equipment were made by himself and were the extra corporeal as these are not like the natural tools such as paws of rabbit to dig with teeth and claws of the lion for tearing the prey, the beaver carpenter's tusk. He does not have thick skins, which could be used as a furry coat to keep him cosy. That is why it compelled him to develop tools to exploits the environment. Slowly the material sciences grew and reach to its pinnacle where men could meet all his requirements and comforts. Hence the material culture that we see today is the products produced by the extra corporeal of man. "It is chiefly threw his equipment that man acts and reacts to the corporeal world, draws sustenance their form and escapes its pearls- in technical language adapts himself to his environment or even adjusts his environment to his need."[1]

Archaeological excavations unfold the material culture of a period or a time. Most of these findings are either buried in the soil or fossilised into rocks or lying under the water or standing as living monuments since ages. Every bit of excavated material whether it is a plant fossil or animal fossil, a coin or terracotta, an inscription or a sculpture it has enormous facts embodied within it and has tremendous resources for reconstruction of a scientific history of India. Archaeological excavations reveal many such truths, such as what were the climatic condition or environment in India thousands of years ago, what kinds of flora and fauna were present at that time and even what caused to the death of man and animal etc. These could be ascertained only from the study of paleobotany, paleontology, paleoenvironmental studies, paleopathological studies of the excavated remains, which are otherwise not explained elsewhere in an explicit manner in literatures, as it should have been, for the reconstruction of the history of science in India. The real cause for the death of Napoleon or Alexander would have been a mystery had there been no DNA analysis of their remains from the grave. Besides, the written history has its own limitations as elucidates only from the period when writing came into existence.

The literally sources or written history contains a very patchy and incomplete record of what mankind had accomplished. Besides most of literary and biographical woks or memoirs are either exaggerated or glorified or composed in symbolic gestures with too many similes and metaphors to please the contemporary kings and leaders. At times the literally works seems to be written to appease certain communities, individual, nations etc. however the archaeological findings do not manifest such kind of bias. Thus these materials are extremely useful for reconstruction of a rational history that can speak about a common man, his feelings, suffering, pain, achievements and accomplishments in various societies he had lived with out being discriminated in the history. He or his society had never been eulogized in Indian history for extraordinary contribution to the mankind. It was only the kings and elites who figured conspicuously in the history. Besides, the written history covers records of human accomplishment of only five thousand or at best eight thousand years' history. Which is only five to eight percent of the overall history of our country? Where as India has long and varied history extending its antiquities to the pre historic times were man endeavoured in

1 Childe. V. Gordon, What happened in history, p.7

the field of science and technology and the period considered to be the genesis of India's scientific accomplishments.

With the change of environment in the beginning of the Holocene period the nomadic life style carne to an end and slowly during the a epipaleolithic period he started making smaller tools called microliths for hunting fast running animals like deer etc. Soon after he realised about domestication of plants and animals in its incipient agriculture state which eventually guided him for a sedentary or a settled life style to protect his plants and animals and there started a new era of agricultural life style. Natural phenomenon such as tides, rainfall, and appearance of the sun, the moon and stellar formations, the ancient Indians constantly observed changes in season, weather patterns and agriculture etc. Vedic literature refers to the condensation of water vapor from seas and oceans due to evaporation the heat of sun and the subsequent formation of clouds and rain. In the earliest scientific text, the *Vaisheshikas* a rudimentary attempt at recording of physical properties of different types of plants and natural substances were done. There was also an attempt at summarizing and classifying the observations made about natural phenomenon.

The most interesting archaeological discovery on ancient agriculture is of course the furrow marks of cultivable land found from the excavations at Kalibangan[1] (Rajasthan). Yet anther find is a terracotta plough unearthed from Banawali[2] (Haryana) which suggests the use of plough and furrow for comparatively tougher land unlike the Harappan or Indus people where a plough or a furrow would not have been required owing to the availability of a fertile land, besides no plough have been reported from the Indus sites in Pakistan.

Besides a large range of agricultural implements were invented such as digging stick hoes crescent shaped sickles, leaf shaped sickles, billhooks and spades. Interestingly iron hoes have also been found in the Painted Grey Ware level (C.1100-1200 BC) at Atranjikhera[3] iron Celts were used as ploughshares. A longitudinal ploughshare of iron has been found at Saradkel[4] in district Ranchi (Jharkhand).

---

1 Kalibangan I- 2900-2700BC., An Encyclopedia of Indian Archaeology, vol. 1, edited by A. Ghosh, p.319

2 Bisht R.S., "Excavations at Banawali, 1974-77," Harappan Civilization: A Contemporary Perspective, ed. G.L. Possehl (New Delhi: Oxford and IBH, 1982), 113-24

3 op. cit., 319

4 An Encyclopedia of Indian Archaeology, vol. 2, edited by A. Ghosh, p.393

Agriculture and irrigation are integral part of the ancient Indian economy. Therefore water harvesting and irrigation like tank irrigation and canal irrigation etc. have also been reported from various excavations. The canal built by Nanda king in Orissa and the dam on the lake *Sudarshan* near Junagadh during the reign of Chandergupta Maurya are interesting example of irrigation and water harvesting. However the water harvesting of Dholavira during the Harappan period is an important landmark of water harvesting in ancient India.

Weights and Measures speak of not only the knowledge of mathematical science but also the devises available at that time for computations. The Weight Weights and Measures were made of chert., semiprecious stone, copper and terracotta-make and were of various sizes and in graduation. Weight rods and a "ruler" made of shell lines drawn 6.7 millimeters apart with a high degree of accuracy. Two of the lines are distinguished by circles and are separated by 33.5 millimeters, or 1.32 inches. This distance is the so-called Indus inch.[1] Smallest weight unit in simple ratio 2, 4, 6, 8, 16 and sets of weight with a regular ratio have also been found Lothal excavations. Binary in smaller weight and than decimal being in the ratios 1,2,1/3 x 8,4,8,16,32,64,1600,200,320,640,1600 have been reported from Mohenjodaro and a set of weight with a regular ratio have also been found Lothal excavations. Binary in smaller weight and than decimal being in the ratios 1,2,1/3 x 8,4,8,16,32,64,1600,200,320,640,1600 have been reported from Mohenjodaro. The Harappan system of weights established the most frequently occurring weight as 13.63 gms.

Ceramic technology of the Indus-Saraswti civilisation was outstandingly remarkable It range from simply fired bricks and plain pottery to finely potted sophisticated wares with pained designs. The Ceramic technology also played a significant role in architecture for the making of terracotta tiles, pipes etc to be used for drains and wells.

Archaeological findings and their scientific investigation as well as interpretations have unfolded many new facts, which show that mathematical astronomy, geometry, algebra and other science arose first in India. Everyone knows that Aryabhatta invented the sign for zero in India about two thousand years ago, much before Aryabhatta the Indus-Saraswati culture knew about the value of Zero as can be seen from the size of the bricks i.e.10: 20:30 and 10:20:40 cms. It is further substantiated by the grid pattern town planning of the Indus people It is much less known that important concepts like algebraić transformation,

---

1 Teresi, p.59, 2002

mathematical logic, abstract language description, binary numbers etc. were known to the Indians in the third millennium B.C.

Mathematics is the science to which Indians have contributed the most. Our decimal system, place notation, numbers 1 through 9, and the ubiquitous 0, are all major Indian contributions to world science. Without them, our modern world of computer sciences, earth-launched satellites, microchips, and artificial intelligence would all have been impossible.[1]

The Indus Civilization introduced the 16 *annas* to a Rupee monetary system, used in south Asian countries until recently before the entry of metric system, seems to stem from base-8 (octal) a gift of Indus mathematics to the world. Other scholars claim that there was a consistent binary system of weights and measures such that the application of this binary measuring system in architectural features such as brick size 10:20:30 or 10:20:40 uniformly and invariably used in all the Indus-Saraswati or Pre Harappan and Harappan sites.

In India, the earliest applications of chemistry took place in the context of medicine, metallurgy, and construction technology and in textile production and dyeing.

The Indians knew knowledge of physical and chemical properties of matter from the time of Indus valley period.The earliest evidence of use of copper in the Indian subcontinent comes from Mehrgarh in Baluchistan in 4000 B.C Iron was also used as an alloy as can be seen from the chemical analysis carried out from the excavated metals of Lothal. Wheeled cart and water wheel were possible due to metal. Advanced agriculture could be feasible only with the knowledge of metal-shares replaced by stone. The crude metal was normally re-melted for refining in clay crucibles, which were subsequently broken up for the recovery of the purified metal. A crucible similar to the one from Lothal with slag sticking to it was recovered at Mohenjo-daro.Channeled crucibles with thick walls of clay mixed sand were used for casting long bars, which were subsequently hammered in to chisels, rods, etc. The thicker bars were cut into thinner ones as indicated by the grooves on some of them obtained from Lothal. Through the Harappans knew casting cire per due.

In the field of architectural science India excelled all other contemporary civilizations in the third millennium BC. The Harappan were considered to be the pioneer it town planning. Scientifically sound fort structure with protective gravity

---

1 Wolpert, 1992, p 194

wall around the town. The thickness of the wall ranges between 7 to 12 meters depending upon the gravity of the annual flood of the river. Mud bricks both sun baked kiln baked domestic structure kept them free from inundation.

The Harappans knew the measurement of time. According to Walter J. Fairservis Jr. the ivory pieces from Mohanjodaro represent the sliding scale from which the day of the lunar calendar followed by Harappans were ascertained[1]. In the 13th century AD the Konark temple in Orissa and the Jantarmantar at Varanasi, Jaipur, Delhi, Ujain and Mathura gave them the systematic and accurate measurement of time, in second, minute, hour, day, week, fortnight, seasons so on and so forth.

The ancient thinkers correctly understood the theory of animal life and particularly of man. The *Brihat Vishnu Purana* states that the aquatic life precedes the monkey life and that the monkey life is the precursor of the human life. The same theory was explained in an interesting way by the *dashavatara* (ten incarnations). But evolution, as everything else, was the manifestation of the supreme spirit (*Atman*) as is testified by *Chandogya* Upanishad. The concept of incarnation or manifestations of Lord in various forms are vividly described in the *vaisnavite* literature. In the *Bhagavadgita* it is stated that the Lord manifests on the earth whenever there is an onslaught over the righteous in order to annihilate the evils and bestow rewards to righteous for the establishment of *dharma* or orders in the world. The concept of incarnation also suggests the evolution of species from aquatic fauna to the modern man (Homo sapiens' sapience) as well as cultural evolution of man. Most interestingly the sequence in which the *avataras* in the *Dasavatara* are depicted is simply amazing. It also shows that how they knew about the human evolution from the aquatic fauna represented as *Matsyavatara* is followed by tortoise, a reptile an animal which dwells in water and on land, which is followed by boar, an animal that paleontologically speaking evolved before the cat family represented as an incarnation in the form of lion which succeeds the boar. After four zoomorphic figures dwarf man is shown as *Vamanavatara*. This is quite interesting as the first evolved Homo sapiens is the Java man who happened to be of dwarf stature. Here after the remaining incarnations represents the intellectual evolution. Finally as you come to the incarnation of Krishna he is the ultimate human possessor of sixteen virtues called *solahkalasamparna*.

The growth of science and the knowledge of the scientific phenomena of the Indians were not only amazing for their acumen and talent but also shows how

---

1 Fairservis W.A.Jr. (1977) Metrology in the Indus Civilization in FIC, 10-16

this continuity of Indian scientific tradition has sustained and eventually enriched the state of science in India. The modern science in India has its genesis in the annals of history. Today India in the field of science and technology is on par with all super powers. In space research, medicine, surgery and super computers Indian scientists rank higher than any other contemporary scientists of the world. However, it is disheartening to note that today's science is not conceived as per the environmental conduciveness. The scientific tempore of contemporary world gave us comforts at the cost of our health and many environmental hazards such as depleting of ozone, pollution and global warming which are threatening the existence of this rare planet earth. It is high time we should learn from the past where our ancestors used to develop scientific devices keeping an eye on the environment.

References

1. Childe. V. Gordon, What happened in history, p.7
2. Kalibangan I- 2900-2700BC., An Encyclopedia of Indian Archaeology, vol. 1, edited by A. Ghosh, p.319
3. Bisht R.S., "Excavations at Banawali, 1974-77," Harappan Civilization: A Contemporary Perspective, ed. G.L. Possehl (New Delhi: Oxford and IBH, 1982), 113-24
4. op. cit., 319
5. An Encyclopedia of Indian Archaeology, vol. 2, edited by A. Ghosh, p.393
6. Teresi, p.59, 2002
7. Wolpert, 1992, p 194
8. Fairservis W.A.Jr. (1977) Metrology in the Indus Civilization in FIC, 10-16

5

Archeological Evidence of Vedic Ostrich - Dadhikravan

ADV. P.V. HOLAY

The theory of Aryan invasion is falsely and maliciously introduced in Indian History by some European scholars when they found that the four Compilations of Vedas and their annexures like Shadangas, Upanishadhas, Brahmanas and Aranyakas are of high quality of literary work composed by brilliant persons in the remotest part of the history of civilizations. Indian civilasation was very much superior to the then existing other civilisations in the world. Study of Dadhikravan in Vedas proves that Aryan invasion theory is wrong.

The word Dadhikravan appears in 3rd, 4th, 7th and 10th Mandal of the Rigveda Nighantu classifies and puts it in the list of horse (vÜo 1.14) and Devata (nsork 5-4-19). While Yaskacharya explains the word as इत्येद् दधात् कमतीति वा दधात् कव्दतीति वा दधादाकारी भवती वा ! Sayanacharya treats it to be horse (). Still then, scholars are not satisfied with these explanations because the description given in Vedas also associates it with a bird. Some scholars think that it is fire (vfXu). Some say that it is the rising Sun and some say that it is a fog covering the rising Sun, which looks like suspended particles of coagulated milk (दधि-क to scatter). All these scholars were not aware that the horse-bird i.e. Ostrich existed in India before 12000 B.P. (Before Present) and had become extinct with the commencement of Holocene period. Dr. Wakankar and his colleagues have discovered thick pieces of Ostrich egg-shells at about 43 places in North -West India. Some of these pieces are archaeological artifacts, because some egg-shell pieces bear designs drawn by man and some are small discs with holes at the center and are found to have been used as necklace pieces. After carbon 14 test for antiquity, they are found to date from 12000 B.P. to 35000 B.P. The description of Dadhikravan given in Vedas totally tallies with the description of full grown Ostrich as it is found in Kalahari desert of South Africa. There are black spots on the pieces of eggshells found in India. The size and density of such spots per sq cm, indicate that both of them belong to the same species.

OSTRICH :

Ostrich is found on the surface of the earth near deserts adjacent to tropic of Cancer and tropic of Capricorn. For its existence, it requires arid climate with a part of plain grasslands with shrubs of short heights. It also requires water for drinking in the vicinity. At present its homeland is Cape Province near Kalahari Desert in South Africa. The Australian Ostrich is called Emu and the American Ostrich is called Riha. The height of the back of the male lin the South Africa is 8 ft with further extension of its neck by about 4 ft. It weighs on an averagae 155 kgs. The female is a bit smaller than the male.

SIMILARITY WITH THE HORSE :

Its legs bear strong hoofs, even stronger than the horse. Its thighs are strong and sturdy. It attacks its adversary by the strokes of its heels by taking a flopping jump. Yuri Dimitriv says that on plain lands it runs with a speed of about 70 kms (44 miles) per hour. He further says that Ostriches can be tamed and used as guards for cattle. Therefore, Ostrich farms were created in Africa in 1896. The domesticated Ostriches in South Africa reached a number of about 2,00,000. In Africa a, Ostriches were used for race course with riders on their backs or with a small vehicle pulled by them. In India at present birds 'Indian Bustard', ' Huma' and 'Maldhok' are wrongly called as Ostrich. They are much small. (Salim Ali page 38/ Picture 75). They cannot be compared with Dadhikravan in Rigveda.

EGGS OF OSTRICH :

The female ostriches lay eggs in a common nest in sand. One nest consists of about 15 to 20 eggs. It is eight inches long with six inches diameter. The wall of empty shell of the egg is thick and sturdy. The Paleolithic man used the eggshell as cups water carrying bottles. Fractions of eggshells are found in India to Archeologists. Some fractions bear decorative designs.

DR. WAKANKAR RESEARCH :

Dr. Wakankar of Ujjain discovered number of rock shelters in Vindhya Hills. There are wall paintings and pictures drawn by the then antique persons.

The pieces of eggshells found by him date back to 12000 B.P. to about 35000 B.P. as revealed by the C-14 test. There is ample evidence in the rock shelter pictures that the relevant Civilisation was religious in nature worshipping natural forces. The pieces of egg-shells are also found in the plains nearby, near water source, e.g. tanks and rivers. Dr. Wakankar concludes that the persons resided in the rock shelters during rainy season and came down to plains after the rainy season was over, occupying temporary hutments.

Till 1990 pieces of ancient Ostrich eggshells are found to archeologist in 44 places in following districts- Banda, Jalgoan, Nagpur, Pune, Bhopal, Raisen,

Hoshangabad, Khargaon, Kota, Bundi, Morena, Jhalabad, Mandasor, Chittorgad, Ajmer, Guna, Kunwargram, Anjar, Nagwada. They are found by Dr. Wakankar, Giriajkumar, Ashok Shani, Chitale, Saxena, Ramesh Kumar Pancholi, Narvare, Sunil Bajpai, Shrinivasan. The thickness of the pieces of egg shell's (2.2. mm) indicates that the Ostrich were full grown in their natural environments (Please see annexure 2)

VEDIC OSTRICH :

We get the description of an animal-cum-bird in hymns Rig. III-20, IV -38, IV-40 and VII-44 Sayanacharya gives us the etymology of the word Dadhikaravan as Dadhi - who bears, who carries his rider and krama - to go (Willson - translation). Nighantu (1 - 14) includes the word in the list of Ashva (horses). This list also includes the word Shyena, which means the bird- hawk.

Yaskacharya defines the word ashva as :

यः क्रचाधवानमश्नुवीतमच: स वचनीयास्यात् - निरूक्त 1.15

This definition of the word 'ashva' includes the word a 'camel-bird' i.e.. Ostrich, Neither Yaskacharya nor Sayanacharya had seen the South African Ostrich. They could not be aware of the discovery made by Dr. Wakankar and his colleagues.

The meaning of the word as predicted by other scholars is as follows :-

Griffith :-

i) Divine horse (note on page 226).

ii) Personified Sun (page 226).

iii) The name is probably derived from "dadhi" - thickened milk and kri- to scatter in allusion to rising sun spreading dew and hoar-frost like milk (page-226).

iv) 'agni' (fire)- as guest of man (page 227)

Ludwig :-

The word does not refer to mythical being but to an actual war horse bearing his name (Griffith's -note on page 226)

Pischel :-

Dadhikaravan was the famous race-horse for the gentlemen of the turf in king Trasadusyus time, What the matchless horse eclipse was in recent days (Griffith Pg. 358)

Mahadev Sastri Joshi :-

From Rig. IV-40-5 it appears that Dadhikrava is some form of fire.

Some are of the opinion that the disc of Sun is Dadhikravan represented as horse.

(Bharatiya Samskruti Kosh Marathi 1967 page 283)

Chitrav Sastri :- Dadhikarava is a divine horse or bird of the Sun; fine drops of water in air which appear at the time of dawn. (Rigved- Marathi 1969 page 300) However the following are the points of description of this Vedic animal-cum-bird in Rigveda which disclose us it's identity.

i) It is a bird resembling hawk (Rik. IV-38-4, IV-40-3)

ii) It is 'patangaro' which means a bid who moves by hopping like grass hopper or a cock. (Rik Iv-40-2)

iii) It carries a rider with rein to control him and move fast- Therefore it has been figuratively called a horse. (Rik. IV-38-6)

iv) When it moves, it creates cloud of dust. (Rik IV-38-6) showing that though it is a bird, it moves on plain ground. It cannot cross mountains and broad rivers.

v) It can be domesticated and used as guard for cattleshed. The Vedic Sage therefore describes him as moving with the cattle (Rik. IV-38-4)

vi) It was used in battles for attacking the groups of adversaries- (Rik. IV-38-4)

vii) It roars like an Ostrich when it attacks its adversaries. (Rig.IV-38-8)

viii) We guess it was worshiped at dawn along with the sun, fire, goddess of morning (Uṣha) because it creates an alarming sound for dawn like cock. (Rick.III-20-1, VII-41-6, VII-44-1, X-101-1)

ix) Sun and fire are separately mentioned in same verse as deities of worship.

(Rik. III-20-5) (Rik. X-101-1). Therefore it cannot be identified either with the sun or fire as done by some scholars.

x) There is a description in Vedas that Dadhikravam moves with chariots.

Yuri Demetriev gives a picture of a light vehicle being drawn by the Ostrich. (Rig. IV -38-6, IV- 40-3). It might have pulled a sledge for Neolithic man.

xi) It moves fast by spreading its wings as described by the sages in (IV-40-3) While running it is in semigliding position.

xi) The description of Dadhikravan given in Vedas is not a poetic description of a horse with wings as given in stories from Arabian Nights or of "Pegasus" as given in Greek Mythology, because, Dadhikravan could not fly at a height in the air, and there is no story attached to it- that of carrying a hero or heroine on its back. It is described as animal - bird for worship.

ICE AGE :

During the last quaternary period of geology there was an Ice age in the Northern Hemisphere (see the map at Annexure 3). It disappeared after the beginning of the Holocene period i.e. about 10 thousand years ago. In Ice age about half of the Northern Europe, Russia, Siberia and Kashmir were covered with ice; part of Mongolia was also covered with ice Layers of ice raised the height of Himalayans ranges. Khyber and Bolan passes in North-Western mountains ranges were also closed being covered with ice. It is worthy to note that though the Northern Hemisphere of the earth was covered with the ice there was arid climate in North western India suitable for the existence of Ostrich. This is probably because Himalayan ranges saved India from the attack of the cold winds from the North. No place of Ostrich egg-shell dating from 10000 B.C. to the end 3000 B.C. are found in North-Western India. Thus of vedic ostrich goes back to at least 10,000 B.C.

CONCLUSION :

The animal-cum-bird-Dadhikravan described in Rigveda was full grown Ostrich, as it is now found in Kalhari desert of South Africa. The Vedic sages saw such Ostriches at least 12000 years ago in Northern India. When Ostriches existed in India there was ice age in Northern Hemisphere it was difficult for any outsider from north to enter Indus- Saravati plane.

Note :- The Rigvedic verses referred in this paper are given in serial order in annexture 1 A & B

REFERENCE BOOKS :

1. Rigveda (Sanskrit) with Sayanasharya - Comentary.
2. Rigveda :- English translation; by Griffith 1973 published by Motilal Banarasidas, Bunglow Road, Javahar Nagar, New Delhi-07

3. Rigveda :- Marathi translation by Chitrav Sastri 1969 published by Bhartiya Charitrakosh Mandal, 1206a/45 Jangali Maharaj Road, Pune -4.
4. Bhartiya Samskritikosh (1967) :- by Mahadev Sastri Joshi published by Bhartiya Samskritikosh Mandal, 413, Shaniwarpeth, Pune.
5. Nighantu
6. Nirukta by Yaskacharya
7. ANIMALS ON PEDESTAL by Yuri Demitriev, RUDGA PUBLICATIONS, Moscow.
8. Ostrich Egg shells And Egg Shell objects in India by Giriraj Kumar, Ashok Shani, Geeta Narvare, articles in MAN and ENVIONMENTS XI (1) - 1990 (pages 20 to 37) published by Department of Archeology, Deccan College Pune.
9. Padmashri - DR WAKANKAR ABHINANDAN GRANTHA published by Bharati Kala Bhawan, Uma Niwas, 29, Varahmihir Marg Madheonagar, Ujjain - 456 001 (M.P.)
10. A BOOK OF INDIAN BIRDS 10th edn. By Salim Ali Bombay Natural History Society, Mumbai.

Annexure B

BOOK III HYMN XX AGNI

1. With lauds at break of morn the priest invoketh Agni, Dawn, Dadhikras and both the Asvins.
   With one consent the Gods whose light is splendid, longing to taste our sacrifice, shall hear us.
2. I call on Savita the God, on Morning, Brhaspati and Dadhikras and Agni, on Varuna and Mitra, on the Asvins, Bhaga, the Vasus, Rudras and Adityas.

BOOK IV HYMN XXXVIII. DADHIKRAS

3. Who gaineth precious booty in the combats and moveth, winning spoil, among the cattle;
   Shown in bright colour, looking on the assemblies beyond the Churl, to worship of the living.
4. Loudly the folk cry after him in battles, as it were a thief who steals away a Garment.
   Speeding to glory, or a herd of cattle, even as a hungry falcon swooping Downward.
5. And, fain to come forth first amid these armies, this way and that with rows of cars he rushes,
   Gay like a bridesman, making him a garland, tossing the dust, champing the rein that holds him.
6. And that strong Steed, victorious and faithful, obedient with his body in the combat,
   speeding straight on amid the swiftly pressing, casts o'ser his brows the dust he tosses upward.

7. And at his thunder, like the roar of heaven, those who attack tremble and and are frighted;
For when he fights against embattled thousands, dread is he in his striving; None may stay him.

8. The people praise the overpowering swiftness of this fleet. Steed who giveth men abundance.
Of him they say when drawing back from battle. Dadhikras hath sped forward with his thousands.

9. Dadhikras hath o'erspread the Fivefold people with vigour, as the Sun Lightens the waters.
May the strong Steed who winneth hundreds, thousands, requite with sweetness these my words and praises.

HYNM XL, DADHIKRAVAN

1. Let us recite the praise of Dadhikravan : may all the Mornings move me to exertin ;
Praise of the Lord of Waters, Dawn and Agni, Brhaspati Son of Arigiras and Surya.

2. Brave, Seeking war and booty, dwelling with the good and with the Swift, may he hasten the food of Dawn.
May he the true, the fleet, the lover of the course, the bird-like Dadhikravan, bring food, strength, and light.

3. His pinion, rapid runner, fans him on his way, as of a bird that hastens Onward to its aim,
And, as it were a falcon's gliding through the air, strikes Dadhikravan's side as he speeds on with might.

4. Bound by the neck and by the flanks and by the mouth, the vigirous Courser lends new swiftness to his speed.
Drawing himself together, as his strength allows, Dadhikras springs along the windings of the paths.

5. The Hamsa homed in light, the Vasu in mid-air the priest beside the altar, in the house the guest.
Dweller in noblest place, mid men, in truth, in sky, born of flood, kine, truth, mountain, he is holy Law.

BOOK VI HYMN XLIV DADHIKRAS

1. I call on Dadhikras, the first, to give you aid, the Asvins, Bhaga, Dawn and Agni kindled well
Indra, and Visnu, Pusan, Brahmanaspati, Adityas, Heaven and Earth, the Waters, and the Light.

4. Foremost is Dadhikravan, vigorous courser; in forefront of the cars, his way he knoweth,
Closely allied with Surya and with Morning, Adityas and Angirases, and Vasus.

6

SCIENCE OF MIND AND CONSCIOUS ELEMENTS UNDERSTANDING THE SĀMKHYAYOGA VIEW

Dr. S.M. MISHRA

In Western Philosophy and Science, mind as such has always been as elusive concept although have been made to underline its exact nature. Data from the metaphysical, behavioural, neuro-biological, biochemical and other disciplines supposed to throw light on the problem of the mind have been mounting day by day. For example the neuro- physiologist maintains that all human behaviour can be explained in terms of stimulus reaction pattern. But then this is only another form of input-output explanation given by the cyber-natician. The behavioural psychologist and some philosophers seek to unite an epitaph to the term mind. On the other hand in the traditional Indian scenario a fundamental inquiry can be noticed into the nature and function of mind and its relation with the soul on one hand and the body on the other. In this respect the watri-lention of Sāṁkhya-yoga, the twin systems, is immense and its influence on the Indian medical science is paramount. The objective of the present paper is to understand at some length the conscious elements in the scientific contents of the Sāṁkhyayoga in the light of the mental phenomena we come across.

1. The fundamental constitution of a living being is out of three factors as held out by Caraka- mind, soul and body- mutual union of which is the foundation of the living world.[1] The same fact is attested by the Sāṁkhyayoga which uses the terms Purāṇa for soul and prakṛti for mind body, One the sattva-predominant prima-material and the other *tamas*-predominent penta-elemental body. It is noteworthy that Caraka, the propounder of Indian medical science, applies Kaṇādian, chemistry for the treatment of the body, while for the treatment of mind, he at once refers to the Sāṁkhya concept of mind.[2] The Sāṁkhyites' mind is a material but supra-chemical energy (know as ojas in Indian bio-chemistry). It is distinct from the chemical energies of the inorganic world in that it is pervaded by the manifestation of a conscient touch (sparśa vijñāna in Indian bio-chemistry).[3]

---

1. Carakasaṁhita, I. 1.34.
   सत्त्वमात्मा शरीरं च त्रयमेतत् त्रिदण्डवत्
   लोकस्तिष्ठति संयोगात् तत्र सर्वं प्रतिष्ठितम्।।
2. Ibid. I 1.41.
3. Ibid. Xi. 38. : तत्र एकं स्पर्शनेन्द्रियम् इन्द्रियाणाम् इन्द्रियव्यापकं चेत:समवायि; स्पर्शनव्याप्तेर्व्यापकमपि च चेत:।

The Sāṁkhya-yoga concept of citta Vṛtti or perceptual conscious act is a particular vṛtti or modification (product or resultant) of the Citta or mental stuff. The mental stuff is a material energy distinguished from all physical energies in that it has the peculiar extra physical capacity of bearing the reflection of a *Puruṣa* (technically called *citicchāyā-patti*) or spiritual conscient self which is *sannikṛṣṭa* or adjacent and not *samaveta* or casualty joined to it.[1] But on that account, again, the material stuff, itself qualified to the effect of a self conscious ego, proceeds to get all its vṛttis or works determined by *puruṣārtha*[2] or self's purposes with in their causal trends possessed by virtue of the mental stuff being otherwise a subtler continuation of *paramāṇus* or physical quantum's. The *citta* or mental stuff is an entity of *samudaya* or relativity between its two dimensions, viz. self consciousness called *sattva* and causality called *tamas*, the relativity itself then rendering it into an energy called rajas. By virtue of its energy-dimension the mind streams in the external world. Its causal dimension then makes it tend to be lost into an external *pariṇāma* or causal transformation or modification to itself while its conscious dimension makes it tend to retrieve itself liquidating the eternality of such of its modification by an in corporation of that within *puruṣārtha* or self's purposes.[3] In the *Sāṁkhya-yoga* view, *citta-vṛtti* or perceptual conscious act, being the output of such a mental energy, should be understood as the product of the relativity dynamism between the two dimensions of the energy viz. the conscious and the causal.

2. In nature the mental body is called Jīva or animal. In the animal the causal tamas-dimension of the mental stuff extends to its extreme in the chemical composition of the body. The corresponding conscious sattva- dimension of the mental energy over there tends to establish *puruṣārtha* or self's purpose in terms of converting the chemical streams into the bio-chemical streams of prāna or life-force. The resultant is the nutritive processes, building the visceral-somatic body integrated into the ego of the animal, instead of the purely causal processes of chemical decomposition and disintegration.

Within the visceral-somatic indriya-tantra or nervous system, again, the causal tamas-aspect of the mental energy is borne by the tendency to generate general random and miscellaneous sense-organ-to muscle sensory-motor impulses and glandular activities in response to external stimuli. The corresponding

1. Sāṁkhyakārikā, 20. तस्मात्रत्संयोगादचेतनं चेतनावदिव लिङ्गम्।
2. Ibid. 31: पुरुषार्थ एव हेतुः।
3. Yogasūtra, 2.21 : तदर्थ एव दृश्यस्यात्मा।

conscious sattva dimension of the mental energy over there tends to establish self's purposes in terms of selecting the particular organically beneficial reactions to the particular stimuli. The resultant is *samskāras* or nervous habits in general and *kleśas* or instincts in particular.

Within the field of instincts, again the causal tamas-aspects of the mental energy is borne by its tendency to keep as *vāsanā* or unconscious in the unmanifest potential state. The corresponding conscious *sattva*-dimensions of the mental energy over there tends to establish self's purposes in terms of unleashing the pent up energy through its respective sense-organ-to muscle course. The resultant is the instinctive sensory-motor response to external stimuli.

3. Within the field of sense-organ-to muscle sensory-motor reaction to environment, again, the causal tamas-aspect of the mental energy is responsible for the whole process being carried out as the unconscious peripheral-to-subcortical spinal reflex. The corresponding conscious sattva-dimension of the mental energy over there tends to establish self's purposes in terms of illuminating all peripheral-subcortical actions with yoga or self-consciousness. The resultant is the emergence or cortical activity to intervene and govern over all the unconscious instinctive and habitual peripheral- subcortical organ-somatic as well as visceral. The perceptual conscious act of our consideration is put up by physiologists as the phenomenon of this cortical activity within the nervous animal. It being so, the following are the elements in the make-up of the perceptual conscious act:

4.1. Yoga and self-consciousness. The basic cortical essence is the nuclear chemical energy, called *rasa* in Indian chemistry, in the infra-chemical free state called *ojas* in Indian bio-chemistry, attained as the last subtlest element-product of the grosser-to-subtler bio-chemical distillation-processes. It is just the finest material essence that manifests the reflection, and is qualified as the conscient matter on account of it, of the adjacent *puruṣa* or conscient self. This conscient cortical essence, which is called *ojas* from the chemical point of view, is called *sparśa vijñāna vāyu* or the energy of conscient touch from the psychological point of view.[1] It streams into the whole chemical body converting the latter into the visceral-and-somatic nervous animal called, *śrotaḥ-puruṣa* or the animal of conscient currents. In any case, this *ojas*-cum-*sparśavijñāna* cortical essence is the buddhi-sattva or mental matter of the Samkhyites showing yoga or self consciousness by putting up a reflection of the objectively projected in the reflection. Yoga or self-consciousness is the basic essence of the perceptual

1. Cf. Footnote. 3.

conscious act in as much as the *puruṣa*-reflecting *buddhi-sattva* or mental matter is the essence or cortical activity underlying consciousness.

4.2 *Aham* or ego.–The cortical essence, which is *buddhi-sattva* or mental matter, capacitated to bear the reflection of its adjacent conscient self to serve for the latter's observation of itself in such an objective projection of it, is in the meantime, and on the same ground, itself become like the conscient self. That is conscient self materially embodied. Self-consciousness of the perceptual conscious is held up by a conscient self that is objectively contained in a material body-medium of it. This refers to the body of the animal and is revealed in the perceptual conscious act as the element of *aham* or ego. The *aham* or ego of the perceptual conscious is the consciently qualified chemical body of the animal. The conscient qualification of the chemical body, which is responsible for the element of *aham* or ego in self-consciousness of the perceptual conscious act, is referred to by Indian physiologists as the *dhāri ojaś*-cum-*sparśa-vijñāna* or nervous conscient energy by which is permeated all the chemical processes of the body turning them into bio-chemical and physiological, visceral and somatic, nervous processes.

4.3 *Manas* or attention. –The ego of which issues the basic self-consciousness of the perceptual conscious act, being materially constituted, proceeds to enter into relations with the external material world. The dynamism of this extension of the ego to the external material world is revealed in the perceptual conscious act as the element of *manas* or attention. The attention-dynamism of the ago is peculiar, that is on account of the ego's double role. The ego is at all put into its dynamism because its material constituent causally demands of it to the effect, which again entails disintegration of the ego as *buddhi- puruṣa* which demands of it contrarily to the effect that the ego should show as an integrated ego to represent the objective projection of the conscient self. This double role of the ego result in putting up its peculiar attention-dynamism whose law is adaptation to environment. Catering to the causal demands it leads to external connections, deferring to the self-conscious demands it resists external connections to degenerate into causal transformation. The compromise is accomplished in the principle of self-adaptation to environment. Indian psychologists refer to it as *bhoga*-cum-*apavarga*, causal connection without transformation. Yoga or self-consciousness basically underlying the perceptual conscious act proves to be a bipolar dynamism of adaptation between the subjective *puruṣa*-side and the objective material side of the same ego. This means that the visceral-somatic nervous ego is differentiated into its subject-side in the central spinal cortical section and object side in the peripheral-subcortical sections, both sides nucleated in *hṛdaya* or the heart, which is the seat of the ego in harmony of adaptation between its two sides the ego as the *mahat* or *buddhi-puruṣa* in pure self-consciousness. The perceptual conscious act is self-consciousness given by the ego in so far as the harmony of self-consciousness is broken of ego at its nuclear *mahat*-seat, which is then differentiated into its spinal-cortical section called, *antaḥkaraṇa* or inner organ of it, on the one side, and the peripherla-subcortical sections called, *vāhya kāraṇas* or outer organs of it, on the other side. The spinal cortical side represents the conscient subject side of the ego in direct reflectional connection with the adjacent *puruṣa* or conscient self, the peripheral- subcortical side represents the objective material side represents the objective material side of the ego in direct causal connection with the external material world. Yoga or self-consciousness of the ego is then represented by the spinal-cortical dynamisms of delivering *puruṣārtha* or self's purposes into the peripheral-subcortical activities. *Manas* or attention revealed of the perceptual conscious act is just this spinal-cortical adaptive dynamism of delivering self's purposes within the peripheral-subcortical causal trends.

4.4 *Ālocana* or sensation. The nuclear ego at ego at *hṛdaya* or the heart puts up spinal cortical cum-peripheral-subcortical trajectories of its *puruṣārtha* carrying *manas*-dynamism at different levels of the animal-being called, *citta-bhumi*-s or mental grounds. Of them the *mūḍha* or dull mental ground would include the bio-chemical visceral ground where the *puruṣārtha*- carrying *manas*-dynamism occurs in terms of changing the chemical process into the bio-chemical distillation processes, leading the grossest *rasa* or amino acids to the subtlest *ojas* or nuclear chemical cohesive fluid in its infra-chemical pure free state. The nuclear chemical cohesive fluid being in the free state as *ojas*, the five kinds of *paramāṇu*-s or physical quantum's, composing the molecules of the animal body, are also released in their infra-chemical free state Indian physiologists hold that the *vāhyendriyas* or peripheral-subcortical outer end-organ are but the particular organs for the particular varieties of the *paramāṇu*-elements of the chemical body to assemble in their free state held by *ojas*. *Ojas* in the meantime bearing the *sparśa-vijñāna vāyu*, energy of conscient touch, which is just the representation of the *manas*-dynamism at the bio-chemical level, lends the conscient quality to the *paramāṇus* as well at the end organs, converting those material end-organs as the sensorty-motor-*buddhīndriyas* or conscient organs. The bio-chemical visceral trajectory completed, the nuclear ego sends up the next sensory-motor somatic trajectories at different levels of higher citta-bhumis or mental grounds beginning with the ground of spinal reflexes. Perceptual conscious is taken as peripheral-cortical-subcortical trajectory of the ego's *manas*-dynamism, beginning at the sensory end-organs in the peripheral region and ending at the muscles in the subcortical region. At both ends, the animal-body, the material container of the self-conscious ego, is in contact with the external physical world inflicting causal demands on the material side of the ego. At the sense-organs those causal connections of the external object are turned into conscient sensations. *Ālocana*[1] or sensation is the external causal connection turned into a non-causal reflectional touch, by the conscient paramānus or physical quantum's assembled at the five respective sense-organ in their free state floating in *ojas*, which are thus responsible for conscient sensation-creation out of causal contact with the external object. To quote Caraka: So said Lord, the son of *Atri* in this context, five are the organs, five are the organic substances, five are the seats of the organs, five are the organic objects, five are the organic knowlwdge[2] this is said in the context of mental organs from the medical point of view. Extra-organic, again, is the mind of the name of *sattva*, which some call *cetas*, and which for its act depends on the object to be judged, inferred etc., and on the *ātman* or self-conscious self, and which is the realiser or the actions of the organs.[3] And governed by the mind the organs become efficient for taking in their objects. And there, eye, ear, nose, tongue and skin-are the five organs.[4] The five organic substances are space, wind, fire, water and earth.[5] The five organic seats are the two eyes, the two ears, the two nostrils, the tongue and the skin.[6] The five organic objects are sound, touch, colour, taste and smell.[7] The five organic knowledge are the visual perception etc.; they are again the products of the adjacence of organ, organic object, mind and the conscient self are momentary, and are giving complete knowledge of objects; this is five for the five organs.[8] And there, although all the organs are built of the chemical combination of the products of all the same five physical elements, yet the fire-element grows predominantly

---

1. The first stimuli in Sāṁkhyain the emergence of perceptual cognition.
2. Carakasamhita, I 8.1-34. Fefle n mceen भगवानात्रेयः। इह खलुपञ्चेन्द्रियाणि, पञ्चोन्द्रिय द्रव्याणि, पञ्चेन्द्रियाधिष्ठानानि, पञ्चोन्द्रियार्थः, पञ्चेन्द्रियबुद्धयो भवन्तीत्युक्तमिन्द्रियधिकारे।
3. Ibid., I 8.4. अतीन्द्रियं पुनर्मनः सत्त्वसंज्ञकं चेत इत्याहुरेके तदर्थात्मसम्पदायत्तचेष्टं चेष्टाप्रत्ययभूतम् इन्द्रियाणाम्। This ake may be an oblique reference to the yoga school.
4. Ibid, I 8.7 मनःपुरःसराणि चेन्द्रियाण्यर्थग्रहणसमर्थानि भवन्ति। तत्र चक्षुः श्रोत्रं घ्राणं रसनं स्पर्शनामति पञ्चोन्द्रियाणि। cp. Sāṁkhyakrāikā.
5. Ibid., I. 8.9. : पञ्चेन्द्रियद्रव्याणि एवं वायुर्ज्योतिरापो भूरिति।
6. Ibid, I. 8.9. पञ्चेन्द्रियाधिष्ठानानि— अक्षिणीकर्णौ नासिके जिह्वा त्वक् चेति।
7. Ibid, I. 8.10. : पञ्चेन्द्रियार्थाः शब्दस्पर्शरूपरसगन्धाः।
8. Ibid, I. 8.11. : पञ्चेन्द्रियबुद्धयःचक्षुर्बुद्धयादिकाः ताः पुनरिन्द्रियेन्द्रियार्थं सत्त्वात्मसन्निकर्षजाः क्षणिका निश्चयात्मिकाश्च इत्येतत् पञ्च-पञ्चकम्।

within the vision-organ, space-element within the auditory organ, earth-element within the olfaction-organ, water-element within the taste-organ, and the wind-element with in the touch-organ.[1] And there, the particular organ in which the particular physical element is predominant would take in the particular object in which the same particular element is predominat; that is because the former possesses the same constituents as the latter and because it is efficient to do so.[2]

4.5 *Saṁkalpa* or cognition.– *Ālocana* or sensation given by the sense-organ is only the undifferentiated external causal contact converted into the undifferentiated conscient reflectional touch of a particular variety corresponding to the particular quantomic variety of the external object concerned the purusārtha-carrying dynamic manas-current takes the sensation up to the respective cortical centre, when its full image and meaning of the external object concerned get differentiated for the ego to be permeated by it Self-consciousness given by the perceptual conscious act proves to be a cognition of *viṣaya* or the external object permeating the ego in the form of a reflectional image and its meaning differentiated out of a sensational touch by the actual external object at the peripheral sense-organ.

4.6 *Prayatna* or Volition.– Self-consciousness given by the perceptual conscious act proves to be a volition in so far as the *puruṣārtha*-carrying *manas*-dynamism is permeating the ego with the reflection of a subcortical impulse for muscular action in respect of a *viṣaya* or external object.[3]

4.7 *Bhāva* or feeling.– Perceptual conscious is the subjective consciousness of the objective self in so far as the latter is fallen to external objective connections, and, keeping itself in its *mahat*-state of subject-object unity as *buddhipuruṣa* or pure mental matter reflecting the adjacent conscient self at the *hṛdaya* or heart to be called the ego, is sending up self-adaptive dynamic trajectory, consisting of peripherla-cortical subcortical sections, propped by the centrifugal-centripetal ambivalent nuclear current to be called manas or attention. The peripheral section provides for the sensation-element, the subcortical section for the volition-element, and the cortical section for the fuller congition-element of the perceptual conscious

---

1. Ibid, I. 8.13. : तत्र अनुमानगम्यानां पञ्चमहाभूतविकारसमुदायात्मकानामपि सतामिन्द्रियाणां तेजश्चक्षुषि एवं श्रोत्रे, घ्राणे क्षितिः आपो रसने, स्पर्शनेऽनिलो विशेषेणोपपद्यते। This view holds true of the Nyāyavaiśeṣika: cp. Nyāysūtra. I. 1.12. घ्राणरसनचक्षुस्त्वक् श्रोत्राणीन्द्रियाणि भूतेभ्यः।
2. Ibid, I. 8.14. : तत्र यद् यदात्मकमिन्द्रियं विशेषात् तत्तदात्मकामेवार्थमनुगृह्णाति, तत्स्वभावाद् विभुत्वाच्च।
3. Volition, as per Nyāya, is a quality in hearing in the ātman, but in Sāṁkhya and Caraka it is a psycho-physical factor.

act. The nuclear connection of the peripheral-cortical-subcortical trajectory is then revealed in the perceptual conscious as the element of *bhāva* or feeling. Let *bhāvas* or feelings and emotions underlying the perceptual conscious act be at once identified as the traces of the *kleśas* or instincts permeating the ego at the nucleus as *vāsanā* or the unconscious, the self-preserving and self-adaptive wishes of it.

5. As matters would have it, *bhāvas* or feeling and emotions formed the special topic of treatment for the Indian *alamkāra śāstra* or science of aesthetics, of which Bharata, the author, of *Nāṭya-śāstra*, is the propounder. In the seventh chapter of *Nāṭya-śāstra* Bharata proposes: 'I shall now narrate the *bhāvas*,'[1] and what he narrates are but the feelings and emotions. The word *bhāva* is taken from *Sāmkhya*.[2]The subtle nucleus of the animal is all permeated with the *bhāvas* or unconscious instincts, just as a piece of cloth, through contact with the fragrant campaka flower, becomes permeated with its perfume, out of that permeation with the feelings it moves into objective differentiation of itself.[3]

Bharata, maintains, the root bhū, with 'I' suffixed, is in the sense of 'doing'. And so bhāvita, vāsita (Saturated), *kṛta* modified are not other than synonyms. In ordinary behaviour too it is well knwon, ' Oh, by this perfume or taste the whole of it is *bhavita'*; and that bears the meaning of *vyāpti* or permeation.[4]

6. Mind in its very essence is the three inter-relative affections of feeling. The mental ego in its very essence is the *puruṣa*-expressing sattva-quality, the *puruṣa*-covering *tamas*-quality and the *rajas*-quality of *sattva-tamas*- conflict-dynamism. Now attraction is pleasure, pleasurable attraction is the *sattva*-quality; repulsion is pain, painful repulsion is the *rajas*-quality; inefficiency is dullness, dull inefficiency is the tamas-quality.[5]

Out of these three fundamental mental dimensions of the ego evolve the five kleśas or instincts viz., *avidyā* or instinct of objective establishment, *asmitā* or egoism, *rāga* or eros, *dveṣa* or aggression, and *abhiniveśa* or fear.

---

1. Nāṭyaśāstra, VII (beginning) : भावानिदानीं व्याख्यास्यामः। अत्राद् भावा इति कस्मात्। किं भवन्तीति भावा:, किं वा भावयन्तीति भावा:। उच्यते वागङ्गसत्त्वोपेतान् काव्यार्थान् भावयन्तीति भावा इति।
2. cf. Sāṁkhyakārikā, 40. : भावैरधिवासितं लिङ्गम्।
3. Vide Tattvakaumudi on Ibidi यथा सुरभिचम्पकसम्पर्कात् वस्त्रं तदामोदवासितं भवति।........ तस्माद् भावैरधिवासितत्वात् संसरति।
4. Nāṭyaśāstra, VII (beginning)
5. Tattvakaumudi or Ṣāṁkhyakārikā, 12.: प्रीतिः सुखं प्रीत्यात्मकः सत्त्वगुणः। अप्रीतिर्दुःखम्, अप्रीत्यात्मको रजोगुण:, विषादो मोहः विषादात्मक स्तमोगुणः।

These five instincts, permeating the nuclear ego and propping up the pcpt cs trajectory consisting of cognitions and volitions, appear in the perceptual conscious act according to Bharata, as the eight *sthāyībhāvas* or permanent feelings and thrity-three *vyabhicāri* or passing feeling. According to Bharata eight are the permanent feelings, thirty three are the passing feelings,, eight are the *sāttvika* feelings.[1]

7. In fine this can be held on available grounds that the ancient Indian photographers -scientists investigated into the mental phenomena in a systematic and practical way and established clinical uses of their knowledge in the practice of medicine and surgery. After all language is the custodian of knowledge and the Śāstraic tradition available today speak of scientific facts which can have bearing on the researchers in the related fields at present.

Bibliography

1. Carakasamhitā	Chaukhamba, Varanasi
2 Nāṭyaśāstra	Bharatiya Vidya Prakashan, Delhi.
3. Nāṭyaśāstra	Ṣaḍdarśanasūtrasaṁgraha, Swami Dwarakadas Sastra, Varanasi.
4 Yogasūtra with Vyāsa omm.	Motilal Banarasidass. Delhi.
5 Sāṁkhyakārikā with Tattvakaumudi	Chaukhamba Varanasi

6 For some modern western views or mind Proceedings on a seminar on 'mind' held at S.V. University. Tirupati in 1970 pub. By the same univ. has been

1. Nāṭyaśāstra, VII. 6 : तत्राष्टौ भावा: स्थायिन:। त्रयस्त्रिंशद् व्यभिचारिण:। अष्टौ सात्त्विका इति। भेदा:।

7

वैदिक अब्विज्ञान

(यजुर्वेद के विशेष सन्दर्भ में)

- डा. वीरेन्द्र कुमार अलंकार

वेदों में विविध विद्याएँ बीज रूप में विद्यमान हैं। यह संयोगमात्र नहीं है कि आधुनिक जो भी विषय आविर्भूत होता है, वैदिक विद्वानों को उसके साक्ष्य वेदों में उपलब्ध होने लगते हैं। अनन्त ज्ञानराशियाँ वेदों में दिखाई देती हैं। वेद-संहिताओं में जैसे काव्य का मूल है, अध्यात्मशास्त्र का उत्कर्ष है और व्याकरण शास्त्र का बीज है, वैसे ही नक्षत्रगति, शरीरविज्ञान, गणित, ग्रह-आकर्षणानुकर्षण आदि विषय भी वेद में मूलरूप में दिखाई देते हैं। आयुर्वेद का स्रोत वेद ही है। विविध रोगों और उनके औषध विज्ञान को वेदों में देखा जा सकता है। वैदिक देवता केवल चेतनता के ही बोधक नहीं हैं, बल्कि प्राकृतिक देवताओं के भी वाचक हैं। कहीं सूर्यविज्ञान, कहीं अग्निविज्ञान, कहीं अश्विन्-विज्ञान, कहीं इन्द्रविज्ञान, कहीं कृषिविज्ञान और कहीं वृष्टि तथा प्रजनन विज्ञान भी वेद में दिखाई देता है। ये विज्ञानपरक सन्दर्भ सभी संहिताओं में विद्यमान हैं। यजुर्वेद में मुख्यतः कर्मवाद का प्रतिपादन प्रायः सभी ने माना है, फिर भी अन्य विषय भी यहाँ कम महत्त्वपूर्ण नहीं हैं। वेदों में जो 'अब्विज्ञान' के सन्दर्भ हैं, उन्हें सम्पूर्ण रूप में प्रतिपादित करना यहाँ सम्भव नहीं है। इसलिए मुख्यतः यजुर्वेद को ही अध्ययन का विषय बनाया गया है। वहाँ भी केवल अब्देवताक याजुष मन्त्रों की चर्चा करनी ही यहाँ अभीष्ट है। यूँ तो यजुर्वेद में वैद्यदेवताक[1] तथा अन्य मन्त्रों में भी यह विज्ञान है, किन्तु यहाँ केवल अब्देवताक[2] सत्ताईस मन्त्रों का अर्थ ही निरूपित किया गया है। इन मन्त्रों के देवता 'आपः' हैं। यहाँ यह भी ध्यातव्य है कि यजुर्वेद में 'जल' शब्द नहीं है, सर्वत्र अप्-शब्द का ही प्रयोग है। अप्-शब्द किन-किन अर्थों में प्रयुक्त हुआ है, यही इस पत्र का विषय है।

यजुर्वेद में अप्-शब्द के अर्थ हैं- जल, स्त्री, विद्वान्, प्राण, वैद्य और प्रजा। इसके अतिरिक्त अपों से सृष्ट्युत्पत्ति तथा रोगविज्ञान का संकेत भी उपलब्ध है। यजुर्वेदीय शतपथ ब्राह्मण में अपों की उत्पत्ति के पश्चात् जगत् की उत्पत्ति कही गई है- सोऽयं पुरुषः प्रजापतिरकामयत... ब्रह्मैव प्रथममसृजत त्रयीमेव विद्याम्।।७।। सोऽपोऽसृजत। वाच एव लोकात्। वागेवास्य साऽसृज्यत। सेदं सर्वमाप्नोद् यदिदं किं च यदाप्नोत् तस्मादापः। यदवृणोत् तस्माद् वा।।८।।[3] आधुनिक विज्ञानी भी जगदुत्पत्तिप्रक्रिया को जलों से सम्बद्ध करते हैं, जिसे वैज्ञानिक

भाषा में नैबूला (Nabulae) कहा जाता है। इसी नैबुला गैस में इलैक्ट्रान आदि की उत्पत्ति मानी गई है।[4] वस्तुतः यहाँ जलों की मूलावस्था निर्दिष्ट की गई है। इसके पश्चात् ज्योति, अर्चि, मरीचि, उदार उत्पन्न हुए, फिर अभ्र और समुद्र रूप बना- 'ज्योतिः अर्चिः, मरीचयः उदाराश्च समुत्पन्नाः, तदनु च तदभ्रमिव समहन्यत तद्वस्तिमभिनत्। स समुद्रोऽभिनत्।..... तद् वा इदमापः सलिलमासीत्।[5] इस सलिलावस्था को धारण कर ये 'आपः' प्रमुख व्यापक हो गए। ये 'आपः' ईश्वरसामर्थ्य से ही उत्पन्न होते हैं- तत्मात्पुरुषात्तप्तादापो जायन्ते।[6] यहाँ यजुर्वेदीय ब्राह्मण में यह स्फुट उद्घोष है कि जगदुत्पत्ति में अब्विज्ञान का बड़ा महत्त्व है। बल्कि 'आपः' से ही जगदुत्पत्ति मानी गई है।

यज्ञविद्या को भी 'आपः' कहा गया है- आपो हि यज्ञः।[7] वृष्टियज्ञ वस्तुतः वृष्टि-विज्ञान है। गृह्यसूत्र आदि में उपलब्ध यज्ञ अग्निहोत्र के ही वाचक नहीं हैं, बल्कि विज्ञान के भी वाचक हैं। वृष्टियज्ञ को धार्मिक अनुष्ठानमात्र नहीं माना जा सकता। अनावृष्टि की स्थिति में वृष्टियज्ञ का आयोजन एक वैज्ञानिक कृत्य ही प्रतीत होता है। यजुर्वेद में 'आपः'का अर्थ वृष्टिजल भी उपलब्ध होता है। इन जलों का रोगों में विशेषतः उपयोग आयुर्विज्ञान की दृष्टि से दिखाई देता है-

'यक्ष्मा' रोग की निवृत्ति में जलचिकित्सा का उल्लेख यजुर्वेद में हुआ है- अपो देवीरुपसृज मधुमतीरयक्ष्माय प्रजाभ्यः।[8] उवट ने इस मन्त्र के भाष्य में कहा है कि ये दिव्य वृष्टि जल मधु से पूरित, रस से आप्लावित, आरोग्यदायक तथा यक्ष्मा के निवारक होते हैं- द्यौर्वा अपो देवीरुपसृज निक्षिप वृष्टिरूपाः मधुमती रसवती आरोग्य - दातृकाः अयक्ष्माय प्रजाभ्यः।[9] महीधर ने भी इन जलों को देवनशील यक्ष्मानिवारक तथा प्रजाओं के सम्पूर्ण आरोग्य को देने वाला कहा है- हे अग्ने यद्वा हे द्यौः, यद्वा हे अध्वर्यो, देवीः देवनशीला अपः उपसृज अस्मिन्खननप्रदेशे आसिञ्च। कीदृशीरपः। मधुमतीः रसवतीः आरोग्यदात्रीः। किमर्थम्- प्रजाभ्यः अयक्ष्माय। यक्ष्मणोऽयक्ष्मं तस्मै प्रजानामारोग्याय।[10] ऋषि दयानन्द भी यहाँ वृष्टि जल से यक्ष्मा आदि रोगों की निवृत्ति का उल्लेख करते हैं- यक्ष्मादिरोगनिवारणाय।[11] यहाँ 'आपः' का अर्थ वृष्टि-जल है। उपर्युक्त याजुष मन्त्र में वृष्टिजल की ही प्रार्थना की गई है। स्मरणीय है कि यजुर्वेद में जलों को अभीष्ट कल्याण करने वाला कहा गया है- शन्नो देवीरभिष्टय आपो भवन्तु पीतये। शंयोरभिस्रवन्तु नः।[12] अथर्ववेद में तो जल को आनुवंशिक रोगों को भी दूर करने वाला कहा गया है- आपो विश्वस्य भेषजीस्तास्त्वा मुञ्चन्तु क्षेत्रियात्।[13] तात्त्विक दृष्टि से देखा जाए तो वनस्पतियों में जो रस है, वह भी जल नहीं, तो और क्या है? इस प्रकार जल ही हमारा पोषण करता है, वह साक्षात् हमारी माता ही है- आपो अस्मान्मातरः।[14]

उपर्युक्त उल्लेख से यह बिल्कुल स्पष्ट है- कि जल चिकित्सा से यक्ष्मा आदि रोगों की निवृत्ति होती है अतः इसका विधिवत् उपाय ढूँढना चाहिए। ध्यातव्य है कि प्राकृतिक चिकित्सा में

रोगी के शिर पर जल धारा प्रपात से दुःसाध्य रोगों की निवृत्ति की जाती है। इसी प्रकार यौगिक आचार्य जलनेती तथा दुग्धनेती से यक्ष्मा (टी.बी.) रोग का इलाज करते हैं, यह मैंने स्वयं देखा है। इस जलचिकित्सा में वृष्टि जल का रोगनिवृत्ति में विशेष उपयोग निर्दिष्ट हुआ है। यह वृष्टिजलविज्ञान आधुनिक विज्ञानियों के लिए एक नई चुनौति है। चरक संहिता के जलवर्ग में जलों का आयुर्वेदीय उपयोग दर्शाया गया है तथा वहाँ वृष्टिजल को सर्वोत्कृष्ट माना गया है।[15]

सोम औषध के निर्माण में यज्ञीय आपः (जलों) का उपयोग दर्शाया गया है- यज्ञाहुतिपरिणामभूता आप उच्यन्ते.....अस्य विशः अस्य सम्बन्धिनं सोमं श्रीणन्ति मिश्रयन्ति।[16] स्मरणीय है कि यज्ञ को 'विशः'। कहा गया है- यज्ञो वै विशः।[17] यहाँ उवट ने यज्ञविद्या अर्थात् यज्ञविज्ञान द्वारा यज्ञीय जलों से सोम मिश्रण की जो चर्चा की है, वह आधुनिक शोधकर्ताओं के लिए एक नई दिशा है। किस प्रकार वह यज्ञीय जल तैयार किया जाता था, तथा वह सोम तत्त्व क्या है- यह विचारणीय है।

वेदों में राजयक्ष्मा रोगनिवारण के लिए यज्ञ का निर्देश भी उपलब्ध है।[18] इस यज्ञ में शुद्धिकृत जलों से राजयक्ष्मा का निवारण होता है। राजयक्ष्मा ही नहीं, बल्कि ज्वर आदि रोगों की चिकित्सा भी जलों से होती है। एलोपेथी के डाक्टर भी तीव्र ज्वर में जलार्द्र वस्त्रपट्टियाँ रखने की सलाह देते हैं, इस पर भी यदि ज्वर शान्त न हो, तो ठण्डे जल में स्नान कराया जाता है। प्राकृतिक चिकित्सक कटिस्नान आदि द्वारा जलों से रोगों की निवृत्ति करते हैं। इसीलिए यजुर्वेद में इन जलों को सुखयुक्त, ज्वरादिरोगरहित और यक्ष्मानिवारक भी कहा गया है- इमा आपः सुशेवाः सुखयुक्ताः, अनमीवाः ज्वरादिरोगरहिताः, अयक्ष्माः यक्ष्मादि-रोगकारकदोषरहिताः स्युः।[19] इन जलों से पाप या दोष मुक्ति होती है तथा इन्हें वाणी से सिद्ध किया जाता है- आप इमाः अंहोमुचः, वाण्या च सिद्धा भवन्ति।[20] यहाँ पर जलों की संस्क्रिया इंगित है। शिवाम्बु चिकित्सा भी आपः चिकित्सा का ही एक रूप है। इस शिवाम्बु चिकित्सा के दो रूप तो अत्यन्त प्रसिद्ध हैं-स्वमूत्र चिकित्सा और गोमूत्र चिकित्सा। भारत के प्रधानमन्त्री स्वर्गीय श्री मोरारजी देसाई स्वमूत्र चिकित्सा के विख्यात उदाहरण थे। कैंसर में भी पर्याप्त लाभ का दावा मूत्रचिकित्सक करते हैं। गोमूत्र चिकित्सा भी कई स्थानों में प्रचलित है। ये चिकित्सक ब्रेन-ट्यूमर, बेहोशी, मिरगी और चर्मरोग जैसे असाध्य रोगों में गोमूत्र का विशेष लाभ मानते हैं। ऐसे प्रयोगों का परीक्षण निष्पक्ष दृष्टि से होना चाहिए। स्वमूत्रचिकित्सा और गोमूत्र चिकित्सा सम्बन्धी अनेक ग्रन्थ देखे जा सकते हैं।

दयानन्द सरस्वती ने यजुर्वेद (३५.११) के एक मन्त्र का देवता 'आपः' माना है **और** वहाँ दुःस्वप्न्य को दूर करने की प्रार्थना की गई है। अथर्ववेद में तो जल द्वारा **स्वप्नदोष की** चिकित्सा भी वर्णित है।[21] आधुनिक प्राकृतिक चिकित्सक भी कटिस्नानादि द्वारा **स्वप्नदोषचिकित्सा** के समर्थक हैं। वैदिक साहित्य में रुद्र को जलचिकित्सक कहा गया है।[22] वस्तुतः **जलों में सब** ओषधियों के गुण हैं।

जल का उपयोग शक्तिवर्धन के लिए भी यजुर्वेद में उपलब्ध है। जल को शक्तिवर्धक (टॉनिक) कहा गया है- देवीरमृता:[23] डाक्टर भी कहते हैं कि प्रतिदिन दस गिलास पानी तो अवश्य ही पीना चाहिए। मेरा जलचिकित्सा में स्वानुभूत प्रयोग है। लगभग २० वर्ष की अवस्था तक (वर्ष १९८२ तक) मुझे कभी-कभी भयंकर ज्वर हो जाता था और डाक्टरों की शरण में जाना मेरी विवशता थी। दूसरा, तब मुझे मूत्र भी सहज रूप में नहीं आता था, एक दम पीला मूत्र। तब अनेक दवाएँ भी लीं, आयुर्वेदीय चिकित्सकों ने चन्दनासव आदि लेने की सलाह दी। तभी एक मित्र की सलाह पर मैंने प्रात: उठते ही तीन-चार गिलास पानी पीना आरम्भ किया तो उसी दिन से मूत्र में मुझे अप्रत्याशित लाभ हुआ और उस रोग से आज २३ वर्ष के बाद भी पूर्णत: मुक्त हूँ। दूसरा अप्रत्याशित लाभ यह हुआ कि तब से आज तक कभी भी भयंकर ज्वर ने आक्रमण नहीं किया। इन २३ वर्षों में मुश्किल से ३,४ बार ही सामान्य ज्वर कभी कभार हुआ है। मैं अब प्रात: उठते ही दो गिलास पानी प्रतिदिन पीता हूँ, जिससे कब्ज की शिकायत भी कभी नहीं रही। सिर-दर्द या उदर में इन्फेक्शन हो जाने पर प्रात: खाली पेट पानी पीकर वमन करने पर भी अपूर्व लाभ होता है। इसी प्रकार जुकाम में लवण मिश्रित गर्म पानी भी अत्यन्त लाभदायी है। कदाचित् इसीलिए यजुर्वेद में जल को पापनाशक और दोषनाशक कहा गया है। जलाभाव में विविध रोग उत्पन्न हो जाते हैं और जलपान से निरन्तर बलवृद्धि होती है, तभी तो यजुर्वेद में जलों को घृत के समान पुष्टिदायक कहा गया है। इससे यह कल्पना की जा सकती है कि वाजीकरण में जल चिकित्सा का विशेष उपयोग है। जल आरोग्यदायक है- इसीलिए पितृतर्पण में जल का विशेष महत्त्व है- हे पुत्रा:! यूयं विविधप्रकारकमुत्तमोत्तमं स्वादिष्टं जलं पितृन् तर्पयत।[25] विशेषत: वृद्धों के लिए जल अमृत है, रोगों में जल ही शक्तिवर्धक है।[23] जल से दीर्घायुष्य प्राप्त होता है, विशेष रूप से वृष्टि जल दीर्घायुष्यदायक है।[27] इसीलिए यजुर्वेद में जलों को दिव्य गुण युक्त कहा गया है- देवीराप इति (१.१२)

वेद में अप्-शब्द से प्राण अर्थ भी ग्राह्य है। इस प्राणविद्या (प्राणायाम विज्ञान) से सम्पूर्ण आरोग्य और दीर्घायुष्य प्राप्त किया जा सकता है। वेदभाष्यकारों ने अर्णव शब्द का अर्थ प्राणविद्या किया है।[28] इसकी पुष्टि शतपथब्राह्मण भी करता है- आपो वै प्राणा:।[29]

वैदिक आचार्य 'आप:' शब्द का अर्थ 'श्रेष्ठ गुणों में वर्तमान स्त्रियाँ' भी करते हैं।[30] वस्तुत: 'आप:' शब्द विशेषण है। जैसे अप्-ग्रहण से शरीर में सर्वत्र शान्ति फैलती है, और सन्तुलन हो जाता है, रोग भी आक्रमण नहीं करते, वैसे ही जो स्त्रियाँ श्रेष्ठ गुणों में रमण करती हैं, वे भी शान्तिकारक होने से तथा पूरे घर में उन्हीं का अधिपत्य होने से 'आप: कहलाती हैं। शतपथ ब्राह्मण में देवियों और सुपत्नियों को 'आप:' कहा गया है- देव्यो आप:,[31] आप: सुपत्न्य:।[32] इस प्रकार 'आप:' शब्द द्वारा गृहस्थाश्रम का निर्देश भी वैदिक साहित्य में उपलब्ध है।

अविद्यारूपी मल के शोधक विद्वान् भी 'आप:' कहलाते हैं[33] क्योंकि जैसे ये आप: (जल) मलों का प्रक्षालन करते हैं, वैसे ही ये आप: (विद्वान्) भी उपदेशवारि से अविद्या मल का प्रक्षालन करते हैं। 'आप:' का अर्थ 'प्रजा' भी है। शतपथ ब्राह्मण में कहा गया है- वीर्यं वा आप:[34] आपो रेत:।[35] यहाँ वैदिक आचार्य प्रजनन विज्ञान की ओर संकेत कर रहे हैं। योषा वाऽआपो वृषाऽग्नि:[36] यहाँ अग्निषोमात्मक जगत् का निर्देश है। आपो ह वा ओषधीनां रस:[37] में रसविज्ञान और अप्सु पृथिवी[38] आपो वा अस्य सर्वस्य प्रतिष्ठा[39] प्रसंगों में सृष्टिविज्ञान की ओर संकेत है। यह ध्यातव्य है कि यजुर्वेद में रुद्र को दिव्य भिषक्[40] और वरुण को भिषक्-पति[41] कहा गया है तथा शतपथ में 'आप:' को वरुण की पत्नियाँ कहा गया है- आपो वरुणस्य पत्न्य आसन्[42] इस प्रकार वरुण और आप: में पति-पत्नी व्यवहार देखा गया है। वरुण चिकित्सक है और उसकी पत्नी हैं- आप:। वस्तुत: वैदिक वाङ्मय में पत्नी शब्द से उस देव की क्रिया शक्तियों का कथन हुआ है। यदि वरुण चिकित्सक है, तो चिकित्साशास्त्र उस वरुण की पत्नी हुआ। यही अभिप्राय यहाँ पत्नी शब्द से ग्राह्य है।

इस प्रकार इस पत्र में यजुर्वेद (वाजसनेयी संहिता) के अब्देवताक, मन्त्रों का अर्थ निरूपण ही किया गया है। इन याजुष संकेतों को बीज मानकर तत्सम्बद्धशास्त्रविशेषज्ञ ही विशेष अध्ययन कर सकते हैं। यजुर्वेद में अग्नि-इन्द्र-वरुणादि देवताक मन्त्रों में भी यह अब्विज्ञान भरा पड़ा है। जगदुत्पति विज्ञान, आयुर्विज्ञान, गृहस्थधर्म आदि अनेक विषय इस वैदिक अब्विज्ञान में ही अन्तर्भूत हैं। अपों के इसी महत्त्व के कारण उन्हें सुमित्र, कल्याणकारी, माता आदि कहा गया है। सुमित्रिया न आप ओषधय: सन्तु,[43] शन्नो देवीरभिष्टय आपो भवन्तु पीतये। शंयोरभिस्रवन्तु न:।[44] आपो हिष्ठा मयोभुवस्ता न ऊर्जे दधातन[45] यो व: शिवतमो रसस्तस्य भाजयतेह न:। उशतीरिव मातर:[46], आपो जनयथा च न:[47] आदि अनेक मन्त्रों में 'आप:' के व्यापक स्वरूप का विवेचन हुआ है। गोपथ-ब्राह्मण[48] में ईश्वर की व्यापनशील शक्तियों को भी 'आप:' कहा गया है।

वेदों में ज्ञान अथवा प्रार्थना की पूर्णता कर्म में मानी गई है। जो कुछ भी प्रार्थनाएं है, उन्हें आचार में लाने से ही वह 'योग' बनता है, उसके अभ्यास से फलप्राप्ति होती है और इन सन्दर्भों के अर्थ की प्राप्ति अथवा ज्ञानक्रिया की समन्विति ही विज्ञान है। वस्तुत: वेद के देवता आध्यात्मिक, आधिभौतिक और पारमार्थिक हैं। उनके विशद अध्ययन के लिए अध्यात्म शास्त्र के साथ साथ भौतिक विज्ञान का अध्ययन भी आवश्यक है।।

सन्दर्भ सूची

1. यजुर्वेद- 12.75,76,77,79,81,83-98,100,101
2. वही, 2.34,3.20,4.2,13,6.10, 13, 17, 23, 27, 10.1, 6,11.38, 50-52,12.55, 13.53, 15. 60, 20.19, 20, 35.11, 12, 36.12, 14-16, 23

3. शतपथब्राह्मण (षष्ठकाण्ड)
4. द्र.-वेदविद्यानिदर्शन-भगवद्दत्त (इतिहास प्रकाशनमण्डल, दिल्ली)
5. द्र.- तैत्तिरीयब्राह्मण- 2.2.9.1
6. शतपथब्राह्मण- 6.1.3.1
7. वही, 3.1.4.15
8. यजुर्वेद - 11.38
9. उवटभाष्य (यजुर्वेद - 11.38)
10. महीधरभाष्य (यजुर्वेद - 11.38)
11. दयानन्दभाष्य (यजुर्वेद - 11.38)
12. यजुर्वेद - 36.12
13. अथर्ववेद - 3.7.5
14. यजुर्वेद - 4.2
15. द्र. चरकसंहिता (सूत्रस्थानम्, अध्याय-27-जलवर्ग)
16. उवटभाष्य (यजुर्वेद-12.55)
17. शतपथब्राह्मण - 8.7.3.21
18. (क) अष्टांगहृदय-5.83: दैवव्यपाश्रयं तत्तद् अथर्वोक्तं च पूजितम्।
    (ख) ऋग्वेद - 10.161.1 (अथर्ववेद - 2.11.1, 20.96.6)
    (ग) तैत्तिरीयसंहिता - 2.3.5.2, 2.5.6.5, काठकसंहिता-11.3, मैत्रायणीसंहिता- 2.2.7
19. द्र. - दयानन्दभाष्य (यजुर्वेद - 4.12)
20. द्र. - वही, 4.13
21. अथर्ववेद - 10.5.24;19.1.10-11
22. ऋग्वेद - 1.43.4 : रुद्रं जलाषभेषजम्।
23. यजुर्वेद - 4.12
24. वही, 4.2
25. दयानन्दभाष्य (यजुर्वेद - 2.34)
26. वही, 3.20
27. अथर्ववेद - 3.31.11 : आ पर्जन्यस्य वृष्ट्या।
28. दयानन्दभाष्य (यजुर्वेद - 13.53, 15.60)
29. शतपथब्राह्मण - 3.8.2.4
30. दयानन्दभाष्य (यजुर्वेद - 6.13, 11.50-52)
31. शतपथब्राह्मण - 1.1.3.7
32. वही, 6.8.2.3
33. दयानन्दभाष्य (यजुर्वेद - 6.17, 20.20)

34. शतपथब्राह्मण - 3.3.4.1
35. वही, 3.8.4.11
36. वही, 1.1.1.1
37. वही, 3.6.1.7
38. जैमिनीयोपनिषद् - 1.10.2
39. शतपथब्राह्मण - 4.5.2.14
40. यजुर्वेद - 16.5 : प्रथमो दैव्यो भिषक् ।
41. वही, 21.40 : वरुणं भिषजां पतिम्।
42. (क) शतपथब्राह्मण - 1.1.1.1, 6.8.2.3
    (ख) तैत्तिरीयसंहिता - 1.1.3.8
43. यजुर्वेद - 35.12
44. वही, 36.12
45. वही, 36.14
46. वही, 36.15
47. वही, 36.16
48. गोपथब्राह्मण (पूर्व)-1.2: तद्यदब्रवीदाभिर्वा अहमिदं सर्वमाप्स्यामि, यदिदं किं चेति तस्मादाप अभवन्, तदपामप्त्वम्।

8

PSYCHOLOGY IN THE ṚGVEDA SAMHITĀ

Professor Ranvir Singh

Psychology is the branch of knowledge which deals with the human mind and knowledge of the mind which one derives from a careful examination of the facts of consciousness and of behaviour. One finds substantial material relating to human mind, psyche, positive and negative facts of consciousness and of behaviour in the *Ṛgveda-Saṁhitā manas* and *cetas* are depicted in whole of the Vedic literature as mind or psyche. In the *Ṛgvedic-Saṁhitā(=Rv.) manas* is mind in its widest sense as applied to all the mental-powers such as intellect, intelligence, understanding, perception, sense, conscience and will.[1] It is, some-times, also joined with *hṛd* or *hṛdaya-* the heart. The spirit and spiritual principle, *i.e.* the word atman has occurred less in comparison to *manas* and has been used for *manas* at some places.[2] *Manas* is also given as *asu* for breath or living soul which escapes from the body at death. In the *mana-āvartana* and three other hymns which follow it in the *Rv.(*X.57-60) *manas* is used in these senses. The *manas* also means thought, imagination, excognition, invention, affection, desire, mood, temper and spirit *etc.*[3]

Cetas is splendour and mind in the *Rv.* and consciousness, intelligence, thinking, soul, heart and mind in the *Vājasaneyi-Saṁhitā* (=*Vs.*XXXIV.3) and the *Atharvaveda* (*Av.*). *Citta* denotes appearance and visible status in the *Rv.*IX.65.12[4] and attending, observing intelligence in the *Rv.*VII.59.8[5] and also thinking, reflecting imagining, thought, intention, aim and wish in the same text at other places.[6]

Manas as a lexica has very extensively been used in *Rv.* which embodies not less than 235 nominal variations/declencials of different case-endings and double the number of compound-words.[7]

This paper concentrates mainly on four-five Ṛgvedic hymns which are exclusively related to *manastattva* along with authentic discussion on some compound and independent usages and other positive thoughts derived from the suggested senses of words added with *sam* at their prior positions. The *Rv* is replete with positive as well as negative aspects of thoughts and one can extensively analyse the different aspects of human psyche on the basis of this material.

In the *Rv.* there is a very interesting episode regarding the re-diversion or fetching the *manas*, *i.e.* vital breath or life or soul of Subandhu-a priest- killed by the other opponent fake priests-*Asuramāyau.* This episode has been covered from the *Rv*.X.57 to X.60 and the seers of these hymns are three brothers -Bandhu, Śrutabandhu and Viprabandhu- the Gaupāyana priests. Subandhu was their fourth brother. The Gaupāyanas were the traditional priests of King Asamāti *Rāthaprau- ṣ haṭan Ikṣvāku* by clan. He appointed his new priests- two deceiver jugglers - *Asuramāyau* namely Kilāta and Ākuli and abandoned Gaupāyanas.[8] To teach a lesson to Asamāti the Gaupāyanas started to perform a *satra* at Khāṇḍavaprastha. This venture generated heat and was tormenting the king Asamāti. His *asuramāyau*- priests-assured him of his welfare and managed to take away the life of Subandhu Gaupāyana. His three brothers intended to redivert and restore his life and chanted the *Viśvedevāḥ* hymn (*Rv*.X.57) preceding by *'mā pra gāma'*-a *gāyatra-svastyayana* of six *ṛcās.*[9] After this they chanted another hymn, *viz. mana-āvartana* (*Rv*.X.58) containing twelve verses (*ṛcās*). This episode has travelled to Kāṭyāyanī, and *Jaiminīya brāhmaṇas, Ṛk-sarvānukramanī* of Kātyāyana and finally to *Bṛhad-devatā* of Śaunaka. In this way it seems to be important historically also. The Ikṣvāku King Asamāti in the company of asuramāyau adopted their food- habits. They live in the huts of *kirātas* and use to cook rice and meat without fire and Asamāti was got enchanted by these juggalaries. Consequently the *Iksṣavākus* were defeated and faced a downfall. These two hymns of Gaupāyana-brothers followed by two others are very important as they narrate the nature and form of *manas* and are having a very strong spirit of a high psyche. All the three Gaupāyanas intends to make their way to the place of Asamāti. When they started from Khāṇḍavdprastha began to enchant the *viśvedevāḥ* hymn to redivert and re-store the life and vital breath of Subandhu. In first *ṛcā* it is prayed that 'they proceed to Asamāti and no enemy would disturb them on their way[10]-'मा प्र गाम पथो वयं मा यज्ञादिन्द्र सोमिनः। मान्तः स्थुर्नो अरातयः।। In continuation narrate with expectation-'may the essential element of the sacrifice (*yajña*), *i.e. Agni* be available at the place of Asamāti (X.57.2). We Gaupāyanas address the psyche (*manas*) with *nārāśaṁsa-soma* and with sacred and spiritual praise of manes.[11] O Subandhu ! your life and psyche may rise again to lead a skilful and vital life and to see the sun for long.[12] O Manes ! may we take back our lost psyche and divine life and enjoy the vital breath.[13] O Soma ! bearing your psyche in our bodily senses and enjoying with off-springs, may we get back the life of Subandhu.[14]

In the next sūkta (X.58) the Gaupāyana-brothers pray to God for *mana-āvartana* which is *devatā* or subject-matter of this hymn. Composed of twelve *ṛcās* this hymn repeats hemistiches - *b,c,d* - of all the *ṛcās* and only the first, *i.e.* hemistich - *a*- is new one and thus...मनो जगाम दूरकम्। तत् त आ वर्तयामसीह क्षयाय जीवसे- is occurring in each *ṛcā*. We call your psyche and life back to your body which has gone far-away to Vaivasvata-yama, so that, you may lead a cheerful living in this world;[15] your psyche which is spread to the heaven (*divam*) and the earth[16]...; your psyche which has been extended in all the directions of surface[17]..(; your psyche which has been diversified in all the four sides[18]...; your psyche which has gone far away in the sea having plenty of water[19]...; your psyche which has entered the ongoing rays of light[20]...; your psyche which has mixed with water and medicinal plants[21]...; your psyche which has touched the sun and dawn[22]...; your psyche which has gone to the giganatic mountains[23]...; your psyche which has met with the whole moving world[24]...; your psyche which has gone far to the farthest one[25]...; your psyche which has surpassed the past and future[26] we call back that-one to lead a cheerful life in this world. In this way the particular nature of mind has been narrated in a very distinct manner in these two hymns.

The third hymn (*Rv*.X.59) of the *Gaupāyana* seers contain two *ṛcās* (fifth and sixth) of *Asunīti* deity, *i.e.* the subject matter. *Asunīti* means one who takes away the *asu, i.e.* life from human body.[27] The fifth *ṛcā* prays *O Asunīte*! please hold our *manas* in us. For a long living expand and increase our span of life. Make us beseated in the broad scene of the sun and grow your body by the consumption of *ghṛta.*[28] Prayers of similar style have been put forth in the sixth *ṛcā*. Eighth *ṛcā* of the next *sūkta* belongs to *jīva* and this hymn has also been seen by the same Gaupāyana's narrates that I hold your *manas* for a safe secure and healthy living in the same manner as the yoke is held by the binding of leathern girth for a proper control of a chariot[29] and the *ṛcā* is ---यथा युगं वरत्र्या नह्यन्ति धारुणाय कम्। एवा दाधार ते मनो जीवातवे न मृत्यवेऽथो अरिष्टतातये-[30] and as the magnanimous earth holds all the vegetation grown on it, says the next *ṛcā*-- यथेयं पृथिवी मही दाधारेमान् वनस्पतीन्। एवा दाधार ते मनो जीवातवे न मृत्यवेऽथो अरिष्टतातये।। The next *ṛcā* narrates the possession taken and restoration of the *manas* of Subandhu from the domain of Vaivasvata yama and the words of *Rv.* are[31] -

यमादहं वैवस्वतात् सुबन्धोर्मन आभरम्।
जीवातवे न मृत्यवेऽथो अरिष्टतातये।।

This narration has again corroborated by Śaunaka, besides *Jaiminīyabrāhmaṇa* and *Sarvānukramaṇī* of Kātyāyana, in his *Bṛhad-devatā* in the following ślokas[32] :-

स्तुतः स राजा सुव्रीस्तस्थौ गौपायनानभि।।
सूक्तेनाऽयस्तुवन्नग्निं द्वैपदेन यथात्रिषु।
अग्निरप्यत्राब्रवीदेतान् अयमन्तःपरिधयसु:।।

सुबन्धोरस्य चेक्ष्वाकोर्मया गुप्तो हितार्थिना।
सुबन्धवे प्रदायासुं जीवेत्युक्त्वा च पावकः।।
स्तुतो गौपायनैः प्रीतो जगाम त्रिदिवं प्रति।
अयं माते (ता) ति हृष्टास्ते सुबन्धोरसुमाह्वयन्।।
शरीरमभिनिर्दिश्य सुबन्धोः पतितं भुवि।
सूक्तशेषं जगुश्चास्य चेतसो धारणाय ते।।
लब्धवासुं चायमत्यस्यां पृथक् पाणिभिरस्पृशन्।

Thus this story has travelled through the tradition and nāṭyāyanī has also been quoted by *Jaiminīya-brāhmaṇa* having narrated this episode. It seems that the opponents, *i.e. Asuramāyau* -Kilāta and Ākuli- were doing the rites without the use of *Agni* and in this way were trying to break the established norms and hence the conflict. Now these type of descriptions reflect the earliest examples of Psychological-Hedonism which is the theory that behaviour is primarily motivated by the desire to avoid pain or to experience pleasure.

Another important hymn of *Rv.*(X,164), of *duḥsvapnanāśana* deity, *i.e.* removal of a nightmare or a bad dream, has been seen by the seer Pracetā Āṅgīrasa. The first *ṛcā* says O Manasaspate ! you please ask the *nirṛti*, *i.e.* misfortune to go away and get out of here forthwith because the *manas* of living person remains full of desire.[33] Continuing these thoughts says that the auspicious persons choose the boon of happiness, yoke the good and gentle horses for their welfare and behold the Vaivasvata with very kind eyes and thus the *manas* of a living person remains associated with many facets.[34] In awakening and sleeping times whatever harm have we imposed on any one by cutting up, by refusing and accusing, these misdeeds of ours, may the *Agni* push them away from us.[35] O Indra ! whatever injury we creat, may the intelligent and thoughtful *Agni* protect us from onset of the weapons of enemy.[36] The last *ṛcā* of this hymn narrates that we have obtained victory today and have got the desired-ones, and moreover have become sinless. Our mental resolutions of awakened and slept-ones, and whatever sin or misdeed we do, all these may go to our enemy or to one who is inimical to us.[37] In this *sūkta*(X.164) one finds for the first time and last time the use of word *saṁkalpa* in fifth *ṛcā*. This word is very important from psychological point of view. This *sūkta* also seems to be unique in its nature as it has been sung for the removal of *duḥsvapna*, *i.e.* nightmare or bad dream. It is very clear from the words used- जाग्रत्स्वप्नः संकल्पः पापः, *i.e.* bad or unwanted resolutions and thoughts of awakened and sleeping positions. One must be free from evil thoughts. This misdeed may be done by his enemies but not by himself. The psyche narrated here seems to be dominated by self-oriented motives. The seer wants that all of his unbecoming thoughts and bad dreams of sin and misdeeds may go away to his enemies. This favour has been demanded in response of the prayers offered through the *ṛcās* as narrated above.

In an another context (*Rv*.X.71.7) the levels of ability of mind has been discussed and it has been depicted that it varies person to person. The friends having same eyes and ears may differ in their *manojava*, *i.e.* the level of capacity of understanding and retaining.[38] *Viśvakarman* an epithet of almighty *paramātman* has been told as swift as the *manas, i.e. -manojuvam-* in performing the deeds of protection.[39] *Pavamāna Soma* has been called *manasaspati* who is having the knowledge of *manas*.[40] In the latter Vedic literature *manas* is said to have been born from *Candramas*. In *Yama-yamī Saṁvāda-sūkta* (*Rv*.X.10) yamī says to yama that you must hold your *manas* into mine, both the *manas* may be in full compromise. The words are[41]-नि ते मनो मनसि धाय्यमस्मे ... Udgītha comments here ... ते तव यमस्य मन: मनसि अस्मे अस्माकं सम्बन्धिनि नि धायि निधीयताम् मम चित्तमनुचित्तं ते अस्तु। *Vajra* is said to be *manojuva*, *i.e.* as swift as the *manas*.[42] *Aśvaḥ* are narrated as rapid as the *manas*.[43] The chariot of Nāsatyau is described as fast as *manas*.[44] *aśva* deity is said to be faster than the *manojava* Indra. The text is[45] - मनोजवा अवर इन्द्र आसीत्।... here Venkaṭa explains- मनोवेग: सूर्यश्च अस्य गमने अवर: भवति। न सममनेन गच्छतीत्यर्थ:। The *cetas* of *Aśva* deity is as swift as the *vāta*,[46] text is- तव चित्तं वात इव ध्रजीमान्।... And this deity is said to be splenderous by his *manas* [47] -देवद्रीचा मनसा दीधयान:। Venkaṭa explains it - मनसा दीप्यमान:...Thus the *cetas* or *manas* is said to be of shining form. Sun comes on its way of daily routine swiftly like the *manas*.[48] The portrayal of *manas* of Indra has been portrayed as *jaitra*, *i.e.* victorious and *nibhṛta*. *i.e.* concealed or submissive.[49] Text is- जैत्रं हीन्द्र निभृतं मनस्तव।...Mudgal the commentator of *Rv*. Explains -हे इन्द्र ! निभृतम् अव्याकुलं तव मन: चित्तम् जैत्रं हि जयनशीलं खलु शत्रून् जित्वास्मभ्यं धानं दातुं समर्थमित्यर्थ:, *i.e.* the *manas* of Indra is stayable and not quivering and thus explains the quality of *manas*. The sense of compound *manīṣa* is understood to be *stuti, i.e.* praise. See the text- *Rv*.X.111.1- मनीषिण: प्रभरधवं मनीषां यथायथा मतय: सन्ति नृणाम्। -and Venkaṭamādhava explains it –हे मनीषिण: प्रभरधवं स्तुतिम् ...। This reference of *Rv*. also suggests that *manīṣā* and *mati* may be separate elements. When Sūryā reached her patigṛha her vehicle was *manas*[50]- मनो अस्या अन आसीद्.. Aśvinau are said to be having a speed as of *manas*[51]-मनोजवसा वृषणा मदच्युता....They are also as penetrating and pinpointing as the *manas*[52]-मनो न येषु हवनेषु तिग्मं- ..and Venkaṭa explains it- यथा तीक्ष्णाग्रं मन: शीघ्रं दूरं गच्छति, एवम्-। *Suparṇa* is said to be as swift as the *manas*[53] - मनोजवा अयमान आयसीमतरत् पुरम्।

According to the above discussion it can be safely assumed that *manas* is an element which is the swiftest one. Any thing which is to be narrated to be rapid has been compared and given simile and resemblance with *manas*. It is *didhyāna*, *i.e.* splendent like a ray, *tigmam*- penetrating or scorching, having the ability to bear good or bad. It is also narrated to have a capacity to be present to the heaven and the earth, to all directions of this world, to the waters of the sea, to the rays of the sun and to the peaks of the mountains. It can penetrate in the veins of medicinal plants and can have knowledge of past and future. It has been expanded to be the vital breath, life or living soul which escapes from body at death, as has been described in *mana-āvartana* and related hymns. It is intellect, perception, conscience or will as has been portrayed in *duḥsvapnanāśanam* hymn. Its vitality remains same in awakened and sleeping positions. Its capacity differs man to man. It has qualities such as submissiveness, aggressiveness, stability and quiveringness *etc*. No where in *Rv.* it has been described or related with the word *mastiṣka*. Actually the word *mastiṣka* occures once in *Rv.* In the context of *yakṣmanāśanam* (X.163) where it has been counted with other organs and the text is[54] अक्षीभ्यां ते नासिकाभ्यां कर्णाभ्यां छुबुकादधि। यक्ष्मं शीर्षण्यं मस्तिष्काज्जिह्वाया वि वृहामि ते। Here Venkaṭamādhava has explained- शिरोऽन्तर्वर्ति मांसं मस्तिष्कः।

The text of the *Rv.* is full of positive thoughts and resolutions. It is evident from the references used with the addition of *sam upasarga*, *e.g.* संदृष्टिः (II.4.4;IV.10.5;VI.16.25); संदृष्टौ (I.144.7;VI.1.4;X.64.11) संधाता and संधिम् (VIII.1.12); संभरः (IV.17.11); संभरणम् (VII.25.2); संभरन्ति (I.162.6); संवननम् (X.93.12) (संवनना (VIII.1.2); संवरणस्य (V.33.10;X.77.6); संवरणात् (VII.3.2); संवरणानि (IX.107.9) (संवरणेषु (IV.21.6); संवसनेषु (IX.86.17); संवसुः (VIII.39.7); संवादाय(VIII.101.4); संवित् (VIII.58.1); संविदम (X.10.14) संविदानः (VII.44.4 *etc.*); संविदानाः (X.30.13); संविदानासः (X.30.14); संविदाने (III.54.6); संवेशने (X.56.1) (many forms of संसद्; संहितम् (I.168.6;VIII.67.21); संहिता (VIII.96.2); संगता (VIII.78.8); संगतानि (I.36.5); संगतासः (VII.76.5); संगतिम् (IV.44.1); संगत्याम् (X.141.1); संगथे (I.91.16 *etc.*); संगमे (I.102.3*etc.*); संगमनः (I.96.6 *etc.*)(संगमनम् (X.14.1); संगमनी (X.125.3); संगमामहै (VIII.91.4); संचक्षे (I.127.11 *etc.*); संचरणे (I.56.2 *etc.*) (संचरणीः (VI.24.4); संजानानाः (I.72.5 *etc.*) संज्ञानम् (X.19.4) and many forms of समन्दृश *etc.* All these references refer to the positive psyche of Ṛgvedic *ṛcās*. We must peep into them.

References :-

1 Monier Williams, '*A Sanskrit-English Dictionary(=Skt.Eng.Dict.)*' MLBD, Delhi, 1986, p.783.

2 *Ṛgvedasaṁhitā (=Rv.)*, I.34.7(-d-) ..आत्मेव वातः स्वसराणि गच्छतम here Skandasvāmin explains-... आत्मशब्देनात्र आत्मसम्बन्धात् मन उच्यते। वात् इत्यपि लुप्तेवशब्द उपमानमेव द्वितीयम् । मन इव च वात इव च यथा वायुमनसी शीघ्रं गच्छतः तद्वत् ।

3 *Skt.Eng.Dict*., p.783.

4 अया चित्तो विपानया (-a-)..; Venkaṭa explains here-- अनया गच्छन्त्याङ्गुल्या ज्ञातः

5 तिरश्चित्तानि वसवो जिघांसति (-b-) here Venkaṭa explains-तिरःकुर्वन् बुद्धीः ...

6 Cf., *Skt.Eng.Dict*., p.395

7 Cf. *Ṛgvedasaṁhitā* in eight parts, with the Coms. Udgītha, Venkaṭamādhava and Mudgala, V.V.R.I., Hoshiarpur -part.VIII, indices, pp.4446-4448.

8 Kātyāyana's *Ṛk-sarvānukramaṇī*, with the Kṛtsna vṛtti of Sadguruśiṣya, Ed. Vijaypal, Savitridevi Bagadia Trust, Calcutta, 1985. X.57 (Kaṇḍikā 57.15-20)- अथ हैक्ष्वाको राजासमातिर गौपायनान् बान्धवादीन् पुरोहितास्त्यक्त्वान्यौ मायाविनौ श्रेष्ठतमौ मत्वा पुरोदधे। तम् इतरे क्रुद्धाः अभिचेरुः। अथ तौ मायाविनौ सुबन्धोः प्राणानाचिक्षिपतुः। अथ हास्य भ्रातरस्त्रयो 'मा प्र गाम' इति षट्कं गायत्रं स्वस्त्ययनं जप्त्वा 'यत्ते यमम्' इति द्वादशर्चम् आनुष्टुभं मन आवर्तनं जेपुः।

9 Jaiminīyam Brāhmaṇam, Crit. ed. By Raghuvira, and Lokeshachandrah, MLBD, Delhi, 1986, III.168- असमतिं राथप्रौष्ठं गौपायना अभ्यदासन् । ते खाण्डवे सत्रमासत । अथ हासमातौ राथप्रौष्ठे किलाताकुली षतुरसुरमायौ । तौ ह स्माऽनग्नावधिधायौदनं पचतोऽनग्नौ मांसम्...तद् वै तच् छश्वद इक्ष्वाकवोऽसुराशनं जग्ध्वा पराभूताः । तमसमातिं राथप्रौष्ठं गौपायनानामाहुतयोऽभ्यतपन् । सोऽब्रवीद्- इमा किराताकुली इमा वै मा गौपायनानामाहुतयोऽभितपन्तीति । तावब्रूतां तस्य वा आवामेव भिषजौ स्व, आवं प्रायश्चित्तिर् आवां तथा करिष्यावो यथा त्वैता नाभितप्स्यन्तीति। तौ परेत्य सुबन्धोर् गौपायनस्य स्वपतः प्रमत्तस्यासुमाहृत्यान्तःपरिधि न्यधत्ताम् । ... Also cf. *Śatapatha-Brāhmaṇam*, with the Coms. of Sāyaṇa and Harisvāmī, Vol.I,1.1.4.14 ... किलाताकुली-इति हासुरब्रह्मावासतुः। Also cf. *Rv*., Ibid., Part-VII,p.3501 where - तत्र शाट्यायनकम् - is added before this reference found in *Jaim.Br*. - III.168.

10 *Rv*.X.57.1- मा प्र गाम पथो वयं मा यज्ञादिन्द्र सोमिनः। मान्तः स्थुर्नो अरातयः ॥

11 *Rv*.X.57.3- मनो न्वा हुवामहे नाराशंसेन सोमेन । पितृणां च मन्मभिः ॥

12 *Rv*.X.57.4- आ त एतु मनः पुनः क्रत्वे दक्षाय जीवसे । ज्योक् च सूर्यं दृशे ॥

13 *Rv*.X.57.5- पुनर्नः पितरो मनो ददातु दैव्यो जनः । जीवं व्रातं सचेमहि ॥

14 *Rv*.X.57.6- वयं सोम व्रते तव मनस्तनूषु बिभ्रतः । प्रजावन्तः सचेमहि ॥

15 *Rv*.X.58.1- यत् ते यमं वैवस्वतं मनो जगाम दूरकम् । तत त आ वर्तयामसीह क्षयाय जीवसे ॥

16 *Rv*.X.58.2(a)- यत् ते दिवं यत् पृथिवीम्-

17 *Rv*.X.58.3(a)- यत् ते भूमिं चतुर्भृष्टिम..-

18 *Rv*.X.58.4(a)- यत् ते चतस्रः प्रदिशो मनो जगाम दूरकम्-

19 *Rv*.X.58.5(a)- यत् ते समुद्रमर्णवं मनो जगाम दूरकम्-

20 *Rv*.X.58.6(a)- यत् ते मरीचीः प्रवतो मनो जगाम दूरकम् -

21 *Rv*.X.58.7(a)- यत् ते अपो यदोषधिर्मनो जगाम दूरकम्.-

22 *Rv*.X.58.8(a)- यत् ते सूर्यं यदुषसं-

23 *Rv*.X.58.9(a)- यत् ते पर्वतान् बृहतो मनो जगाम दूरकम...

24 *Rv*.X.58.10(a)- यत् ते विश्वमिद्रं जगन्मनो जगाम दूरकम् ।...

25 *Rv*.X.58.11(a)- यत् ते परा: परावतो मनो जगाम दूरकम्...

26 *Rv*.X..58.12(a)- यत् ते भूतं च भव्यं च मनो जगाम दूरकम् ।
तत् त आ वर्तयामसीह क्षयाय जीवसे ॥

27 *Rv*.Viśvabandhu, X.59.5-Venkaṭa explains *asunīti* -या मनुष्याणाम् असुं शरीरतो नयति सा असुनीतिर्नाम ।

28 *Rv*.X.59.5- असुनीते मनो अस्मासु धारय जीवातवे सु प्र तिरा न आयु: ।
रारन्धि न: सूर्यस्य संदृशि घृतेन त्वं तन्वं वर्धयस्व ॥

29 *Rv*.X.60.8- यथा युगं वरत्रया नह्यन्ति धरुणाय कम् ।
एवा दाधार ते मनो जीवातवे न मृत्यवेऽथो अरिष्टतातये ॥

30 *Rv*.X.60.9

31 *Rv*.X.60.10

32 *Bṛhad-devatā* of Śaunaka, Ed. Ram Kumar Rai, Chaukhamba Sanskrit Sansthan, Varanasi, 1989, VII.97-102.

33 *Rv*.X.164.1.- अपेहि मनसस्पतेऽप क्राम परश्चर ।
परो निर्ऋत्या आचक्ष्व बहुधा जीवतो मन: ॥

34 *Rv*.X.164.2- भद्रं वै वरं वृणते भद्रं युंजन्ति दक्षिणम् ।
भद्रं वैवस्वते चक्षुर्बहुत्र जीवतो मन: ॥

35 *Rv*.X.164.3- यदाशसा नि:शसाभिशसोपारिम जाग्रतो यत् स्वपन्त: ।
अग्निर्विश्वान्यप दुष्कृतान्यजुष्टान्यारे अस्मद् दधातु ॥

36 *Rv*.X.164.4- यदिन्द्र ब्रह्मणस्पतेऽभिद्रोहं चरामसि ।
प्रचेता न आङ्गिरसो द्विषतां पात्वंहस: ॥

37 *Rv*.X.164.5- अजैष्माद्यासनाम चाभूमानागसो वयम् ।
जाग्रत्स्वप्न: संकल्प: पापो यं द्विष्मस्तं स ऋच्छतु यो नो द्वेष्टि तमृच्छतु ॥

38 *Rv*.X.71.7- अक्षण्वन्त: कर्णवन्त: सखायो मनोजवेष्वसमा बभूवु: ।
आदधनास: उपकक्षास उ त्वे ह्रदा इव स्नात्वा उ त्वे ददृश्रे ॥

39 *Rv*.X.81.7(a & b)- वाचस्पतिं विश्वकर्माणमूतये मनोजुवं वाजे अद्या हुवेम ।

40 *Rv*.IX.11.8(c)-...मनश्चिन्मनसस्पति: ।

41 *Rv*.X..10.3. (c)

42 *Rv*.VI.22.6

43 *Rv*.VI.62.3(b&c)-..उग्रेत्था धिय ऊहथु: शश्वदश्वै:। मनोजवेभिरिषिरै शयध्यै- Skanda explains it- अश्वै:। कीदृशै: । मनोजवेभि: मनस: इव जवो येषां ते मनोजवा अत्यन्तशीघ्रगतय:,...

44 *Rv*.VI.63.7(c)-'प्र वां रथो मनोजवा असर्जीष:.., Skanda explains it-..वाम् युवयो: स्वभूत: रथ: अपि मनोजवा: मनस इव जवो वेगो यस्य स मनोजवा । अत्यन्त शीघ्रगतिरित्यर्थ:।..,

45 *Rv*.I.163.9(b)

46 *Rv*.I.163.11

47 *Rv*.I.163.12(b)

48 *Rv*.I.71.9(a+b)- मनो न योऽध्वनः सद्य एत्येकः सत्र सूरो विश्व ईशे । Mudgala comments - दृष्टान्तः। मनः न यथा मनः शीघ्रं गच्छति तद्वत् ।

49 *Rv*.I.102.5(d)

50 *Rv*.X.85.10(a)

51 *Rv*.VIII.22.6(a)

52 *Rv*.VIII.100.8(a+b)

53 *Rv*.X.61.30(a)

54 *Rv*.X.163.1

9

योगायुर्वेदयोः प्राकृतिक चिकित्सा

डॉ० रामेश्वर दत्त शर्मा

पातञ्जल योगशास्त्र में योग की परिभाषा करते हुए महर्षि पतञ्जलि कहते हैं "**योगश्चित्तवृत्तिनिरोधः**" अर्थात् चित्त की वृत्तियों का निरोध, नियंत्रण, उपशमन करना ही योग कहलाता है। वस्तुतः प्रत्यक्ष में अनुभव होता है कि मानव मानसिक तनाव, घबराहट, दबाव और अशान्ति से कितना बेचैन और अशान्त रहता है। इस तनाव से विशेषकर आज के युग में हृदय, रोग, उच्च रक्तचाप, सिरदर्द और पेट की अव्यवस्था आदि अनेक रोग मनुष्यों को बुरी तरह से दुःखी किये हुए हैं।

पाश्चात्य देशों में जहाँ भौतिकवादी वातावरण अपनी चरम सीमा पर पहुँचा हुआ है वहाँ कितनी अशान्ति और मानसिक असन्तुलन है, इसका परिणाम हम प्रतिदिन देखते हैं, कितने लोग आत्महत्या जैसे दुस्साहसिक कार्य करते हैं। इसी कारण से उन देशों में योग को अपनाय जा रहा है। योग शिविर लगाये जाते हैं। धड़ाधड़ योगाश्रम खोले जा रहे हैं। प्राकृतिक चिकित्सा की यौगिक पद्धति के तीन चरण हैं— (१) उचित आहार, (२) उचित योगाभ्यास, (३) व्यक्ति के जीवन से सम्बद्ध बातों का समुचित ज्ञान।

चिकित्सात्मक योग के अन्तर्गत आने वाले प्रायः सभी रोगियों के आहार के पदार्थों में काफी समानता है। निरामिष भोजन में प्राकृतिक रूप से पके फल, सलाद, हरी पत्ती वाली तरकारियाँ, अंकुरित चने, मूंग, गेहूँ की रोटियाँ, दालें आदि ऐसी अनेक वस्तुएँ हैं, जिन्हें व्यक्ति दैनिक जीवन में यदि प्रयोग में लाता रहता है तो न केवल वह स्वस्थ रहता है अपितु अनेक रोगों से छुटकारा भी पा लेता है। मनुष्य जन्म से ही दूध, दही, लस्सी स्वच्छ जल पर आश्रित रहता है। सात्विक भोजन पर वह स्वस्थ रहता है, जबकि गर्म मसाले, तली हुई वस्तुएं उसके आमाशय को विकृत, दूषित और खराब कर देती हैं। इसी प्रकार तम्बाकू, शराब तथा अन्य नशीली वस्तुएँ मानव मस्तिष्क को अशान्त कर देती हैं। मस्तिष्क में ज्ञान तन्तुएँ इतने सूक्ष्म होते हैं, कि तनिक से नशे से वे निष्क्रिय हो जाते हैं और मूढ़ मनुष्य उसे आनन्द समझता है। योग शास्त्र में ध्यान, समाधि, प्रेक्षाध्यान आदि अनेक ऐसे साधन हैं, जिनके प्रयोग से व्यक्ति उस आनन्दानुभूति को प्राप्त कर लेता है, जहाँ लोभ, लालच, क्रोध, मोह, मद, मात्सर्य, ईर्ष्या, द्वेष, हिंसक भावना आदि आवेगों से भी सर्वथा मुक्त हो जाता है। आम साधारण व्यवहार में लोग कहते रहते हैं, जैसा खाओ अन्न, वैसा होता है मन। संसार में मनुष्य के अतिरिक्त लाखों अन्य जीवों-जन्तुओं की योनियाँ हैं। वे सभी बीमार नहीं होते। क्योंकि उनका निश्चित और नियमित आहार होता है। रोगों का मूल कारण आहार और विहार ही होते हैं। माधव निदान में सर्वप्रथम यही कहा है:—

मिथ्याहार विहाराभ्यां दोषा ह्यामाशयाश्रया:।

बहिर्निरस्य कोष्ठाग्निं ज्वरदास्यूरसानुगा:।।

अर्थात् जब व्यक्ति के आहार और विहार दूषित हो जाते हैं, असन्तुलित हो जाते हैं, तब वायु-पित्त-कफ सम्बन्धी दोष आमाशय में स्थित होते हुए बाहर आते हुए कोष्ठाग्नि को बाहर निकालते हुए ज्वर-बुखार को पैदा कर देते हैं। स्मरण रहे वात, पित्त और कफ ये आयुर्वेद के अनुसार धातु भी हैं, अर्थात् शरीर को धारण शक्ति प्रदान करते हैं। शरीर को हृष्ट-पुष्ट और स्वस्थ बनाते हैं, किन्तु जब ये मिथ्या आहार-विहार से दूषित हो जाते हैं, तब ये रोग पैदा करते हैं। आहार के विषय में एक सुन्दर कहावत प्रचलित है कि कोई एक कौवा वैद्य जी के मकान की मुंडेर पर बैठकर वैद्य जी पूछने लगा:— कोऽरुक्, कोऽरुक्, कोऽरुक्, अर्थात् निरोग कौन, निरोग कौन, निरोग कौन? वैद्य जी ने उत्तर दिया—कालभुक्, हितभुक् मितभुक्। वास्तव में देखा जाए तो स्वस्थ रहने के ये तीन मानदण्ड प्राकृतिक चिकित्सा के सर्वश्रेष्ठ साधन हैं। शरीर एक मशीन के समान है, एक इंजन के समान है इसे खुराक चाहिए, खुराक से इसमें शक्ति आती है, इसे ऊर्जा मिलती है। परन्तु कब, जब इसे खुराक की आवश्यकता हो तब भोजन का समय निश्चित होना चाहिए। आयुर्वेद में लिखा है कि **''याम मध्ये न भोक्तव्यं यामद्वयं न लंघयेत्''** अर्थात् प्रहर के बीच में भोजन न करें और दोपहर से अधिक समय का लंघन न करें। संसार में प्रत्येक मनुष्य को भिन्न-भिन्न कार्य करने पड़ते हैं, उनके स्थान, कहीं ठण्डे हैं, कहीं गर्म हैं, कहीं पर्वतीय हैं, कहीं जलीय हैं, इसी प्रकार काम भी कहीं शारीरिक हैं, कहीं बौद्धिक हैं, फिर प्रत्येक व्यक्ति की शारीरिक स्थिति मानसिक प्रवृत्ति भिन्न-भिन्न होती है, अत: सभी के लिए एक सा नियम कारगर सिद्ध नहीं हो सकता। अत: जिस व्यक्ति का शरीर जब खुराक माँगे तभी देना— यह सिद्धान्त अधिक लाभप्रद होगा। इसीलिए कालभुक् कहा गया है, जिसको जिस समय भोजन की आवश्यकता हो,तभी भोजन दिया जाए। भोजन कैसा हो? आयुर्वेद में तीन प्रकार के भोजन बताये गए हैं— सात्त्विक, राजसिक, तामसिक। सात्त्विक भोजन में बहुत ही कम मसाले होते हैं। भोजन ताजा, आकर्षक तथा पुष्टिकारक हो, इसे सरल ढंग से तैयार किया जाए। राजसिक भोजन में कुछ भुने हुए, कुछ तले हुए, रसदार तथा बहुत मसालेदार अनेक प्रकार के मिष्ठान्न तथा पेय पदार्थ होते हैं। तामसी भोजन में अत्यन्त उत्तेजक पदार्थ होते हैं— गले सड़े बासी पदार्थ इस भोजन में होते हैं।

वास्तव में यदि देखा जाए तो कोई भी भोजन चाहे वह निरामिष हो या सामिष हो, अपने आप में न सात्विक होता है और न राजसिक होता है, न तामसिक होता है, उसके बनाने के ढ़ंग से ही उसे उस वर्ग में रखा जाता है, भोजन का कोई अपना वर्ग नहीं होता। लोगों की यह सामान्य धारणा होती है कि सामिष भोजन तामसी होता है और निरामिष भोजन सात्विक होता है, किन्तु यह धारण सर्वथा निर्मूल, निराधार और गलत है। आलू या फूलगोभी को भी तामसी के रूप में तैयार किया जा सकता है तथा मांस या मुर्गे को सात्विक रूप में बनाया जा सकता है। यह व्यक्ति की रुचि पर निर्भर करता है।

दूसरी बात यह है कि योग में कैलोरी की गणना के आधार पर भोजन का मूल्यांकन नहीं किया जाता। इसके विपरीत, भोजन के गुण तथा खाने के ढंग पर विशेष ध्यान दिया जाता है। हितभुक् से अभिप्राय है कि जिस व्यक्ति के शरीर के लिए जो पदार्थ गुणकारी हो, लाभप्रद हो, शक्तिदायक हो, वही पदार्थ उसके लिए भोज्य है। किसी के लिए दही उपयुक्त है तो किसी के लिए दूध। कुछ ऐसा प्रदेश हैं जहाँ दूध दही होता ही नहीं, लाहोलस्पीति ऐसे स्थान हैं, जहाँ भेड़ के मांस के अतिरिक्त कुछ खाने को है ही नहीं। बंगाल, मद्रास, उड़ीसा, आन्ध्रप्रदेश, केरल, कर्नाटक आदि ऐसे प्रदेश हैं, जहाँ मछली का भोजन सर्वोत्तम होता है। हरियाणा, राजस्थान, पंजाब, ऐसे प्रदेश हैं, जहाँ दही-दूध-घी का भोजन सर्वोत्तम होता है, अत: प्रत्येक व्यक्ति के लिए जो हितकारक भोजन हो उसे ही वह व्यक्ति लेवे।

मितभुक्—इसका अभिप्राय यह है कि ८५% भोजन लिया जाए। जल वायु की भी मात्रा पर्याप्त रूप में रहे। भोजन को चबा-चबा कर ग्रहण किया जाए। सयाने लोग कहते हैं दाँत का और आंत का चोली-दामन का सम्बन्ध होता है। जो भोजन धीरे-धीरे चबा कर खाया जाएगा, वह शीघ्र पच जाता है। अन्न न पचने के कारण 'माधवनिदान' में निम्नलिखित बतलाए हैं:—

अत्यम्बुपानाद् विषमाशनाच्च सन्धारणात् स्वप्नविपर्ययाच्च।
कालेऽपि सात्म्यं लघु चापि भुक्तमन्नं न पाकं भजते नरस्य॥

अधिक जल पीने से, असमान भोजन करने से, वेगों-मूत्र-मल आदि के रोकने से, नींद के पूरा न लेने से, समय पर अपने अनुकूल और हल्का भोजन भी पचता नहीं।

विहार—शरीर प्रतिदिन व्यायाम चाहता है। यह व्यायाम अनेक प्रकार का हो सकता है। योग में आसनों के महत्त्व को बहुत बढ़-चढ़ कर बताया गया है। प्राणायाम इसमें मुख्य साधन होता है। प्राणायाम सांस के व्यायाम का विशेष रूप है। इसके अनेक रूप हैं— (१) रेचक (सांस छोड़ना), (२) पूरक (सांस खींचना), (३) कुम्भक (सांस रोकना)।

आसन– शरीर को विविध मुद्राओं में मोड़ना आसन कहलाते हैं। आसनों के अनेक भेद हैं—वे भेद भिन्न-भिन्न प्रकार के रोगों को समूल नष्ट करते हैं। इन आसनों के कुछ प्रमुख भेद ये हैं—

(१) उत्तानपाद आसन– फर्श पर पीठ के सहारे लेट कर छत्त की ओर देखा जाता है। दोनों हाथों को शरीर के साथ फैलाएँ, जिससे हथेलियां फर्श का स्पर्श करें। दोनों पैरों को सीधा रख कर एड़ियों तथा अंगूठों को मिलाएँ रखें। स्वाभाविक रूप से सांस लें । इस आसन से पेट की सभी मांसपेशियों का व्यायाम भीतरी तथा बाहरी दोनों रूपों में हो जाता है। यह आसन पेंक्रियाज की गड़बड़ियों को रोकता है एवं कब्ज, वायुविकार, अपच तथा आंत की अव्यवस्थाओं को ठीक कर देता है।

(२) पवनमुक्तासन– एक घुटने को सीने की ओर उठाएँ, उसी ओर के हाथ को टखने पर रख कर दूसरे हाथ को घुटने पर रखें। इसी प्रकार दूसरे घुटने को सीने की ओर उठाएँ उसी ओर के हाथ को टखने पर रखकर दूसरे हाथ को घुटने पर रखें।

लाभ-लगभग उत्तानपाद वाले आसन के लाभ ही इस आसन के लाभ हैं।

(३) भुजंगासन– पेट के सहारे लेट जाएँ। सिर को किसी भी कनपटी के सहारे रहने दें। हथेलियों को दोनों तरफ कंधों के नीचे रखें। अंगुलियों के छोर कंधों के आखिरी छोर पर रहें। कुहनियाँ शरीर से सटाकर मुड़ी हुई रखें। एड़ियों को सटाकर पैरों के अंगूठों को जमीन पर सपाट रखें।

लाभ–पाचन संस्थान को सुदृढ़ बनाता है।

(४) सूर्यनमस्कार आसन– दोनों पैरों को करीब दो फीट की दूरी पर रखते हुए खड़े हो जाएँ। दोनों हाथों को बगल में लटकने दें। सिर सीधा रखें। सामने देखें।

लाभ–इस आसन के अनेक लाभ हैं। यह नलिकाहीन संस्थान की सभी ग्रन्थियों को हल्के ढंग से सक्रिय करता है।

(५) धनुरासन– पेट के सहारे लेट जाएँ। दोनों ओर हाथों को फैला लें फिर सिर को किसी कनपटी के सहारे जमीन पर टिका दें। पैरों और एड़ियों को सटाकर रखें।

लाभ– अधिवृक्क (ऐड्रिनल) चुल्लिका (वायराडय) उपचुल्लिका (पाराथायरायड) पीयूष (पिट्यूटरी) तथा यौन ग्रन्थियों को स्वस्थ एवं नीरोग बनाता है।

इनके अतिरिक्त अनेक आसन हैं। किन्तु ध्यान रहे इन आसनों के करने में योग्य गुरु के सान्निध्य और उनके निर्देशन की परम आवश्यकता है।

विहार से भ्रमण करना भी अभीष्ट है। सौ दवाई एक घुमाई बराबर होती है। प्रात: सायं भ्रमण करना प्राकृतिक चिकित्सा का सर्वोत्तम साधन है। (बसन्ते भ्रमणं पथ्यम्। बसन्त ऋतु में भ्रमण अत्यन्त लाभप्रद होता है)।

आयुर्वेद–आयुर्वेद का अर्थ है आयु को जानना, आयु को समझना। यह शब्द स्वयं प्राकृतिक चिकित्सा को इंगित करता है। आयुर्वेद में यद्यपि चरक, सुश्रुत, अष्टांगहृदय, भाव प्रकाश, माधवनिदान, निघण्टु आदि अनेक ग्रन्थ हैं। परन्तु उन सभी ग्रन्थों में, स्वेदन, तेल, स्नान, स्नेहन, लंघन आदि अनेक ऐसे चिकित्सा के साधन हैं, जिन्हें हम प्राकृतिक संसाधनों में गिनते हैं। उदाहरण के लिए लस्सी छाछ तक्र से अतिसार संग्रहणी, पेचिश वगैरह रोग स्वत: ही शान्त हो जाते हैं–

न तक्रसेवी व्यथते कदाचिन्न तक्रदग्धा: प्रभवन्ति रोगा:।
यथा सुराणाममृतं सुखाय, तथा नराणां भुवि तक्रमाहु:।।

तक्र लस्सी का सेवन करने वाला व्यक्ति कभी पीड़ा को प्राप्त नहीं करता है। तक्र से नष्ट हुए रोग पुन: उत्पन्न नहीं होते। जैसे स्वर्ग में देवताओं के लिए अमृत है वैसे ही पृथिवी पर मनुष्यों के लिए तक्र है।

OUR CONTRIBUTORS

1. Agrawal Dr. Ashok
Prof. & Head
Department of Botony
Kurukshetra University,
Kurukshetra (136119)

2. Agrawal Dr. (Mrs) Vibha
Principal, D.N. College for Girls
Kurukshetra (136118)

3. Alamkar, Dr. Veerendra Kumar
Head, Deptt. of Sanskrit
Punjab University
Chandigarh

4. Arya, Dr. Ravi Prakash
901, N.C.H. D.R.D.O. Complex
Lucknow Road.
Timarpur, Delhi (110054)

5. Bhartiya, Dr. M.C
M.M.H. College
R-6, Bharat deep Market,
Flat 102, Raj Nagar,
Ghaziabad, U.P

6. Deshpande, Mr. Deepak
'Anand-Vilas' Chitali Road
Dhantoli, Nagpur (440012)

7. Didolkar, Prof. V.K.
Deptt of Metallurgy
Visveshvaraiya National Institute
of Technology, Nagpur

8. Dongre, Prof. Dr. N.G.
B-62, Brij, Enclave
Varanasi (221005)

9. Goyal, Dr. V.K.
Reader & Head,
Deptt. of Physics
University College
Kurukshetra

10. Gupta, Dr. Aayam
Dabar Dhanvantary Hospital
Shri Dhanvantary Ayurvedic
College, Sec. 46B
Chandigarh

11. Gupta, Dr. S.P.
Reader, Deptt. of physics
Kurukshetra University.

12. Gupta, Dr. (Mrs) Vinay
Lecturer in Sanskrit
D.N. Mahila Mahavidyalaya
Kurukshetra

13. Holay, Adv. P.V.
West Park Road
Dhantoli, Nagpur (Maharashtra)
440012

14. Jain, Dr (Mrs) Anubha
Lecturer in Sanskrit
Khalsa College
Yamuna Nagar (Haryana)

15. Kaushal, Dr. C.D.S
Lecturer in Sanskrit,
University college,
Kurukshetra University Kurukshetra

16. Kawadkar. M.K
Annapurna Apartments
18, Pande Lay-Out,
Khamla Road, Nagpur- 440025

17. Kumar, Dr. Vikram
Professor, Deptt. of Sanskrit,
University of Punjab, Chandigarh

18. Mishra, Dr. Prafulla Kumar
Prof. P.G. Deptt. of Sanskrit
Utkal University, Bhubaneshwar

19. Mishra, Prof. Rajendra
Ex-Head, Sanskrit Deptt. & Ex-Dean, H.P. University, Shimla
& Ex-Vice-Chancellor,
Sampoornanand Sanskrit
Vishvavidyalaya, Varanasi

20. Mishra Dr. Surendra Mohan
Reader, Deptt, of Sanskrit, Pali &
Prakrit, Kurukshetra University,
Kurukshetra -136119

21. Mishra, Dr. Rajeshwar
Reader, Deptt, of Sanskrit, Pali &
Prakrit, Kurukshetra University,
Kurukshetra-136119)

22. Panda, Dr. N.C
Senior Research officer
Indira Gandhi National Centre for
Arts, B 1/3- E-7, Lane II,
Ravindrapuri Extension,
Varanasi-221005

23. Prabhu, Mr. C.S.R.
Deputy Director General
NIC A block, B.R.K.R. Bhawan
Tank Bond Road
Hyderabad (500063)

24. Purohit, Mr. Rajesh
Curator, Shrikrishna Museum
Kurukshetra (136118)

25. Rajput, Prof. Dr. B.S
Ex-V.C. Kumaun University
Ex- V.C. Garhwal University
1-11 Gamma-2, Greater Noida,
Distt. Gautama Buddha Nagar
(U.P.)

26. Saini, Dr. (Mrs.) Suman
Department of Sanskrit
Kurukshetra- University,
Kurukshetra (136119)

27. Sharma, Dr (Mrs) Indu
Professor, Institute of Skt. and
Indological Studies,
Kurukshetra- University,
Kurukshetra (136119)

28. Sharma, (Dr.) Rameshwar Datt,
Director Haryana Sanskrit
Academy, Kothi No. 169, Sec-12
Panchkula (Haryana)

29. Shastri, Dr. J.L.N
Dabar Dhanvantary Hospital
Shri Danvantary Ayurvedic
College, Sec. 46B,
Chandigarh

30. Singh, Dr. Ramvir
Director, Institute of Sanskrit
Indological Studies
Kurukshetra- University,
Kurukshetra (136119)

31. Sinha, Prof. Himmat Singh
Retd. Chairperson, Deptt. of
Philosophy, K U.
H.No. 304/11, Mohala Mahadcv,
Kurukshetra (136119)

32. Trivedi, Dr. Chandra Prakash
Principal, Govt. Girls, College,
Ratlam- 457001

33. Varnekar, Dr. Chandragupta
Cumins Girls Engineering College,
Karve Nagar
Pune-411052

34. Vartak, Dr. P.V.
Founder & Chairman of Veda
Vijnana Mandal, 521, Shaniwar
Petha, Pune (411030)

35. Dr. Vasant Kumar
Lecturer in Sanskrit
Dayal Singh College
Karnal (Haryana)

36. Verma Dr. Madan Lal
68, Sumangal, New Colony,
Kurukshetra (136118)

37. Vidyalankar, Dr. Kailash Chandra
767/13 Urban Estate
Kurukshetra (136118)

38. Dr. (Mrs) Vijaya Rani
Prof. and Chairperson,
Deptt. of Sanskrit, Pali & Prakrit,
Kurukshetra University
Kurukshetra

39. Wadekar, Prof. M.L.
Director, Oriental Institute
M.S. University of Baroda,
Near Palace Gate, Palace Road
Vadodara